T0074860

GENERATORS OF MARKOV CHAINS

Elementary treatments of Markov chains, especially those devoted to discrete-time and finite state-space theory, leave the impression that everything is smooth and easy to understand. This exposition of the works of Kolmogorov, Feller, Chung, Kato, and other mathematical luminaries, which focuses on time-continuous chains but is not so far from being elementary itself, reminds us again that the impression is false: an infinite, but denumerable, state-space is where the fun begins.

If you have not heard of Blackwell's example (in which all states are instantaneous), do not understand what the minimal process is, or do not know what happens after explosion, dive right in. But beware lest you are enchanted: 'there are more spells than your commonplace magicians ever dreamed of.'

Adam Bobrowski is Professor and Chairman of the Department of Mathematics at Lublin University of Technology, Poland. He is a pure mathematician who uses the language of operator semigroups to describe stochastic processes. He has authored and coauthored nearly 70 papers on the subject and five books, including *Functional Analysis for Probability and Stochastic Processes* (2005) and *Convergence of One-Parameter Operator Semigroups* (2016).

CAMBRIDGE STUDIES IN ADVANCED MATHEMATICS

All the titles listed below can be obtained from good booksellers or from Cambridge University Press. For a complete series listing, visit www.cambridge.org/mathematics.

Already Published

150 P. Mattila *Fourier Analysis and Hausdorff Dimension*
151 M. Viana & K. Oliveira *Foundations of Ergodic Theory*
152 V. I. Paulsen & M. Raghupathi *An Introduction to the Theory of Reproducing Kernel Hilbert Spaces*
153 R. Beals & R. Wong *Special Functions and Orthogonal Polynomials*
154 V. Jurdjevic *Optimal Control and Geometry: Integrable Systems*
155 G. Pisier *Martingales in Banach Spaces*
156 C. T. C. Wall *Differential Topology*
157 J. C. Robinson, J. L. Rodrigo & W. Sadowski *The Three-Dimensional Navier–Stokes Equations*
158 D. Huybrechts *Lectures on K3 Surfaces*
159 H. Matsumoto & S. Taniguchi *Stochastic Analysis*
160 A. Borodin & G. Olshanski *Representations of the Infinite Symmetric Group*
161 P. Webb *Finite Group Representations for the Pure Mathematician*
162 C. J. Bishop & Y. Peres *Fractals in Probability and Analysis*
163 A. Bovier *Gaussian Processes on Trees*
164 P. Schneider *Galois Representations and (ϕ, Γ) Modules*
165 P. Gille & T. Szamuely *Central Simple Algebras and Galois Cohomology* (*2nd Edition*)
166 D. Li & H. Queffelec *Introduction to Banach Spaces, I*
167 D. Li & H. Queffelec *Introduction to Banach Spaces, II*
168 J. Carlson, S. Müller-Stach & C. Peters *Period Mappings and Period Domains* (*2nd Edition*)
169 J. M. Landsberg *Geometry and Complexity Theory*
170 J. S. Milne *Algebraic Groups*
171 J. Gough & J. Kupsch *Quantum Fields and Processes*
172 T. Ceccherini-Silberstein, F. Scarabotti & F. Tolli *Discrete Harmonic Analysis*
173 P. Garrett *Modern Analysis of Automorphic Forms by Example, I*
174 P. Garrett *Modern Analysis of Automorphic Forms by Example, II*
175 G. Navarro *Character Theory and the McKay Conjecture*
176 P. Fleig, H. P. A. Gustafsson, A. Kleinschmidt & D. Persson *Eisenstein Series and Automorphic Representations*
177 E. Peterson *Formal Geometry and Bordism Operators*
178 A. Ogus *Lectures on Logarithmic Algebraic Geometry*
179 N. Nikolski *Hardy Spaces*
180 D.-C. Cisinski *Higher Categories and Homotopical Algebra*
181 A. Agrachev, D. Barilari & U. Boscain *A Comprehensive Introduction to Sub-Riemannian Geometry*
182 N. Nikolski *Toeplitz Matrices and Operators*
183 A. Yekutieli *Derived Categories*
184 C. Demeter *Fourier Restriction, Decoupling and Applications*
185 D. Barnes & C. Roitzheim *Foundations of Stable Homotopy Theory*
186 V. Vasyunin & A. Volberg *The Bellman Function Technique in Harmonic Analysis*
187 M. Geck & G. Malle *The Character Theory of Finite Groups of Lie Type*
188 B. Richter *Category Theory for Homotopy Theory*
189 R. Willett & G. Yu *Higher Index Theory*

'Fascinating Markov Chains' by Marek Bobrowski. In a different version of this cartoon by W. Chojnacki, the role of the anonymous examiner is played by an antagonist of A. A. Markov, that is, Pavel Niekrasov, who says, 'Contrary to my beliefs, these chains are really fascinating!' See [50] for the entire story.

Generators of Markov Chains

From a Walk in the Interior to a Dance on the Boundary

ADAM BOBROWSKI
Lublin University of Technology

CAMBRIDGE
UNIVERSITY PRESS

University Printing House, Cambridge CB2 8BS, United Kingdom

One Liberty Plaza, 20th Floor, New York, NY 10006, USA

477 Williamstown Road, Port Melbourne, VIC 3207, Australia

314–321, 3rd Floor, Plot 3, Splendor Forum, Jasola District Centre, New Delhi – 110025, India

79 Anson Road, #06–04/06, Singapore 079906

Cambridge University Press is part of the University of Cambridge.
It furthers the University's mission by disseminating knowledge in the pursuit of education, learning, and research at the highest international levels of excellence.

www.cambridge.org
Information on this title: www.cambridge.org/9781108495790
DOI: 10.1017/9781108863070

First published 2021

A catalogue record for this publication is available from the British Library.

ISBN 978-1-108-49579-0 Hardback

To my dancing Beatka

I met a man once . . . to whom Heyst exclaimed, in no connection with anything in particular (it was in the billiard-room of the club): 'I am enchanted with these islands!' He shot it out suddenly, a propos des bottes, as the French say, and while chalking his cue. And perhaps it was some sort of enchantment. There are more spells than your commonplace magicians ever dreamed of.

J. Conrad (J. T. K. Korzeniowski), *Victory*

Nie sprawiłeś mi zawodu, synu. Przeciwnie, zadziwiłeś mnie. Zdołałeś dać z siebie więcej, niżem od ciebie oczekiwał.

Teodor Parnicki, *Srebrne orły*

Markov chains merely walk in their regular state space, but on the cliffs of their boundaries, they dance.

Johann Gottfried von Spacerniak

Contents

Preface

Application to business is the root of prosperity, but those who ask questions that do not concern them are steering the ship of folly towards the rock of indigence.
Arsheesh the greedy fisherman in C. S. Lewis's *The Horse and His Boy*

The theory of Markov chains, whether time discrete or time continuous, is one of the integral parts of the theory of stochastic processes. This book, however, is not devoted to the popular part of this rich theory, so the reader will not learn here about recurrent and transient states, ergodic theorems, or convergence to equilibrium. (In Arsheesh's words, thus, we will ask questions that do not concern us, having no business in mind, in the hope that we will somehow reach Narnia and the North.) Instead, we focus on the equally intriguing question of how a continuous-time Markov chain may be described by means of its Kolmogorov (intensity) matrix or its generator, and we study the interplay between the notions just named. We argue in particular that, despite their popularity, Kolmogorov (intensity) matrices are less suitable for such description than generators. Whereas, in their relative simplicity, they allow an intuitive formulation of processes, in general, they fail to describe more delicate phenomena.

Therefore, in Chapter 2, we compare these two notions in the light of two examples due to Kolmogorov, Kendall, and Reuter. These examples show that whereas the intensity matrix determines in a sense the way the generator acts, it may not determine the generator's domain, and without information on the shape of the domain, a Markov chain is not completely specified. Furthermore, in Chapters 3 and 4, devoted to boundary theory, we show that an explosive intensity matrix characterizes the chain only locally, up to a time of explosion. Put otherwise, the matrix characterizes merely the minimal chain, which after explosion is undefined. There are, however, infinitely many postexplosion processes that dominate the minimal chain. Their generators may differ from the generator of the minimal chain by extra terms and may have different domains; both the domain and the terms contain crucial information on the postexplosion process; by nature, this information cannot be found in the intensity matrix. Chapter 5 presents a similar view from the dual perspective.

* * *

The way information on boundary behavior of a Markov chain is reflected in its generator is a recurring subject in this book. It transpires that in l^1, the exit boundary introduces additional terms in the generator, whereas the entrance boundary affects its domain. But in l^∞, these things are turned upside down: roughly, the exit boundary always perturbs the domain, but the entrance boundary may either introduce new terms or perturb the domain. The fact that in passing from a space to its dual a perturbation of the way a generator acts may become a perturbation of the domain, and vice versa, has been observed also in other contexts (see, e.g., [17] or Chapter 50 in [16]), but in the case of Markov chains, this phenomenon seems to be particularly intriguing and spectacular.

* * *

Another idea, perhaps borrowed from [39], that permeates the book is that instead of describing extraordinary Markov chains by rigorous, but involved, stochastic constructions, stochastic intuitions may be developed by approximating this chain's semigroup of transition probabilities by a sequence of semigroups of transition probabilities related to some finite-space chains. (The theory of semigroups of operators and suitable convergence theorems are presented in Chapter 1.) Since the latter chains have considerably simpler, well-understood structure, this idea works well. This is in particular how an insight is gained into Blackwell's and both Kolmogorov–Kendall–Reuter examples. This is also how the discrete boundary for an explosive Markov chain is introduced.

* * *

Yet another characteristic of the book is that we (the reader and I) allow ourselves the comfort of discovering new results gradually, step by step, not trying to reach the mountain peak immediately via the most efficient route. The proof of an introductory result may thus be more complicated than that of a general theorem, presented later; in proving the former, we are simply yet not so clear about the general view. Neither are we afraid to spend some time looking at an illustrative example, which perhaps involves more calculations than one could wish to go through, before the idea of a general theorem following it dawns on us.

For instance, it takes several pages of investigating a particular (pure birth) Markov chain before formula (3.36) is discovered, and yet another couple of pages before an analogous formula (3.39) is derived. It is only then that the much more general formula (3.55), built by analogy to (3.36) and (3.39), is discussed, and the proof that the operator in (3.55) is a Markov generator is less than one page long (see also Section 3.5.5). Similarly, the two-pages-long proof of the master theorem of Section 3.7.9 is preceded by the more than four-pages-long discussion of a particular 'two infinite ladder' example of a Markov chain (see Sections 3.5.6–3.5.9).

I still think the intuitive should go before the abstract and that discovering is more fun than learning: Sections 3.5.5 and 3.7.9 occupy a special place in this book and are worth being understood thoroughly. Besides, the general idea is to enjoy ourselves as much as possible.

* * *

A couple of words are due on the way boundary theory for Markov chains is presented in this book. The primary sources of information on this theory are W. Feller [41, 42] and K. L. Chung [22]. My presentation is closer to Feller's, and draws on his heavily, but the two still differ significantly. First of all, Feller focuses on the Laplace transform of transition probabilities (of postexplosion processes), that is, on the resolvents of generators, whereas – at least to my taste – description of generators themselves is more appealing. In characterizing the latter, certain functionals show up naturally as building blocks for the generators, and thus it is the set of these functionals that is defined as the boundary. It is only later, as a sort of afterthought, that I identify these functionals as sojourn solutions for the related minimal chain. For Feller it was the latter that were the starting point of his investigations.

* * *

Besides one place in Chapter 5 (Section 5.8.4), the text is self-contained, but some basic knowledge of real analysis (including Laplace transform), probability theory, and functional analysis is needed for its understanding. I assumed the reader is familiar with random variables, Banach spaces, linear operators and their norms, and so on. Nevertheless, only elementary properties of these objects are used without explanation: I have made a definite effort to make the text as reader-friendly as possible.

* * *

I would like to thank the members of the Lublin University of Technology mathematical research seminar (Adam Gregosiewicz, Andrzej Łagodowski, Iwona Malinowska, Małgorzata Murat, Ernest Nieznaj, Elżbieta Ratajczyk, Renata Rososzczuk, Katarzyna Steliga, and Łukasz Stępień) for their careful study of early drafts of this book, in their various forms, and for spotting a number of places that needed smaller and greater corrections. Tomasz Szarek also read through the entire manuscript, in just two months, and suggested several improvements; I am grateful to him for this. A number of invaluable corrections I owe, furthermore, to Marta Tyran–Kamińska. Special thanks go to Jacek Banasiak, Wojciech Chojnacki, Ryszard Rudnicki, and Yuri Tomilov for their encouragement and advice. It was by reading N. H. Bingham's interview with David Kendall [12], suggested to me by Yura, that I became aware of Kendall and Reuter's work on Markov chains, dots started to connect, and the idea of the present book dawned on me. I am also indebted to my physicians, in particular, to Drs. Andrzej Kwiecień, Krzysztof Paprota, and Justyna Biegańska-Siczek for keeping me alive (and kicking) while I was writing this book. Finally, I would like to thank Clare Dennison and Roger Astley from Cambridge University Press for their professional support and advice before and during the process of writing and producing this book.

A Nontechnical Introduction

Markov chains are one of the simplest stochastic processes. In the archetypical example of a Markov chain, that is, in the (symmetric) simple random walk on the set of integers, a traveler or a particle starting at i makes a step to the right, to $i + 1$, with probability $\frac{1}{2}$ or a step to the left, to $i - 1$, with the same probability. At its new position (either $i - 1$ or $i + 1$), it continues in the same fashion, forgetting the past and the way it reached this position.

Amazingly, many quantitative results may be proved about such apparently completely chaotic movement. For example, a particle starting at any point will surely reach 0 at a certain time in the future. What is probably less obvious is that the same is true about an analogous random walk in two dimensions, where a particle may go to the left, to the right, up, or down. However, in three and more directions, the situation is quite different: the probability of reaching 0 from a nonzero state i is smaller than 1, and so is the probability of reaching any state different from i (see, e.g., [35]).

What distinguishes Markov chains from other, more complex stochastic processes is the state-space, that is, the set of possible states. For Markov chains, the state-space is by definition denumerable: it is either finite or there is a one-to-one correspondence between its elements and the elements of the set of positive integers. For example, the set of integers may be arranged in a sequence, and so may be the state-space of the simple random walk in any dimension.

Of course, as a result of such a rearrangement, natural neighbors of a point will probably lie quite a distance apart from each other and from the point, and the description of the walk will become less intuitive. But we are naturally led to a more general object: in a chain that is more general than the simple random walk, a particle starting at i may go to any other point j of the state-space (or stay in i), and it does that with probability characteristic to that point. In fact, that probability, denoted $p_{i,j}$ (and termed 'transition probability'), in general depends on both i and j. The only restriction is that in the honest Markov chains, the sum of $p_{i,j}$ over all j is 1 for all i.

In this much more general setting, still much may be said of the nature of the process: for example, under reasonable assumptions on the probabilities $p_{i,j}$, existence of stationary distributions may be proved. (Informally speaking, stationary distributions are arrangements of collections of a large number of particles with the property that a random displacement of these particles according to the rules provided by $p_{i,j}$'s, though changing positions of many or all individual particles, does not change the arrangement as a whole.) Conditions are also known under which any other arrangement of particles, with displacement rules hidden in $p_{i,j}$, will in time become more and more close to the stationary distribution, even to the point of being practically indistinguishable, and we can even estimate the speed with which this happens [72, 78].

Everything said above concerns Markov chains in discrete time: in a discrete-time Markov chain, as described above, all changes occur at multiples of a unit time. However, there are also time-continuous Markov chains. Whereas in a discrete-time Markov chain, everything hinges on the probabilities $p_{i,j}$, to describe a continuous-time Markov chain, one needs intensities of jumps. In the simplest case of right-continuous processes with lefthand limits, if the intensity of a jump from a state i is $q_i > 0$, the process stays at i for an exponential time T so that $\mathbb{P}(T > t) = \mathrm{e}^{-q_i t}, t \geq 0$. (If $q_i = 0$, the process stays at i forever.) Then, it jumps to a different state j, and the probability that a particular j is chosen is $p_{i,j} = q_{i,j}/q_i$, where $q_{i,j}$ is a (given) intensity of jump from i to j. The condition that the sum of $p_{i,j}$ over all j should be 1 tells us that we should have $\sum_{j \neq i} q_{i,j} = q_i$. At j, the process forgets its past and continues the same procedure, with q_i replaced by q_j.

Even from this picture, it is somewhat clear that continuous-time Markov chains are in many aspects similar to discrete-time Markov chains. In the case where the state-space is finite, they indeed are much the same, and the description in terms of intensities is both appealing and useful. If the state-space is infinite (but denumerable), however, unexpected difficulties and curious phenomena might occur. First of all, some and even all states might be instantaneous, which is to say that all intensities might be infinite. Upon arriving at such a state, the process leaves 'immediately,' and the description given above does not apply, or at least does not tell the entire story. Moreover, if worst comes to worst and the times before consecutive jumps are dramatically shorter and shorter, the definition in terms of jumps may turn out inadequate, leaving the process undefined after a finite random time (being equal to the infinite sum of shorter and shorter times before consecutive jumps). In other words, intensities of jumps do not contain the information on what happens with the process after such 'explosion.' On top of that, there can be many ways such explosion occurs and many, trivial and nontrivial, ways the process returns to the state-space.[1]

In such cases, to describe the process in full, instead of the set, or a matrix, of intensities, one needs a more complex object: a closed operator in the Banach space of absolutely summable sequences, the generator of a related semigroup of transition matrices in this space. In this book, we discuss when and in what sense such generators may be identified with matrices of jump intensities and how they can be constructed and employed when intensity matrices are seen to be less useful. In particular, we explain how the information on postexplosion behavior is built into the generators. Roughly speaking, this is done either by modifying the way the generator of the minimal chain acts (the minimal chain is the chain undefined after explosion) or by modifying the generator's domain.

[1] Each way explosion occurs is a set (or, better, an equivalence class of sets) where the process 'sojourns,' and each such sojourn set may be seen as an additional point of the state-space for the process, a point of its exit boundary. Similarly, each nontrivial way the process may return to the state-space is a point of its entry boundary.

1

A Guided Tour through the Land of Operator Semigroups

1.1 Semigroups and Generators

This introductory chapter is meant as a gentle, short introduction to the theory of operator semigroups. From the beginning, this theory has been devised and seen, especially by W. Feller and E. B. Dynkin, as an efficient tool for describing Markov processes. Today – notwithstanding advances of stochastic analysis we have witnessed over the last several decades – its usefulness is still undeniable.

We present all the relevant facts of the theory, needed for understanding the main body of the book, but – for brevity – refrain from presenting detailed proofs of the main theorems on generation, perturbation, approximation and convergence of semigroups. (The almost self-contained Section 1.5, devoted to sun-dual semigroups, is an exception to this rule.) Instead, we explain all the needed notions and illustrate the theory with simple, but illuminating examples. We hope that the reader who has not come across the theory of semigroups of operators yet, and decides to take this guided tour will be well prepared for studying the rest of the book, and, if necessary, to study more advanced books on semigroups. A list of monographs containing all the missing proofs – and much more – is given in Section 1.7.

Here and there in this chapter, we will be guided by intuitions from stochastic processes and Markov chains in particular, even though the latter will not be defined before Chapter 2.

1.1.1 *Strongly continuous semigroups of operators*

A C_0 semigroup or a **strongly continuous semigroup** in a Banach space $\mathbb{X}$ is a family $T = \{T(t), t \geq 0\}$ (written also as $(T(t))_{t\geq 0}$ or $\{T_t, t \geq 0\}$) of bounded linear operators such that

(a) $T(t)T(s) = T(t+s)$, $t, s \geq 0$,
(b) $T(0) = I_{\mathbb{X}}$ (the identity operator in $\mathbb{X}$),
(c) $\lim_{t \to 0} T(t)x = x$, $x \in \mathbb{X}$,

with the last limit in the norm of $\mathbb{X}$. Condition (a) is termed the **semigroup property**. If merely the first two conditions are satisfied, the family T is said to be a **semigroup**.

The following intuition is hidden behind this notion. Throughout large parts of this book, $\mathbb{X}$ will be the space

$$l^1 = l^1(\mathbb{I})$$

of absolutely summable sequences $x = (\xi_i)_{i \in \mathbb{I}}$ indexed by elements of a countable set $\mathbb{I}$. (Recall that the norm in this space is $\|x\| = \sum_{i \in \mathbb{I}} |\xi_i|$.) Nonnegative elements of l^1 with coordinates summing to 1 can be thought of as initial distributions of an underlying Markov chain, and then, in many cases of interest to us, $T(t)$ can be interpreted as mapping an initial distribution of the chain to its distribution at time t. Then point (a) expresses Markovian nature of the chain, and points (b)–(c) are an assumption of continuous dependence on initial data (see Chapter 2 for details).

1.1.2 *An example of a strongly continuous semigroup*

Let r_i, $i \geq 2$ be positive numbers. For any $t \geq 0$, the formula

$$T(t)\,(\xi_i)_{i \geq 1} = (\xi_1 + \sum_{j=2}^{\infty} \xi_j (1 - \mathrm{e}^{-r_j t}), \xi_2 \mathrm{e}^{-r_2 t}, \xi_3 \mathrm{e}^{-r_3 t}, \ldots) \tag{1.1}$$

defines a linear operator in $l^1 := l^1(\mathbb{N})$. This operator is bounded with norm not exceeding 1, since

$$\|T(t)x\| \leq |\xi_1| + \sum_{j=2}^{\infty} |\xi_j|(1 - \mathrm{e}^{-r_j t}) + \sum_{i=2}^{\infty} |\xi_i| \mathrm{e}^{-r_i t} = \sum_{i=1}^{\infty} |\xi_i| = \|x\|.$$

This norm in fact equals 1, because for nonnegative $x \in l^1$ the inequality in the calculation presented above may be replaced by the equality. It is also easy to see, and the reader should check it, that the semigroup property holds.

We claim that $\{T(t), t \geq 0\}$ is a strongly continuous semigroup. Indeed, for any $x \in l^1$,

$$\|T(t)x - x\| \leq 2 \sum_{j=2}^{\infty} |\xi_j|(1 - \mathrm{e}^{-r_j t}),$$

and the right-hand side converges to 0, as $t \to 0+$ by the Lebesgue Dominated Convergence Theorem (see, e.g., Section 1.6.1). Alternatively, for a nonzero x, and given $\epsilon > 0$, one may find $i_0 \in \mathbb{N}$ such that $\sum_{i=i_0+1}^{\infty} |\xi_i| < \frac{\epsilon}{4}$, and then a $t_0 > 0$ such that $\sup_{i=1,\dots,i_0}(1 - \mathrm{e}^{-r_i t}) < \frac{\epsilon}{4\|x\|}$ for all $t \in [0, t_0]$. For such t,

$$2\sum_{j=2}^{\infty} |\xi_j|(1 - \mathrm{e}^{-r_j t}) \le 2\sum_{j=2}^{i_0} |\xi_j|(1 - \mathrm{e}^{-r_j t}) + 2\sum_{j=i_0+1}^{\infty} |\xi_j|(1 - \mathrm{e}^{-r_j t})$$

$$\le \frac{\epsilon}{2\|x\|}\sum_{j=2}^{i_0} |\xi_j| + 2\sum_{j=i_0+1}^{\infty} |\xi_j| \le \frac{\epsilon}{2} + \frac{\epsilon}{2} = \epsilon.$$

This proves the claim.

To gain some intuition about what the semigroup defined above describes, consider the following stochastic movement with state-space $\mathbb{N}$: a traveler starting at an $i \ge 2$ stays there for an exponential time with parameter r_i (so that the probability that the traveler is still at i at time t is $\mathrm{e}^{-r_i t}$) and then jumps to the state 1 to stay there for ever. A traveler starting at 1 simply stays there for ever. If ξ_i is the probability that a traveler starts at i, then the ith coordinate of the vector $T(t)\,(\xi_i)_{i\ge 1}$ is the probability that the traveler will be at i at time $t \ge 0$.

1.1.3 *A semigroup that is not strongly continuous*

Let (1.1) be modified so that the first coordinate of the vector on the right-hand side is ξ_1. Then, $T(t)$ is again a linear operator with norm 1 (corresponding to stochastic movement in which a traveler starting at $i \ge 2$ after an exponential time with parameter r_i disappears). The so-modified family $\{T(t), t \ge 0\}$ may be checked to be a strongly continuous semigroup.

The same (modified) formula defines also a semigroup of operators of norm 1 in the space l^∞ of bounded sequences $(\xi_i)_{i\ge 1}$. (Here, the norm is given by $\|x\| = \sup_{i\in\mathbb{I}} |\xi_i|$.) However, if $\sup_{i\ge 2} r_i = \infty$, this semigroup is not strongly continuous. To see this, consider the vector $x = (1, 1, 1, \dots)$. Then, for each $t > 0$,

$$\|T(t)x - x\| = \|(0, 1 - \mathrm{e}^{-r_2 t}, 1 - \mathrm{e}^{-r_3 t}, \dots)\| = \sup_{i\ge 2} |1 - \mathrm{e}^{-r_i t}| = 1.$$

Hence, $T(t)x$ cannot converge to x in the norm of l^∞, as $t \to 0$, showing that $\{T(t), t \ge 0\}$ is not a strongly continuous semigroup in l^∞.

1.1.4 *Continuity*

A closer inspection of Example 1.1.2 shows that the map $[0, \infty) \ni t \mapsto T(t)x$ is continuous in the sense of the norm in l^1 for all $x \in l^1$. This is not a coincidence: it can be shown that this is the property of all C_0 semigroups. In other words, assumptions (a)–(c) of the definition of a strongly continuous semigroup imply that the functions $[0, \infty) \ni t \mapsto T(t)x$, sometimes termed trajectories of the semigroup, are continuous for all $x \in \mathbb{X}$.

1.1.5 *The generator*

Closed form expressions for semigroups of operators are seldom available. But, fortunately, the semigroups can be fully described by their 'derivates,' that is, certain operators, called generators. The situation is somewhat similar to the fact that even if we know initial position $s(0)$ and an exact formula for velocity $v(t)$ of a moving object at any time $t \geq 0$ in terms of elementary functions, a closed form for its position $s(t)$ at time $t \geq 0$ in terms of such functions may be hard to find, or in fact may not exist (take, e.g., $v(t) = \mathrm{e}^{-t^2}$). This is despite the fact that the relation between velocity and position is very well known and simple: $s(t) = s(0) + \int_0^t v(t')\,\mathrm{d}t'$. In terms of Markov chains, the semigroup contains all the information on transition probabilities, whereas its generator gathers the information on transition rates (and more).

Formally, the infinitesimal generator (or simply: the generator) of $(T(t))_{t\geq 0}$ is defined by

$$\boxed{Ax = \lim_{t\to 0+} t^{-1}(T(t)x - x)}$$

for those x for which the limit exists in the sense of the norm in $\mathbb{X}$.

The second part of the previous sentence is important: in situations of interest, the limit presented above rarely exists for all $x \in \mathbb{X}$. If it does, A turns out to be bounded, and the semigroup $\{T(t), t \geq 0\}$ can be recovered from A by means of the following formula:

$$T(t) = \sum_{n=0}^{\infty} \frac{t^n A^n}{n!}, \qquad t \geq 0; \tag{1.2}$$

the series here converges in the operator norm. The semigroup generated by a bounded operator A will be denoted $\{\mathrm{e}^{tA}, t \geq 0\}$, and referred to as an exponent of A.

Let's look at some examples; we will come back to our discussion of generators in Section 1.1.11.

1.1.6 *An example of a generator*

Bounded linear operators in $\mathbb{R}^3$ may be identified with 3×3 matrices. In this sense, the formula

$$T(t) = \frac{1}{5}\begin{bmatrix} 5 & 0 & 0 \\ 5 - 4e^{-t} - e^{-6t} & 2e^{-t} + 3e^{-6t} & 2e^{-t} - 2e^{-6t} \\ 5 - 6e^{-t} + e^{-6t} & 3e^{-t} - 3e^{-6t} & 3e^{-t} + 2e^{-6t} \end{bmatrix}$$

defines a semigroup of operators in $\mathbb{R}^3$; and this is regardless of whether $T(t)x$ is the product of the matrix $T(t)$ and a three dimensional column vector x or the product of a three dimensional row vector x and the matrix $T(t)$. A direct calculation shows that the limit $\lim_{t \to 0+} t^{-1}(T(t)x - x)$ exists and equals Ax, where

$$A = \begin{bmatrix} 0 & 0 & 0 \\ 2 & -4 & 2 \\ 0 & 3 & -3 \end{bmatrix},$$

for all $x \in \mathbb{R}^3$. Therefore, the generator here may be identified with the matrix A.

1.1.7 *An example of application of* (1.2)

Take $a \geq 0$ and $b \geq 0$ such that $a + b > 0$, and let $\mathbb{X}$ be $\mathbb{R}^2$ equipped with any of the equivalent norms. The space $\mathcal{L}(\mathbb{X})$ may be identified with the space of 2×2 matrices. Let's find e^{tA} for $A = \begin{pmatrix} -a, & a \\ b, & -b \end{pmatrix}$, using (1.2). We claim that

$$e^{tA} = \frac{1}{a+b}\begin{pmatrix} b + ae^{-(a+b)t}, & a - ae^{-(a+b)t} \\ b - be^{-(a+b)t}, & a + be^{-(a+b)t} \end{pmatrix}. \tag{1.3}$$

To prove this, we first introduce

$$B := A + (a+b)I_{\mathbb{X}} = \begin{pmatrix} b, & a \\ b, & a \end{pmatrix}.$$

Since $B^2 = (a+b)B$, we have, by induction, $B^n = (a+b)^{n-1}B$. It follows that

$$e^{tB} = I_{\mathbb{X}} + \frac{1}{a+b}\sum_{n=1}^{\infty} \frac{t^n (a+b)^n}{n!} B = I_{\mathbb{X}} + \frac{1}{a+b}(e^{(a+b)t} - 1)B.$$

Next, we note that, similarly as for complex numbers, $e^{t(B+C)} = e^{tB}e^{tC}$ provided $BC = CB$. Since $I_{\mathbb{X}}$ commutes with any operator (matrix), by the definition of B, we obtain

$$\begin{aligned} e^{tA} &= e^{-(a+b)t} e^{tB} = e^{-(a+b)t} I_{\mathbb{X}} + \frac{1}{a+b}(1 - e^{-(a+b)t})B \\ &= \frac{1}{a+b}(B - e^{-(a+b)t} A), \end{aligned}$$

completing the proof.

1.1.8 *Another example of application of* (1.2)

Let $\mathbb{X} = C[0, \infty]$ be the space of continuous functions on $[0, \infty)$ with limits at ∞. Equipped with the supremum norm, $\mathbb{X}$ is a Banach space, and for any $a > 0$, the operator defined by $Ax(p) = a[x(p+1) - x(p)]$, $p \geq 0$ is bounded, since $\|Ax\| \leq 2a\|x\|$. To compute e^{tA} for $t \geq 0$, we use the definition and the fact that $B := A + aI_{\mathbb{X}}$ is a scalar multiple of the shift operator: $Bx(p) = ax(p+1)$, $p \geq 0$ so that $B^n x(p) = a^n x(p+n)$, $p \geq 0$. Therefore,

$$\begin{aligned} e^{tA} x(p) &= e^{-at} e^{t(A + aI_{\mathbb{X}})} x(p) = \sum_{n=0}^{\infty} e^{-at} \frac{a^n t^n}{n!} x(p+n) \\ &= \mathbb{E}\, x(p + N(t)), \qquad t \geq 0, \end{aligned}$$

where $\mathbb{E}$ denotes expected value. In the last line $N(t)$ is a Poisson-distributed random variable with parameter at:

$$\mathbb{P}(N(t) = k) = e^{-at} \frac{(at)^k}{k!}.$$

In other words, the exponent $\{e^{tA}, t \geq 0\}$ describes a Poisson process. In this process, if the starting point is p then after time $t \geq 0$ with probability $\mathbb{P}(N(t) = k)$ the process is at $p + k$.

1.1.9 *Example: The generator of the semigroup of Section 1.1.2*

Let us come back to Example 1.1.2, and, to focus our attention, let's agree that $r_i = i, i \geq 2$. To find the generator of the semigroup (1.1), we note first that convergence in the sense of l^1 norm implies convergence of all coordinates. Therefore, if $x = (\xi_i)_{i \geq 1}$ is in the domain $\mathcal{D}(A)$ of the generator, then the ith coordinate of Ax must be $\lim_{t \to 0+} \frac{e^{-it} - 1}{t} \xi_i = -i\xi_i, i \geq 2$. Since Ax, being the limit of elements of l^1, belongs to l^1, we see that a necessary condition for x to belong to $\mathcal{D}(A)$ is that

$$\sum_{i=2}^{\infty} i|\xi_i| < \infty. \tag{1.4}$$

We will show that all x's satisfying this condition belong to $\mathcal{D}(A)$ and that for such x, Ax is equal to

$$y := \Big(\sum_{i=2}^{\infty} i\xi_i, -2\xi_2, -3\xi_3, \cdots\Big).$$

For, under assumption (1.4), y belongs to l^1. Moreover, the ith coordinate of $t^{-1}(T(t)x - x) - y$ is

$$\begin{cases} \left[\frac{e^{-it}-1}{t} + i\right]\xi_i, & i \ge 2, \\ -\sum_{j=2}^{\infty} \xi_j \left[\frac{e^{-jt}-1}{t} + j\right], & i = 1, \end{cases}$$

and we have $\frac{e^{-it}-1}{t} + i = \frac{i}{t}\int_0^t (1 - e^{-is})\,ds$. Therefore,

$$\|t^{-1}(T(t)x - x) - y\| \le 2\sum_{i=2}^{\infty} i|\xi_i|\frac{1}{t}\int_0^t (1 - e^{-is})\,ds.$$

Since $\lim_{t\to 0+} \frac{1}{t}\int_0^t (1 - e^{-is})\,ds = 0$ and $0 \le \frac{1}{t}\int_0^t (1 - e^{-is})\,ds \le 1$, $i \ge 2$, the right-hand side above converges to 0 by assumption (1.4) and the Lebesgue Dominated Convergence Theorem. (Alternatively, one may argue as in Example 1.1.2.) This shows that $x \in \mathcal{D}(A)$ and $Ax = y$.

1.1.10 *Example: Isomorphic semigroups*

Suppose that Banach spaces $\mathbb{X}$ and $\mathbb{Y}$ are isomorphic: there is a bounded linear map $I: \mathbb{Y} \to \mathbb{X}$ with bounded left and right inverse $I^{-1}: \mathbb{X} \to \mathbb{Y}$ so that $II^{-1} = I_{\mathbb{X}}$ and $I^{-1}I = I_{\mathbb{Y}}$. Suppose also that $\{T(t), t \ge 0\}$ is a strongly continuous semigroup in $\mathbb{X}$ with generator A. Then the operators

$$I^{-1}T(t)I, \quad t \ge 0$$

form a strongly continuous semigroup in $\mathbb{Y}$. Since the limit

$$\lim_{t\to 0+} \frac{I^{-1}T(t)Iy - y}{t}$$

exists iff so does

$$\lim_{t\to 0+} \frac{T(t)Iy - Iy}{t},$$

a y belongs to the domain of the infinitesimal generator, say, $A^{\diamond}$, of $\{I^{-1}T(t)I, t \ge 0\}$ iff Iy belongs to $\mathcal{D}(A)$, and then

$$A^{\diamond}y = I^{-1}AIy.$$

The semigroups $\{I^{-1}T(t)I, t \geq 0\}$ and $\{T(t), t \geq 0\}$, which of course play symmetric roles, are said to be **isomorphic** to each other, and so are said to be the generators A and $A^{\diamond}$; other authors prefer to speak of **similar semigroups**. Sometimes, one of the isomorphic semigroups/generators is easier to handle and knowing the isomorphism involved aids the analysis of the other. See, for example, Exercise 1.1.18 or Section 3.3.7.

1.1.11 *Generators are densely defined*

As already stated, the limit defining the generator rarely exists for all $x \in \mathbb{X}$; $\mathcal{D}(A)$ is rarely equal to the entire $\mathbb{X}$. In fact, $\mathcal{D}(A) = \mathbb{X}$ iff the generator is bounded, and this happens iff

$$\lim_{t \to 0+} \|T(t) - I_{\mathbb{X}}\| = 0,$$

which is a much stronger condition than assumption (c) of the definition of Section 1.1.1.

All we can say at the first inspection of $\mathcal{D}(A)$ is that it is a linear subset of $\mathbb{X}$. Fortunately, $\mathcal{D}(A)$ turns out to be always dense in $\mathbb{X}$: any $x \in \mathbb{X}$ may be approximated by elements of $\mathcal{D}(A)$. In particular, for generators of semigroups we need to search in the class of densely defined operators.

The reason for $\mathcal{D}(A)$ to be dense is as follows. Since the trajectories $t \mapsto T(t)x$ are continuous functions of t, one may think of Riemann-type integrals $\int_a^b T(t)x \, \mathrm{d}t$ for any $0 \leq a < b$. These are defined in the same way as integrals of real-valued functions, as limits of approximating Riemann sums, and possess similar properties. For example, continuity of a function $f : [a, b] \to \mathbb{X}$ guarantees that the integral $\int_a^b f(t) \, \mathrm{d}t$ is well defined, provided $\mathbb{X}$ is a Banach space (see, e.g., [14], pp. 60–63), and

$$\left\| \int_a^b f(t) \, \mathrm{d}t \right\| \leq \int_a^b \|f(t)\| \, \mathrm{d}t. \tag{1.5}$$

(This estimate will be found useful later.) Therefore, one may think of $\int_0^h T(t)x \, \mathrm{d}t$ for any $h > 0$ and $x \in \mathbb{X}$, and it can be seen that such elements belong to $\mathcal{D}(A)$ (with $A \int_0^h T(t)x \, \mathrm{d}t = T(t)x - x$). Moreover, it is a well-known property of Riemann integrals that for a continuous function f on an interval $[a, b]$,

$$\lim_{h \to 0+} h^{-1} \int_s^{s+h} f(t) \, \mathrm{d}t = f(s), \qquad s \in [a, b),$$

and Riemann integrals of vector-valued functions also possess this property. Applying this to $f(t) = T(t)x$ and $s = 0$, we obtain

$$\lim_{h\to 0+} h^{-1}\int_0^h T(t)x\,\mathrm{d}t = x.$$

This however means that any $x \in \mathbb{X}$ may be approximated by elements of $\mathcal{D}(A)$.

1.1.12 *Generators are closed*

Generators of semigroups, though generally unbounded, are closed. By definition this means that conditions

$$x_n \in \mathcal{D}(A), n \geq 1, \quad \lim_{n\to\infty} x_n = x \quad \text{and} \quad \lim_{n\to\infty} Ax_n = y$$

imply that x belongs to $\mathcal{D}(A)$ and $Ax = y$. In other words, the graph of A, defined as the following subset of the Banach space $\mathbb{X} \times \mathbb{X}$:

$$\mathbb{G}_A = \{(x, y) \in \mathbb{X} \times \mathbb{X}; x \in \mathcal{D}(A), y = Ax\}$$

is closed in $\mathbb{X} \times \mathbb{X}$.

For example, it may be checked directly that the generator we calculated in Section 1.1.9 is closed. That the latter operator is not bounded (in the sense that there is no M such that $\|Ax\| \leq M\|x\|$ for $x \in \mathcal{D}(A)$) may be seen by considering the vectors $e_i \in l^1$ which at the ith coordinate have 1 and are composed of 0's otherwise: since $Ae_i = -ie_i$, which implies $\|Ae_i\| = i\|e_i\|$, the hypothetical constant M would need to be larger than all $i \in \mathbb{N}$.

Even though the notion of closedness is very important, in searching for generators of semigroups one rarely needs to prove directly that a candidate for a generator is closed. This is because, even if A is already proved to be densely defined and closed, before one can claim that A is a generator, another condition, discussed in the next section (Section 1.2), needs to be checked, and operators satisfying this condition are automatically closed.

1.1.13 *Closability*

Analysis of generators of Markov chains often involves closable operators. To recall, a linear operator A in a Banach space $\mathbb{X}$ is said to be closable if conditions

$$x_n \in \mathcal{D}(A), n \geq 1 \quad \lim_{n\to\infty} x_n = 0 \quad \text{and} \quad \lim_{n\to\infty} Ax_n = y$$

imply $y = 0$. When combined with linearity this condition means that the closure of the graph of A in the norm of $\mathbb{X} \times \mathbb{X}$ is still a graph (of another operator). The latter, 'larger' operator, termed the closure of A and denoted $\overline{A}$, is defined by

$$\overline{A}x = \lim_{n\to\infty} Ax_n$$

on the domain composed of $x \in \mathbb{X}$ for which there are $x_n \in \mathcal{D}(A)$ such that $\lim_{n\to\infty} x_n = x$ and the limit $\lim_{n\to\infty} Ax_n$ exists; the definition of $\overline{A}x$ does not depend on the choice of $(x_n)_{n\ge1}$ precisely because A is closable (cf. point 2 of the proof presented in Section 5.4.7).

For example, suppose that there is a closed operator B such that

$$\mathbb{G}_A \subset \mathbb{G}_B.$$

Then A is automatically closable, and $\overline{A}$ is simply the restriction of B to $\mathcal{D}(\overline{A})$. In fact, $\overline{A}$ (if it exists) is the smallest (in the sense of inclusion of graphs) closed operator extending A.

1.1.14 *Cores*

Sometimes one faces a situation that is in a sense 'inverse' to the one discussed above: a closed operator A and a subset $\mathcal{D}$ of its domain are given, and the question is whether for each $x \in \mathcal{D}(A)$ there is a sequence $(x_n)_{n\ge1}$ of elements of $\mathcal{D}$ such that $\lim_{n\to\infty} x_n = x$ and $\lim_{n\to\infty} Ax_n = Ax$. If this is the case, $\mathcal{D}$ is said to be a core for A. Since the graph of $A_{|\mathcal{D}}$ (of the operator A restricted to $\mathcal{D}$) is a subset of $\mathbb{G}_A$, closability of $A_{|\mathcal{D}}$ is not an issue here: instead, we would like to know whether $\mathcal{D}$ is 'large enough' so that A may be 'recovered' from $A_{|\mathcal{D}}$. Such information is vital in the situations in which it is hard to describe the entire $\mathcal{D}(A)$ but a manageable description of $\mathcal{D}$ is available. As an easy exercise, the reader may check to see that the set of linear combinations of basis vectors is a core for the generator A of Section 1.1.9.

Before closing this section, we need to say a word about connection between semigroups and Cauchy problems.

1.1.15 *Semigroups and Cauchy problems*

As commented in 1.1.4, the assumption of continuity of $t \mapsto T(t)x$, at $t = 0$ for all $x \in \mathbb{X}$ implies, via the semigroup property, that $t \mapsto T(t)x$ is continuous at all $t \ge 0$. Similarly, if x belongs to $\mathcal{D}(A)$ then, by definition, $t \mapsto T(t)x$ is differentiable at $t = 0$ with (right-hand) derivative equal Ax. The semigroup property allows extending this attribute of $t \mapsto T(t)x$ to the entire half-line: it can be shown that this map is continuously differentiable there, that $T(t)x$ belongs to $\mathcal{D}(A)$ for all $t \ge 0$, and that

$$\frac{\mathrm{d}}{\mathrm{d}t}(T(t)x) = AT(t)x = T(t)Ax, \qquad t \ge 0.$$

In other words, for $x \in \mathcal{D}(A)$, the abstract Cauchy problem

$$u'(t) = Au(t), \qquad t \geq 0, u(0) = x$$

is well posed with solution $u(t) = T(t)x$ which, furthermore, may be proved to be unique.

For example, for any $(\xi_1, \xi_2) \in \mathbb{R}^2$, the pair

$$(u_1(t), u_2(t)) := (\xi_1, \xi_2) \cdot \mathrm{e}^{tA},$$

where e^{tA} is defined in (1.3), solves the Cauchy problem

$$\begin{aligned} u_1'(t) &= -au_1(t) + bu_2(t), \\ u_2'(t) &= au_1(t) - bu_2(t), \qquad t \geq 0, \end{aligned}$$

with initial condition $u_1(0) = \xi_1, u_2(0) = \xi_2$. Similarly, see 1.1.8, for any $x \in C[0, \infty]$, $u(t, p) := \mathbb{E}\, x(p + N(t))$ solves

$$\frac{\partial u(t, p)}{\partial t} = a[u(t, p + 1) - u(t, p)], \qquad p, t \geq 0$$

with initial condition $u(0, p) = x(p)$.

1.1.16 Exercise

Check directly that the generator calculated in Section 1.1.9 is closed.

1.1.17 Exercise

Let $\{T(t), t \geq 0\}$ be a strongly continuous semigroup with generator A. For a given number $a > 0$, let $S(t) = T(at)$. Check that $\{S(t), t \geq 0\}$ is a strongly continuous semigroup, and that its generator, say, B, is related to A as follows: $x \in \mathcal{D}(B)$ iff $x \in \mathcal{D}(A)$ and $Bx = aAx$. For a far-reaching generalization of this result see the Volkonskii Formula (Sections 5.5.6–5.5.9).

1.1.18 Exercise

(Example of isomorphic semigroups.)

(a) Let $\mathbb{X} = C[0, \infty]$ be the space of continuous functions on $[0, \infty)$ with limits at infinity, equipped with the usual supremum norm. Check that

$$T(t)x(\tau) = x(\tau + t)$$

defines a strongly continuous semigroup of operators in $\mathbb{X}$, that the domain of its generator A is composed of continuously differentiable functions x such that $x' \in \mathbb{X}$ (deduce that $\lim_{\tau\to\infty} x'(\tau) = 0$) and that $Ax = x'$.

(b) Check that the space $\mathbb{Y} = C[0, 1]$ of continuous functions on $[0, 1]$, equipped with the usual supremum norm is isomorphic to $C[0, \infty]$ with $I\colon C[0, 1] \to C[0, \infty]$ given by

$$Ix(\tau) = x(\mathrm{e}^{-\tau}), \qquad \tau \geq 0.$$

Check also that the isomorphic image of the semigroup of point (a) is

$$I^{-1}T(t)Ix(\tau) = x(\tau\mathrm{e}^{-t})$$

and characterize the generator of this image.

1.1.19 Exercise

Suppose that $\mathbb{X}_1$ is a subspace of a Banach space $\mathbb{X}$. Let $\{T_t, t \geq 0\}$ be a strongly continuous semigroup of linear operators with generator A, such that $T_t\mathbb{X}_1 \subset \mathbb{X}_1$ (this is to say that $\mathbb{X}_1$ is invariant for $\{T_t, t \geq 0\}$: $T_tx \in \mathbb{X}_1$ provided $x \in \mathbb{X}_1$). Prove that $\{S_t, t \geq 0\}$ where $S_t = (T_t)_{|\mathbb{X}_1}$ is the restriction of $\{T_t, t \geq 0\}$ to $\mathbb{X}_1$ is a strongly continuous semigroup of operators in the Banach space $\mathbb{X}_1$, with the generator B given by

$$\mathcal{D}(B) = \mathcal{D}(A) \cap \mathbb{X}_1, \qquad Bx = Ax, x \in \mathcal{D}(B).$$

(The so-defined B is called the **part of** A in $\mathbb{X}_1$.)

1.2 The Hille–Yosida Theorem

1.2.1 *Resolvent equation*

It can be shown that for a strongly continuous semigroup $\{T(t), t \geq 0\}$ there are constants $\omega \in \mathbb{R}$ and $M \geq 1$ such that $\|T(t)\| \leq M\mathrm{e}^{\omega t}$, $t \geq 0$. Since nearly all semigroups considered in this book are composed of contractions:

$$\|T(t)\| \leq 1, \qquad t \geq 0, \tag{1.6}$$

from that time on we focus on such semigroups. This will allow us to avoid unnecessary technical complications.

The condition that really decides whether a densely defined operator A is a generator, is related to the so-called **resolvent equation** (for A):

$$\boxed{\lambda x - Ax = y.} \tag{1.7}$$

In this equation, $\lambda > 0$ and $y \in \mathbb{X}$ are given, whereas $x \in \mathcal{D}(A)$ is to be found. It can be shown that, if A is the generator of a contraction semigroup then the resolvent equation has precisely one solution $x \in \mathcal{D}(A)$ for any $y \in \mathbb{X}$ and $\lambda > 0$.

If $\{T(t), t \geq 0\}$ is known, the solution may be found explicitly:

$$x = \int_0^\infty \mathrm{e}^{-\lambda t} T(t) y \,\mathrm{d}t, \tag{1.8}$$

that is, x is the Laplace transform of the trajectory of the semigroup. Technically, on the right-hand side we have an improper Riemann integral. Since trajectories of the semigroup are continuous, so are the maps $t \mapsto \mathrm{e}^{-\lambda t} T(t) y$, and thus the integrals $\int_0^\tau \mathrm{e}^{-\lambda t} T(t) y \,\mathrm{d}t$ are well defined for all $\tau > 0$. Then, for $\tau < \sigma$,

$$\left\| \int_0^\sigma \mathrm{e}^{-\lambda t} T(t) y \,\mathrm{d}t - \int_0^\tau \mathrm{e}^{-\lambda t} T(t) y \,\mathrm{d}t \right\| = \left\| \int_\tau^\sigma \mathrm{e}^{-\lambda t} T(t) y \,\mathrm{d}t \right\|$$

and this, by (1.5) and (1.6), does not exceed $\int_\tau^\sigma \mathrm{e}^{-\lambda t} \,\mathrm{d}t \, \|y\|$. Since

$$\lim_{\sigma, \tau \to \infty} \int_\tau^\sigma \mathrm{e}^{-\lambda t} \,\mathrm{d}t = 0,$$

the limit $\lim_{\tau \to \infty} \int_0^\tau \mathrm{e}^{-\lambda t} T(t) y \,\mathrm{d}t$ exists and this is precisely how the right-hand side of (1.8) is defined.

As a by-product of (1.8), we obtain also

$$\begin{aligned} \|\lambda x\| &= \left\| \lambda \lim_{\tau \to \infty} \int_0^\tau \mathrm{e}^{-\lambda t} T(t) y \,\mathrm{d}t \right\| = \lambda \lim_{\tau \to \infty} \left\| \int_0^\tau \mathrm{e}^{-\lambda t} T(t) y \,\mathrm{d}t \right\| \\ &\leq \lim_{\tau \to \infty} \int_0^\tau \lambda \mathrm{e}^{-\lambda t} \,\mathrm{d}t \, \|y\| = \|y\|. \end{aligned}$$

This is a crucial discovery: $\lambda \|x\| \leq \|y\|$, that is,

the map $y \mapsto \lambda x$ is a contraction.

1.2.2 *The Hille–Yosida Theorem (version I)*

The Hille–Yosida Theorem is in a sense a converse to the findings of the previous section: it says that if A is a densely defined operator such that for all $\lambda > 0$ and all $y \in \mathbb{X}$ the resolvent equation has a unique solution x, and for all $\lambda > 0$ the map $y \mapsto \lambda x$ is a contraction, then A is the generator of a semigroup of contractions.

Thus, the question of generation is reduced to that of studying solutions of (1.7).

1.2.3 *Resolvent operators*

It is often convenient to have the Hille–Yosida Theorem expressed in slightly different terms.

To this end, fix a $\lambda > 0$ and think of the operator $\lambda - A : x \mapsto \lambda x - Ax$ with domain $\mathcal{D}(A)$ and the range in $\mathbb{X}$. The fact that the resolvent equation has a solution for each $y \in \mathbb{X}$ means that the range of $\lambda - A$ is the entire $\mathbb{X}$. In other words, $\lambda - A$ has a right inverse. Moreover, the fact that the solution to the resolvent equation is unique means that $\lambda - A$ has a left inverse. These inverses then must coincide and thus we may think of x as the value of $(\lambda - A)^{-1}$ on y.

The operator $(\lambda - A)^{-1}$, often denoted R_λ, is known as the resolvent operator (we also frequently say simply 'the resolvent' and usually this does not lead to misunderstandings). To repeat, by definition, $(\lambda - A)^{-1}$ is the inverse of $\lambda - A$, both a right and the left inverse.

1.2.4 *The Hille–Yosida Theorem (version II)*

In terms of inverses of $\lambda - A$, the Hille–Yosida Theorem may be phrased as follows.

A densely defined operator A is the generator of a strongly continuous contraction semigroup $\{T(t), t \ge 0\}$ iff

- for all $\lambda > 0$, $\lambda - A$ has a left and right inverse $(\lambda - A)^{-1}$,
- $\|\lambda(\lambda - A)^{-1}\| \le 1, \lambda > 0$.

1.2.5 *Comments and notation*

(a) Since $R_\lambda = (\lambda - A)^{-1}$, the family $R_\lambda, \lambda > 0$ satisfies the **Hilbert equation**:

$$R_\lambda - R_\mu = (\mu - \lambda) R_\mu R_\lambda, \qquad \lambda, \mu > 0, \tag{1.9}$$

which could also be obtained from the fact that R_λ is the Laplace transform of the semigroup.

(b) In terms of $(\lambda - A)^{-1}$, relation (1.8) takes the form $(\lambda - A)^{-1} = \int_0^\infty e^{-\lambda t} T(t)\, dt$; the resolvent of the generator is the Laplace transform of the semigroup. A uniqueness theorem for the Laplace transform then implies in particular that an operator must not generate two different semigroups. We will write

$$\{e^{tA}, t \ge 0\} \qquad \text{or} \qquad \{T_A(t), t \ge 0\}$$

for the semigroup generated by A.

(c) Yosida's and Hille's proofs of the Hille–Yosida Theorem provide two different approximations of the semigroup generated by an operator A. The Hille approximation is

$$e^{tA}x = \lim_{n\to\infty}\left(\frac{n}{t}R_{\frac{n}{t}}\right)^n x, \qquad x \in \mathbb{X}. \tag{1.10}$$

The Yosida approximation involves the operators $A_\lambda := \lambda^2 R_\lambda - \lambda I_{\mathbb{X}}$. These are bounded linear operators, and it turns out that their exponents e^{tA_λ} are composed of contractions. Moreover, for $x \in \mathbb{X}$,

$$e^{tA}x = \lim_{\lambda\to\infty} e^{tA_\lambda}x = \lim_{\lambda\to\infty}\sum_{n=0}^{\infty}\frac{(tA_\lambda)^n}{n!} = \lim_{\lambda\to\infty} e^{-\lambda t}\sum_{n=0}^{\infty}\frac{(\lambda^2 R_\lambda)^n}{n!}. \tag{1.11}$$

(d) As a by-product of the approximations described above, it may be shown that to check that A generates a contraction semigroup it suffices to check that $(\lambda - A)^{-1}$ exists for all sufficiently large $\lambda > 0$ and that for such λ, $\|\lambda(\lambda - A)^{-1}\| \le 1$.

We complete this section with two examples of application of the Hille–Yosida Theorem. For these, we need to recall the notion of a Markov operator.

1.2.6 *Markov operators*

Suppose $L^1 := L^1(\Omega, \mathcal{F}, \mu)$ is the space of (equivalence classes of) absolutely integrable functions on a measure space $(\Omega, \mathcal{F}, \mu)$. A linear map $P\colon L^1 \to L^1$ such that

(a) $Px \ge 0$ for $x \ge 0$,
(b) $\int_\Omega Px\,d\mu = \int_\Omega x\,d\mu$ for $x \ge 0$

is called a **Markov operator**. If condition (b) is replaced by $\int_\Omega Px\,d\mu \le \int_\Omega x\,d\mu$ for $x \ge 0$, P is said to be a **sub-Markov operator**.

We claim that each sub-Markov operator is a contraction: to show this, for $x \in L^1$, let us write $x = x^+ - x^-$, where $x^+ = \max(x, 0)$ and $x^- = \max(-x, 0)$. We have $P(x^+) - Px \ge 0$, since $x^+ - x \ge 0$. Thus,

$$(Px)^+ = \max(Px, 0) \le P(x^+).$$

Since $x^- = (-x)^+$, we also have

$$(Px)^- = (-Px)^+ = [P(-x)]^+ \le P[(-x)^+] = P(x^-).$$

Therefore,

$$\int_\Omega |Px| \, d\mu = \int_\Omega [(Px)^+ + (Px)^-] \, d\mu \le \int_\Omega [P(x^+) + P(x^-)] \, d\mu$$
$$\le \int_\Omega [x^+ + x^-] \, d\mu = \int_\Omega |x| \, d\mu.$$

This shows that $\|Px\| \le \|x\|$, as claimed.

We also note that Markov operators preserve integral: this means that $\int_\Omega Px \, d\mu = \int_\Omega x \, d\mu$ for all $x \in L^1$. For,

$$\int_\Omega Px \, d\mu = \int_\Omega P(x^+ - x^-) \, d\mu = \int_\Omega Px^+ \, d\mu - \int_\Omega Px^- \, d\mu$$
$$= \int_\Omega x^+ \, d\mu - \int_\Omega x^- \, d\mu = \int_\Omega x \, d\mu.$$

Finally, referring to the Hille–Yosida Theorem, we note that if $\lambda R_\lambda = \lambda(\lambda - A)^{-1}, \lambda > 0$ are Markov operators, then so are $e^{tA}, t \ge 0$. This can be seen directly either from Hille's or Yosida's approximations, because the limit of a sequence of Markov operators is clearly a Markov operator. (In Section 2.4.3 we provide a detailed proof of this statement based on Yosida's approximation. An argument based on Hille's approximation is even simpler.)

1.2.7 Definition

The generator of a semigroup of Markov operators will be said to be a **Markov generator**.

1.2.8 Example

If P is a Markov operator in L^1, then, as we have just seen, $\|Px\| \le \|x\|, x \in L^1$. Moreover, for nonnegative x, $\|Px\| = \int_\Omega Px \, d\mu = \int_\Omega x \, d\mu = \|x\|$. Hence, $\|P\| = 1$. Is it true that a linear map $P: L^1 \to L^1$ of norm 1 such that $x \ge 0$ implies $Px \ge 0$ is automatically a Markov operator?

To see that the answer is in the negative (see also Exercise 1.2.12), consider $L^1 := L^1[0,1]$, the space of (equivalence classes of) Lebesgue integrable functions on the unit interval $[0,1]$, and let $Px(s) = sx(s), s \in [0,1], x \in L^1$. Clearly, $Px \ge 0$ provided $x \ge 0$, but P is not a Markov operator because for $x(s) = s, s \in [0,1]$, we have $x \ge 0$, $\int_0^1 x(s) \, ds = \frac{1}{2}$ and $\int_0^1 Px(s) \, ds = \frac{1}{3}$. Nevertheless, $\|P\| = 1$. Indeed, the calculation

$$\|Px\| = \int_0^1 s|x(s)| \, ds \le \int_0^1 |x(s)| \, ds = \|x\|$$

shows that $\|P\| \le 1$. Also, defining, for $n \ge 1$, $x_n(s)$ to be equal $\alpha_n^{-1}s$ for s between $1 - n^{-1}$ and 1, and 0 otherwise, where $\alpha_n = \int_{1-n^{-1}}^1 s\,\mathrm{d}s$, we see that $\|x_n\| = 1$. Since $\lim_{n\to\infty} \alpha_n^{-1} \int_{1-n^{-1}}^1 s^2\,\mathrm{d}s = 1$, it follows that $\|P\| \ge \sup_{n\ge 1} \|Px_n\| = 1$, completing the argument.

1.2.9 *Example: Hille's and Yosida's approximations of the semigroup* (1.1)

Let us use the Hille–Yosida Theorem to see that the operator of Section 1.1.9 is a generator and recover formula (1.1) for the semigroup it generates from Hille's and Yosida's approximations.

So, let $\mathcal{D}(A) \subset l^1$ be composed of $x = (\xi_i)_{i\ge 1}$ such that $\sum_{i=2}^{\infty} i|\xi_i| < \infty$, and let $Ax = (\sum_{i=2}^{\infty} i\xi_i, -2\xi_2, -3\xi_3, \dots)$. In coordinates, the resolvent equation for A reads

$$\lambda\xi_1 - \sum_{i=2}^{\infty} i\xi_i = \eta_1,$$
$$\lambda\xi_i + i\xi_i = \eta_i, \quad i \ge 2,$$

where $y = (\eta_i)_{i\ge 1}$. Thus, its only possible solution is

$$\xi_1 = \lambda^{-1}\left(\eta_1 + \sum_{i=2}^{\infty} \frac{i\eta_i}{\lambda + i}\right), \tag{1.12}$$
$$\xi_i = \frac{\eta_i}{\lambda + i}, \quad i \ge 2.$$

It is easy to see that the so-defined $(\xi_i)_{i\ge 1}$ belongs to $\mathcal{D}(A)$:

$$\sum_{i=2}^{\infty} i|\xi_i| = \sum_{i=2}^{\infty} \frac{i|\eta_i|}{\lambda + i} \le \sum_{i=2}^{\infty} |\eta_i| < \infty.$$

Moreover, $(\eta_i)_{i\ge 1} \mapsto \lambda\,(\xi_i)_{i\ge 1}$ is a Markov operator. For, ξ_i's are clearly nonnegative provided so are η_i's, and then

$$\lambda \sum_{i=1}^{\infty} i\xi_i = \eta_1 + \sum_{i=2}^{\infty} \frac{i\eta_i}{\lambda + i} + \sum_{i=2}^{\infty} \frac{\lambda\eta_i}{\lambda + i} = \sum_{i=1}^{\infty} \eta_i.$$

Since Markov operators are contractions, and A is obviously densely defined, the Hille–Yosida Theorem shows that A is the generator of a contraction semigroup. Moreover, since $\lambda\,(\lambda - A)^{-1}, \lambda > 0$ are Markov operators, so are the $\mathrm{e}^{tA}, t \ge 0$.

It remains to find e^{tA}. Let's use the Hille approximation first. Relation (1.12) says that the ith coordinate of $\lambda R_\lambda (\eta_i)_{i\geq 1}$ is $\frac{\lambda}{\lambda+i}\eta_i$, $i \geq 2$. It follows that the ith coordinate of $(\lambda R_\lambda)^n (\eta_i)_{i\geq 1}$ is $\left(\frac{\lambda}{\lambda+i}\right)^n \eta_i$. We know that $\left(\frac{n}{t} R_{\frac{n}{t}}\right)^n (\eta_i)_{i\geq 1}$ converges to $e^{tA} (\eta_i)_{i\geq 1}$. On the other hand

$$\lim_{n\to\infty} \left(\frac{\frac{n}{t}}{\frac{n}{t}+i}\right)^n \eta_i = \lim_{n\to\infty} \frac{\eta_i}{(1+\frac{it}{n})^n} = e^{-it}\eta_i.$$

Since convergence in the norm of l^1 implies convergence in coordinates, the ith coordinate of $e^{tA} (\eta_i)_{i\geq 1}$ is $e^{-it}\eta_i$, $i \geq 2$.

A closed form of the first coordinate of $(\lambda R_\lambda)^n (\eta_i)_{i\geq 1}$ is available (and the reader who wants to check his skills in using induction arguments is encouraged to find it), but complicated, and thus we will take an easier route. We know that e^{tA} is a Markov operator. Since Markov operators preserve the integral, the sum of coordinates of $e^{tA} (\eta_i)_{i\geq 1}$ is the same as the sum of coordinates of $(\eta_i)_{i\geq 1}$. It follows that the first coordinate of $e^{tA} (\eta_i)_{i\geq 1}$ is $\eta_1 + \sum_{j=2}^{\infty} \eta_j (1 - e^{-it})$, as in (1.1).

An argument based on Yosida's approximation is similar. Since, as already remarked, the ith coordinate of $(\lambda R_\lambda)^n (\eta_i)_{i\geq 1}$ is $\left(\frac{\lambda}{\lambda+i}\right)^n \eta_i$, the ith coordinate of $e^{t\lambda^2 R_\lambda} (\eta_i)_{i\geq 1}$ is $e^{t\frac{\lambda^2}{\lambda+i}}\eta_i$, and the ith coordinate of $e^{tA_\lambda} (\eta_i)_{i\geq 1} = e^{-\lambda t} e^{t\lambda^2 R_\lambda} (\eta_i)_{i\geq 1}$ is $e^{t(\frac{\lambda^2}{\lambda+i}-\lambda)}\eta_i = e^{-\frac{\lambda i}{\lambda+i}t}\eta_i$, $i \geq 2$. Since $\lim_{\lambda\to\infty} e^{tA_\lambda} (\eta_i)_{i\geq 1} = e^{tA} (\eta_i)_{i\geq 1}$ in the l^1 norm (and this convergence implies convergence in coordinates), it follows that the ith coordinate of $e^{tA} (\eta_i)_{i\geq 1}$ is $e^{-it}\eta_i$, $i \geq 2$. The first coordinate is calculated as with the Hille approximation.

It should perhaps be stressed here again that, despite (1.10) and (1.11), closed forms for semigroups are rarely given. Our example was not meant to suggest otherwise; its sole goal is to illustrate, in a simple case, how Hille's and Yosida's approximations work in practice.

1.2.10 *Example: The generator of uniform motion to the right with jump from the boundary*

Let $\mathbb{X} = L^1([0, 1], \mu)$ be the space of (equivalence classes) of functions $x\colon [0, 1] \to \mathbb{R}$ which are absolutely integrable with respect to the Borel measure

$$\mu := leb + \delta_1,$$

where δ_1 is the Dirac delta measure at $\tau = 1$, and leb is the Lebesgue measure on $[0, 1]$. In other words, a class in $\mathbb{X}$ is composed of absolutely integrable functions that differ on a set of Lebesgue measure but at $\tau = 1$ have the same finite value. The norm in $\mathbb{X}$ is, as customary,

$$\|x\| = \int_0^1 |x| \, \mathrm{d}\, leb + |x(1)|.$$

A class x belongs to the domain of the operator of interest to us if one of its representatives, also denoted x, is absolutely continuous on $[0, 1)$ with x' absolutely integrable with respect to the Lebesgue measure, and such that

$$ax(1) = x(0), \tag{1.13}$$

where $a > 0$ is a given parameter. This is to say that for such a representative, there is an absolutely integrable $z(= x')$ on $[0, 1]$ such that

$$x(\tau) = ax(1) + \int_0^\tau z(\sigma) \, \mathrm{d}\sigma, \qquad \tau \in [0, 1).$$

In particular, it is meaningful to speak of $x(1-) := ax(1) + \int_0^1 z(\sigma) \, \mathrm{d}\sigma$; this value, however, may differ from $x(1)$. For such x, $Ax \in \mathbb{X}$ is defined by specifying one of its representatives, also denoted Ax, by

$$Ax(\tau) = \begin{cases} -x'(\tau), & \tau \in [0, 1), \\ x(1-) - ax(1), & \tau = 1. \end{cases}$$

We claim that A is a Markov generator in $\mathbb{X}$. For the first step of the proof, we note that the resolvent equation for A may be written as the following system:

$$\begin{aligned} \lambda x(\tau) + x'(\tau) &= y(\tau), \qquad \tau \in [0, 1), \\ \lambda x(1) + ax(1) - x(1-) &= y(1), \end{aligned} \tag{1.14}$$

where $\lambda > 0$ and $y \in \mathbb{X}$ are given whereas $x \in \mathcal{D}(A)$ is searched-for (in fact, here we think of representatives of x and y). An x satisfies the first of these two equations if $x_1(\tau) := \mathrm{e}^{\lambda t} x(\tau)$ satisfies $x_1'(\tau) = \mathrm{e}^{\lambda\tau} y(\tau)$, and so x_1 must be absolutely continuous with $x_1(\tau) = x_1(0) + \int_0^\tau \mathrm{e}^{\lambda\sigma} y(\sigma) \, \mathrm{d}\sigma$, $\tau \in [0, 1)$; hence, so must x and

$$x(\tau) = x(0)\mathrm{e}^{-\lambda\tau} + \mathrm{e}^{-\lambda\tau} \int_0^\tau \mathrm{e}^{\lambda\sigma} y(\sigma) \, \mathrm{d}\sigma, \qquad \tau \in [0, 1). \tag{1.15}$$

The second equation in (1.14) yields $x(1) = \frac{y(1)+x(1-)}{\lambda+a}$ and thus the necessary condition (1.13) for x to belong to $\mathcal{D}(A)$ forces

$$x(0) = a\frac{y(1) + x(0)\mathrm{e}^{-\lambda} + C_\lambda(y)}{\lambda + a}$$

where $C_\lambda(y) = \mathrm{e}^{-\lambda}\int_0^1 \mathrm{e}^{\lambda\sigma} y(\sigma)\,\mathrm{d}\sigma$. It follows that there is precisely one solution to (1.14), and it is given by (1.15) supplemented with

$$x(0) = a\frac{y(1) + C_\lambda(y)}{\lambda + a(1 - \mathrm{e}^{-\lambda})}, \qquad x(1) = \frac{y(1) + C_\lambda(y)}{\lambda + a(1 - \mathrm{e}^{-\lambda})}. \tag{1.16}$$

We have established that the resolvent equation has a unique solution. It is also clear from the analysis that $x \geq 0$ provided $y \geq 0$. We claim next that for such y

$$\lambda \int_{[0,1]} x\,\mathrm{d}\mu = \int_{[0,1]} y\,\mathrm{d}\mu.$$

To see this, observe that for $x \in \mathcal{D}(A)$, we have

$$\int_{[0,1]} Ax\,\mathrm{d}\mu = -\int_0^1 x'(\tau)\,\mathrm{d}\tau + x(1-) - ax(1) = x(0) - ax(1) = 0$$

because x, being a member of $\mathcal{D}(A)$ (more precisely: its class belongs to $\mathcal{D}(A)$), satisfies (1.13). Thus, the claim may be obtained by integrating both sides of the resolvent equation $\lambda x - Ax = y$.

Our analysis shows that the map

$$\lambda R_\lambda := \lambda\,(\lambda - A)^{-1}$$

is a Markov operator, and because such operators have norm equal to one, we obtain $\|\lambda R_\lambda\| = 1$. Since $\mathcal{D}(A)$ may be seen to be dense in $\mathbb{X}$, by the Hille–Yosida Theorem, A is the generator of a contraction semigroup. Moreover, as discussed at the end of Section 1.2.6, the semigroup generated by A is composed of Markov operators.

This semigroup governs the evolution of distributions of the following Markov process with state-space $[0, 1]$. A particle starting at $\tau \in [0, 1)$ moves to the right with constant velocity $v = 1$. Once it reaches $\tau = 1$, it stays there for an exponential time with parameter $a > 0$, and later jumps to $\tau = 0$, where it starts moving to the right again. If the initial distribution x of this process is absolutely continuous with respect to the Lebesgue measure, except perhaps for a probability mass at $\tau = 1$, then at time $t > 0$ the distribution has the same property and is given by $\mathrm{e}^{tA}x$. It is somewhat clear that a closed form for e^{tA} will be rather hard to find for large t.

1.2.11 Exercise

Let $\mathbb{X} = c_0$, the space of sequences $(\xi_i)_{i\geq 1}$ converging to zero. Use the Hille–Yosida Theorem to show that the operator A defined by $A\,(\xi_i)_{i\geq 1} = (-i\xi_i)_{i\geq 1}$

on sequences $(\xi_i)_{i\ge 1} \in c_0$ such that $(i\xi_i)_{i\ge 1} \in c_0$ is the generator of a contraction semigroup. Check to see that $\mathrm{e}^{tA}\,(\xi_i)_{i\ge 1} = \left(\mathrm{e}^{-it}\xi_i\right)_{i\ge 1}, t \ge 0$.

1.2.12 Exercise

Let $l^1 := l^1(\mathbb{N})$ be the space of absolutely summable sequences $(\xi_i)_{i\ge 1}$. Check to see that P mapping $(\xi_i)_{i\ge 1}$ to $(\eta_i)_{i\ge 1}$ where $\eta_i = (1-\frac{1}{i})\xi_i, i \ge 1$, has norm 1 and is strictly sub-Markov in the sense that $\eta_i \ge 0$ and $\sum_{i=1}^\infty \eta_i < \sum_{i=1}^\infty \xi_i$ provided $(\xi_i)_{i\ge 1} \ge 0$ is nonzero.

1.2.13 Exercise

Let P be a Markov operator in a space L^1. Check to see that the semigroup generated by $Q = P - I_{L^1}$ is composed of Markov operators. If P is sub-Markov, Q generates a semigroup of sub-Markov operators.

1.3 Perturbation Theorems

1.3.1 *Phillips's Perturbation Theorem*

Even with the help of the Hille–Yosida Theorem, proving that a given operator A is a contraction semigroup generator is often not an easy task. One of the tricks of the trade is to write A as the sum of two operators

$$A = A_0 + B$$

where A_0 is usually 'simpler' than A (but has the same domain), and B is bounded. The Phillips Perturbation Theorem says, then, that if A_0 is the generator of a strongly continuous semigroup, then so is A. Moreover,

$$\mathrm{e}^{tA} = \sum_{n=0}^{\infty} S_n(t), \tag{1.17}$$

where $S_0(t) = \mathrm{e}^{tA_0}$ and $S_{n+1}(t) = \int_0^t \mathrm{e}^{(t-s)A} B S_n(s)\,\mathrm{d}s, n \ge 0$. The series (1.17) is known as the Dyson–Phillips series.

1.3.2 *Trotter's Product Formula*

It should be noted that the semigroup $\{\mathrm{e}^{tA}, t \ge 0\}$ given by (1.17) need not be composed of contractions even if $\{\mathrm{e}^{tA_0}, t \ge 0\}$ is. However, if both $\{\mathrm{e}^{tA_0}, t \ge 0\}$ and $\{\mathrm{e}^{tB}, t \ge 0\}$ (the latter existing as an exponent of B) are composed of contractions then so is $\{e^{tA}, t \geq 0\}$. This is a direct consequence of Trotter's Product Formula, which says that (even under less restrictive assumptions than those given above)

$$e^{tA}x = \lim_{n\to\infty} \left(e^{\frac{t}{n}A_0}e^{\frac{t}{n}B}\right)^n x;$$

as a limit of contractions, e^{tA} is a contraction. Similarly, if $\{e^{tA_0}, t \geq 0\}$ and $\{e^{tB}, t \geq 0\}$ are both composed of Markov operators then so is the semigroup $\{e^{tA}, t \geq 0\}$.

1.3.3 Example

Here is a typical example of application of Trotter's Product Formula. Let $q_{i,j}, i, j \geq 1$ be given numbers with the following properties (a) $\sum_{j=1}^{\infty} q_{i,j} = 0$, for all $i \geq 1$, (b) $q_{i,j} \geq 0, i \neq j$, and (c) $\sup_{i\geq 1}(-q_{ii}) < \infty$ (we thus assume that $q_{i,j}$'s form a bounded Kolmogorov matrix; see Section 2.2.8).

In the space l^1 of absolutely converging sequences $(\xi_i)_{i\geq 1}$, we define an operator A by

$$A\,(\xi_i)_{i\geq 1} = \left(\sum_{j=2}^{\infty} j\xi_j + \sum_{j=1}^{\infty} q_{j,1}\xi_j,\ -2\xi_2 + \sum_{j=1}^{\infty} q_{j,2}\xi_j,\ -3\xi_3 + \sum_{j=1}^{\infty} q_{j,3}\xi_j, \cdots\right)$$

on the domain $\mathcal{D}(A)$ composed of $(\xi_i)_{i\geq 1}$ such that $\sum_{j=1}^{\infty} j|\xi_j| < \infty$. We claim that A is a Markov generator.

To this end, we write

$$A\,(\xi_i)_{i\geq 1} = A_0\,(\xi_i)_{i\geq 1} + B\,(\xi_i)_{i\geq 1}$$

where

$$A_0\,(\xi_i)_{i\geq 1} = \left(\sum_{j=2}^{\infty} j\xi_j,\ -2\xi_2,\ -3\xi_3, \cdots\right) \quad \text{and}$$

$$B\,(\xi_i)_{i\geq 1} = \left(\sum_{j=1}^{\infty} q_{j,i}\xi_j\right)_{i\geq 1}.$$

Then, we note that whereas the formula for $A_0\,(\xi_i)_{i\geq 1}$ makes sense only for $(\xi_i)_{i\geq 1} \in \mathcal{D}(A)$, the formula for $B\,(\xi_i)_{i\geq 1}$ makes sense for all $(\xi_i)_{i\geq 1} \in l^1$. Moreover, for $x = (\xi_i)_{i\geq 1}$,

$$\begin{aligned}
\|B\,(\xi_i)_{i\geq 1}\| &\leq \sum_{i=1}^{\infty}\left|\sum_{j=1}^{\infty} q_{j,i}\xi_j\right| \leq \sum_{i=1}^{\infty}\left(\sum_{j\neq i}^{\infty} q_{j,i}|\xi_j| - q_{i,i}|\xi_i|\right) \\
&\leq q\|x\| + \sum_{j=1}^{\infty}|\xi_j|\sum_{i\neq j}^{\infty} q_{j,i} \leq q\|x\| + \sum_{j=1}^{\infty}(-q_{j,j})|\xi_j| \\
&\leq 2q\|x\|,
\end{aligned}$$

where

$$q = \sup_{i\geq 1}(-q_{i,i}),$$

proving that B is a bounded operator. Since in A_0 we recognize the generator of the semigroup of Section 1.1.2 (see Section 1.1.9), this shows, by the Phillips Perturbation Theorem, that A is a generator. Furthermore, since the semigroup generated by A_0 is composed of Markov operators, we will be able to conclude, by Trotter's Product Formula, that A is a Markov generator once we show that B is.

Let $B_0 := B + qI_{l^1}$ and $P := q^{-1}B_0$. Since

$$B_0\,(\xi_i)_{i\geq 1} = \left((q+q_{i,i})\xi_i + \sum_{j\neq i} q_{j,i}\xi_j\right)_{i\geq 1}$$

and $q + q_{i,i} \geq 0$, it is clear that $B\,(\xi_i)_{i\geq 1}$ is nonnegative provided so is $(\xi_i)_{i\geq 1}$. Moreover, for such $(\xi_i)_{i\geq 1}$, the sum of coordinates of $B_0\,(\xi_i)_{i\geq 1}$ equals

$$\sum_{i=1}^{\infty} q\xi_i + \sum_{i=1}^{\infty}\sum_{j=1}^{\infty} q_{j,i}\xi_j = q\sum_{i=1}^{\infty}\xi_i + \sum_{j=1}^{\infty}\xi_j\sum_{i=1}^{\infty} q_{j,i} = q\sum_{i=1}^{\infty}\xi_i.$$

It follows that P is a Markov operator, and thus $B = q(P - I_{l^1})$ is a Markov generator by combined Exercises 1.2.13 and 1.1.17.

1.4 Approximation and Convergence Theorems

In this section we discuss Trotter–Kato and Sova–Kurtz Approximation and Convergence Theorems for semigroups of operators. Even though these theorems occupy a special place in this book, we refrain from giving examples of their applications here: they will be found throughout the book. Plenty of other examples may be found in [16].

1.4.1 *Approximation theorems*

The Trotter–Kato Approximation Theorem says that for a sequence $\{e^{tA_n}, t \geq 0\}, n \geq 1$ of contraction semigroups in a Banach space $\mathbb{X}$ to converge to a semigroup $\{e^{tA}, t \geq 0\}$ in the sense that

$$\lim_{n\to\infty} e^{tA_n}x = e^{tA}x, \qquad t \geq 0, x \in \mathbb{X}, \tag{1.18}$$

it is necessary and sufficient that

$$\lim_{n\to\infty} (\lambda - A_n)^{-1} x = (\lambda - A)^{-1} x, \qquad \lambda > 0, x \in \mathbb{X}. \tag{1.19}$$

Moreover, if the latter condition is satisfied, the limit in (1.18) is uniform in compact subsets of $[0, \infty)$: for any $t_0 > 0$,

$$\lim_{n\to\infty} \max_{t\in[0,t_0]} \|e^{tA_n}x - e^{tA}x\| = 0, \qquad t \geq 0, x \in \mathbb{X}. \tag{1.20}$$

The Sova–Kurtz version of this theorem [61, 81] says that (1.20) (or, equivalently, (1.18)) holds iff for any $x \in \mathcal{D}(A)$ there is a sequence $(x_n)_{n\geq 1}$ of elements of $\mathbb{X}$ such that

(a) $x_n \in \mathcal{D}(A_n), n \geq 1$,
(b) $\lim_{n\to\infty} x_n = x$, and
(c) $\lim_{n\to\infty} A_n x_n = Ax$.

The latter version is often easier to use since it does not require knowledge of the resolvents $(\lambda - A_n)^{-1}$, which may not be available in a manageable form.

1.4.2 *Convergence theorems*

Often the limit semigroup (i.e., $\{e^{tA}, t \geq 0\}$) is not *a priori* given, and even its existence is not granted: we have a sequence of contraction semigroups $\{e^{tA_n}, t \geq 0\}, n \geq 1$, and we want to know whether the limit

$$\lim_{n\to\infty} e^{tA_n}x \tag{1.21}$$

exists for all $t \geq 0$ and $x \in \mathbb{X}$. Perhaps surprisingly, in this case the sole existence of

$$R_\lambda x := \lim_{n\to\infty} (\lambda - A_n)^{-1} x \tag{1.22}$$

does not guarantee convergence (1.21). There are examples showing that resolvents may converge whereas the semigroups converge only for the null vector $x = 0$. As it transpires, nevertheless, using $R_\lambda, \lambda > 0$ one may characterize the subspace, say,

$$\mathbb{X}_0,$$

sometimes termed the **regularity space**, of x's in $\mathbb{X}$ such that the limit

$$\lim_{n\to\infty} \mathrm{e}^{tA_n}x$$

exists uniformly in compact subsets of $[0, \infty)$.

To explain, the operators $\lambda R_\lambda, \lambda > 0$ are contractions because so are $\lambda(\lambda - A_n)^{-1}, \lambda > 0$ for each n. Similarly, $R_\lambda, \lambda > 0$ satisfy the Hilbert equation (1.9), since so do $(\lambda - A_n)^{-1}, \lambda > 0$ for each n. It follows that $R_\lambda, \lambda > 0$ have a common kernel and range. It can be proved that

$$\mathbb{X}_0 \text{ is the closure of the common range of } R_\lambda, \lambda > 0, \tag{1.23}$$

and this is the first part of the Trotter–Kato version of convergence theorem. The other part says that on $\mathbb{X}_0$ there is a strongly continuous semigroup of operators $\{T(t), t \ge 0\}$ such that

$$T(t)x = \lim_{n\to\infty} \mathrm{e}^{tA_n}x, \qquad t \ge 0, x \in \mathbb{X}_0,$$

and the convergence is again uniform on compact subsets of $[0, \infty)$.

The generator of $\{T(t), t \ge 0\}$ can also be characterized: when restricted to $\mathbb{X}_0$, the operators $R_\lambda, \lambda > 0$ have trivial kernel (i.e., the kernel is composed merely of the null vector $x = 0$), and thus there exists inverses $(R_\lambda)^{-1}_{|\mathbb{X}_0}, \lambda > 0$. By the Hilbert equation

$$A := \lambda I_{\mathbb{X}_0} - (R_\lambda)^{-1}_{|\mathbb{X}_0} \tag{1.24}$$

defined on the common range of $(R_\lambda)_{|\mathbb{X}_0}$, does not depend on $\lambda > 0$; this operator is the generator of $\{T(t), t \ge 0\}$.

In particular, if $\mathbb{X}_0 = \mathbb{X}$, that is, if the range of $R_\lambda, \lambda > 0$ is dense in $\mathbb{X}$,

$$\mathcal{D}(A) \text{ is the range of } R_\lambda, \lambda > 0. \tag{1.25}$$

1.4.3 *Sova–Kurtz version*

In practice, formula (1.24) is rarely useful in characterizing the generator of the limit semigroup: inverting $(R_\lambda)_{|\mathbb{X}_0}$ is not an easy task. The Sova–Kurtz version is more informative. It says that A can be characterized by means of the so-called **extended limit** of $A_n, n \ge 1$.

The domain $\mathcal{D}(A_{\mathrm{ex}})$ of this extended limit is composed of $x \in \mathbb{X}$ such that there is a sequence $(x_n)_{n\ge 1}$ with the following properties (cf. Section 1.4.1):

(a) $x_n \in \mathcal{D}(A_n), n \ge 1$,
(b) $\lim_{n\to\infty} x_n = x$, and
(c) $\lim_{n\to\infty} A_n x_n$ exists.

Unless we want to deal with multi-valued operators, $A_{ex}x$ cannot be simply defined as $\lim_{n\to\infty} A_n x_n$, because this limit may depend on the choice of $(x_n)_{n\geq 1}$. This notwithstanding, if the limit (1.22) exists,

$$\mathbb{X}_0 \text{ is the closure of } \mathcal{D}(A_{ex}).$$

Consider also the subset, say, $\mathcal{D}$, of $x \in \mathcal{D}(A_{ex})$ such that for some $(x_n)_{n\geq 1}$ of the definition of A_{ex}, $\lim_{n\to\infty} A_n x_n$ belongs to $\mathbb{X}_0$. It can be seen that for any other $\left(x'_n\right)_{n\geq 1}$ of the definition, $\lim_{n\to\infty} A_n x'_n = \lim_{n\to\infty} A_n x_n$. In other words, A_{ex} as restricted to $\mathcal{D}$ is single-valued: it has only one possible value in $\mathbb{X}_0$. Moreover,

$$(A_{ex})_{|\mathcal{D}} \text{ is the generator of the limit semigroup.}$$

Finally, we note that existence of the limit (1.22) may be checked in terms of A_{ex}: this limit exists for all $\lambda > 0$ iff there is a $\lambda > 0$ such that for all $y \in \mathbb{X}$ there is an $x \in \mathcal{D}(A_{ex})$ such that $y = \lambda x - A_{ex}x$.

1.4.4 *Kurtz's Singular Perturbation Theorem*

The Trotter–Kato and Sova–Kurtz Convergence Theorems settle the question of convergence that is uniform in compact subintervals of $[0, \infty)$. However, there are examples of semigroups that converge in a less regular way; it often happens, especially in singular perturbation theory, that the semigroups converge also outside of the regularity space $\mathbb{X}_0$. By nature (and definition of $\mathbb{X}_0$) this convergence is not uniform around $t = 0$: it is uniform in compact subintervals of $(0, \infty)$.

For a simple, somewhat typical instance, think of a Markov generator A_0 in a certain space $\mathbb{X}$ of absolutely integrable functions, and of a bounded operator B that is also a Markov generator, and let $(r_n)_{n\geq 1}$ be a sequence of positive numbers converging to ∞. Since $r_n B, n \geq 1$ are bounded and are Markov generators,

$$A_n := A_0 + r_n B, \qquad n \geq 1 \tag{1.26}$$

are also Markov generators (by the combined Phillips Perturbation Theorem and Trotter's Product Formula). The latter operators form a singular perturbation of A_0, since the influence of the operator B in the limit is 'much stronger' than that of A_0.

A simple case of Kurtz's Singular Perturbation Theorem says that the semigroups generated by $A_n, n \geq 1$ converge provided the following two conditions are met:

(i) $\lim_{t\to\infty} e^{tB}x =: Px$ exists for all $x \in \mathbb{X}$.
(ii) PA_0 with domain $\mathcal{D}(A_0) \cap \mathbb{X}'$ is a generator in $\mathbb{X}'$, where $\mathbb{X}' := \mathit{Range}\ P$.

Then $\mathbb{X}'$ is the regularity space for the semigroups $\{e^{tA_n}, t \ge 0\}, n \ge 1$:

$$\lim_{n\to\infty} e^{tA_n}x = e^{tPA_0}x, \qquad x \in \mathbb{X}', t \ge 0$$

and the limit is uniform in compact subintervals of $[0, \infty)$. Moreover,

$$\lim_{n\to\infty} e^{tA_n}x = e^{tPA_0}Px, \qquad x \notin \mathbb{X}', t > 0$$

and the limit is uniform in compact subintervals of $(0, \infty)$.

See our Section 4.2.5 for an illustrative example of this theorem (this section is also the only place where this theorem is used in this book).

1.5 Dual and Sun-Dual Semigroups

The material presented in this section will be needed only in Chapter 5, and thus may be omitted at the first reading. We start by recalling the definition of the dual for an operator.

1.5.1 *The dual of an operator*

Let $\mathbb{X}$ and $\mathbb{Y}$ be Banach spaces, and let $A : \mathbb{X} \to \mathbb{Y}$ be a bounded linear operator. For any functional f on $\mathbb{Y}$, the map $f \circ A$ is a linear functional on $\mathbb{X}$. Since $\|f \circ A\|_{\mathbb{X}^*} \le \|f\|_{\mathbb{Y}^*}\|A\|_{\mathcal{L}(\mathbb{X},\mathbb{Y})}$, the map $f \mapsto f \circ A$, denoted A^*, is a bounded linear map from $\mathbb{Y}^*$ to $\mathbb{X}^*$ and $\|A^*\| \le \|A\|$. The operator A^* is called the **dual operator** or the **adjoint operator** of A. Since, as a consequence of the Hahn–Banach Theorem, for any $x \in \mathbb{X}$,

$$\|Ax\| = \sup_{f\in\mathbb{Y}^*, \|f\|=1} |f(Ax)| = \sup_{f\in\mathbb{Y}^*, \|f\|=1} |(A^*f)x| \le \|A^*\|\|x\|,$$

we see that $\|A\| \le \|A^*\|$, and so $\|A\| = \|A^*\|$.

1.5.2 *Trouble with continuity of the dual semigroup*

Let $\{T_t, t \ge 0\}$ be a strongly continuous semigroup in a Banach space $\mathbb{X}$. Then the family $\left\{T_t^*, t \ge 0\right\}$ is also a semigroup since

$$T_t^* T_s^* = [T_s T_t]^* = T_{s+t}^*;$$

we refer to $\left\{T_t^*, t \ge 0\right\}$ as the **dual semigroup**. By 1.5.1, if $\{T_t, t \ge 0\}$ is composed of contractions, then so is $\left\{T_t^*, t \ge 0\right\}$.

Interestingly, though, $\left\{T_t^*, t \ge 0\right\}$ need not be strongly continuous. For example, if $\mathbb{X}$ is the space $C_0(\mathbb{R})$ of continuous functions x on $\mathbb{R}$ that vanish at both $-\infty$ and ∞, and $T_t x(\tau) = x(\tau + t)$, $\tau \in \mathbb{R}, t \ge 0$, then $\mathbb{X}^*$ is the space of Borel signed measures, and T_t^* maps a measure μ to its translate μ_t given by

$$\mu_t(B) = \mu(B_{-t})$$

where, for a Borel set B, $B_{-t} := \{\tau | \tau + t \in B\}$. In particular, for Dirac delta measures we have $T_t^* \delta_a = \delta_{a+t}, a \in \mathbb{R}$, and, since $\|\delta_a - \delta_b\| = 2$ provided $a \neq b$, there is no hope for $\lim_{t \to 0} \|T_t^* \delta_a - \delta_a\| = 0$.

On the other hand, if $\mathbb{X} = c_0$ and

$$T_t \, (\xi_i)_{i \geq 1} = \left(\mathrm{e}^{-it} \xi_i\right)_{i \geq 1}, \qquad (\xi_i)_{i \geq 1} \in c_0$$

(see Exercise 1.2.11) then $\mathbb{X}_0^* = l^1$ and

$$T_t^* \, (\eta_i)_{i \geq 1} = \left(\mathrm{e}^{-it} \eta_i\right)_{i \geq 1}, \qquad (\eta_i)_{i \geq 1} \in l^1,$$

and this is a strongly continuous semigroup.

1.5.3 *The sun-dual semigroup*

The two seemingly contradicting examples from the previous section may be reconciled by noting that there is always a subspace of $\mathbb{X}^*$ where the dual semigroup is strongly continuous; if worst comes to worst, this subspace may be trivial, that is, composed of the zero functional. We denote this subspace by $\mathbb{X}^{\odot}$:

$$\mathbb{X}^{\odot} := \{f \in \mathbb{X}^* | \lim_{t \to 0+} \|T_t^* f - f\| = 0\}.$$

In the second example, $\mathbb{X}^{\odot}$ just happens to be equal to the entire $\mathbb{X}^*$. In the first example, it contains the subspace of measures that are absolutely continuous with respect to the Lebesgue measure, which may be identified with $L^1(\mathbb{R})$. For, in the latter space the dual semigroup is the familiar semigroup of translations to the right:

$$T_t^* \phi(\tau) = \phi(\tau - t), \qquad t \geq 0, \text{ almost all } \tau \in \mathbb{R}.$$

It is quite easy to see that $\mathbb{X}^{\odot}$ is closed and invariant for the dual semigroup, that is, $x \in \mathbb{X}^{\odot}$ implies $T_t^* x \in \mathbb{X}^{\odot}$ for all $t \geq 0$. It follows that (cf. Exercise 1.1.19) $\{T_t^*, t \geq 0\}$ as restricted to $\mathbb{X}^{\odot}$ is a strongly continuous semigroup in the Banach space $\mathbb{X}^{\odot}$. This semigroup is termed **the sun-dual** of $\{T_t, t \geq 0\}$ and denoted $\left\{T_t^{\odot}, t \geq 0\right\}$. Thus, by definition,

$$T_t^{\odot} = (T_t^*)_{|\mathbb{X}^{\odot}}, \qquad t \geq 0.$$

The remainder of this section is devoted to characterizing $\mathbb{X}^{\odot}$ and the generator of the sun-dual semigroup. As we shall see, a natural path to such a characterization leads through the notion of a dual of the generator of the original semigroup.

1.5.4 *The dual of a densely defined operator*

Let $\mathbb{X}$ be a Banach space and $A\colon \mathbb{X} \supset \mathcal{D}(A) \to \mathbb{X}$ be a densely defined operator (think of A as of a generator). We define $\mathcal{D}(A^*)$ as the set of $f \in \mathbb{X}^*$ such that

$$|f(Ax)| \le C\|x\|, \qquad x \in \mathcal{D}(A) \tag{1.27}$$

for some constant $C = C(f)$. Then, $\mathcal{D}(A)$ being dense in $\mathbb{X}$, the map

$$x \mapsto f(Ax), \qquad x \in \mathcal{D}(A)$$

may be uniquely extended to a continuous functional on $\mathbb{X}$. In other words, (1.27) is equivalent to assuming that there is a (uniquely determined) functional $g \in \mathbb{X}^*$ such that

$$g(x) = f(Ax), \qquad x \in \mathcal{D}(A).$$

Thus, for $f \in \mathcal{D}(A^*)$ it makes sense to agree that $A^*f = g$.

The so-defined operator A^*, with domain $\mathcal{D}(A^*)$ is termed **the dual of** A. It is clear that in the case where A is a bounded linear operator, the definition just given is consistent with the one introduced in 1.5.1.

1.5.5 Lemma

Suppose A is a densely defined operator and $\lambda \in \rho(A)$. Then $\lambda \in \rho(A^*)$ and

$$\left(\lambda - A^*\right)^{-1} = R_\lambda^*$$

where R_λ is the usual shorthand for $(\lambda - A)^{-1}$.

Proof To clarify, we assume that there is $R_\lambda \in \mathcal{L}(\mathbb{X})$ (where $\mathbb{X}$ is the space where A is defined) such that

(a) for any $x \in \mathbb{X}$, $R_\lambda x$ belongs to $\mathcal{D}(A)$ and $\lambda R_\lambda x - AR_\lambda x = x$,
(b) for $x \in \mathcal{D}(A)$, $\lambda R_\lambda x - R_\lambda Ax = x$,

and claim that R_λ^* plays a similar role for A^*.

The calculation

$$R_\lambda^*[\lambda f - A^*f](x) = (\lambda f - A^*f)(R_\lambda x) = f(\lambda R_\lambda x) - f(AR_\lambda x) = f(x),$$

which makes sense for $f \in \mathcal{D}(A^*)$ and $x \in \mathbb{X}$, where we used assumption (a), shows that the map $f \mapsto \lambda f - A^* f$ is injective, and that R_λ^* is the left inverse for this map. Also, for any $f \in \mathbb{X}^*$ we have

$$R_\lambda^* f(Ax) = f(R_\lambda Ax) = f(\lambda R_\lambda x) - f(x), \qquad x \in \mathcal{D}(A).$$

This reveals that $x \mapsto R_\lambda^* f(Ax)$ coincides on $\mathcal{D}(A)$ with the bounded linear functional $\lambda R_\lambda^* f - f$. It follows that $R_\lambda^* f$ is a member of $\mathcal{D}(A^*)$ and $A^* R_\lambda^* f = \lambda R_\lambda^* f - f$. Thus, R_λ^* is seen to be a right inverse for $\lambda - A^*$. Since R_λ^* is both the left and right inverse for $\lambda - A^*$, we are done. □

When combined with 1.5.1, the lemma presented above says that if A is a generator then A^* is a so-called Hille–Yosida operator; by definition, a Hille–Yosida operator is short of being a generator only because it is perhaps not densely defined. However, it can be proved (see, e.g., [14], Section 8.2.3) that the part of A^* in $cl\mathcal{D}(A^*)$, which we denote A_{p}^*, generates a semigroup in $cl\mathcal{D}(A^*)$. (To recall, the domain of A_{p}^* is the set of $f \in \mathcal{D}(A^*)$ such that $A^* f$ belongs to $cl\mathcal{D}(A^*)$, and $A_{\mathrm{p}}^* f = A^* f$ for f in this domain.) As we shall see next, $cl\mathcal{D}(A^*)$ coincides with $\mathbb{X}^\odot$ and the semigroup generated by A_{p}^* is the sun-dual of $\{T_t, t \ge 0\}$.

1.5.6 Theorem

We have

$$\boxed{\mathbb{X}^\odot = cl\,\mathcal{D}(A^*),}$$

where cl denotes closure in the norm of $\mathbb{X}^*$. Moreover, for the semigroup $\{\mathrm{e}^{tA_{\mathrm{p}}^*}, t \ge 0\}$ in $\mathbb{X}^\odot$ generated by A_{p}^* we have

$$\mathrm{e}^{tA_{\mathrm{p}}^*} = T_t^\odot, \qquad t \ge 0.$$

Proof For $f \in cl\mathcal{D}(A^*)$, the function $t \mapsto \mathrm{e}^{tA_{\mathrm{p}}^*} f$ is continuous and

$$\int_0^\infty \mathrm{e}^{-\lambda t} \mathrm{e}^{tA_{\mathrm{p}}^*} f \,\mathrm{d}t = \left(\lambda - A_{\mathrm{p}}^*\right)^{-1} f = \left(\lambda - A^*\right)^{-1} f = R_\lambda^* f, \qquad \lambda > 0.$$

In particular, for any $x \in \mathbb{X}$,

$$\int_0^\infty \mathrm{e}^{-\lambda t} \mathrm{e}^{tA_{\mathrm{p}}^*} f(x) \,\mathrm{d}t = \left(\int_0^\infty \mathrm{e}^{-\lambda t} \mathrm{e}^{tA_{\mathrm{p}}^*} f \,\mathrm{d}t\right)(x) = f(R_\lambda x).$$

On the other hand, for any $x \in \mathbb{X}$, $t \mapsto (T_t^* f)(x) = f(T_t x)$ is also continuous, and

$$\int_0^\infty \mathrm{e}^{-\lambda t} (T_t^* f)(x) \,\mathrm{d}t = \int_0^\infty \mathrm{e}^{-\lambda t} f(T_t x) \,\mathrm{d}t = f\left(\int_0^\infty \mathrm{e}^{-\lambda t} T_t x \,\mathrm{d}t\right)$$
$$= f(R_\lambda x).$$

Since the Laplace transform, restricted to continuous functions, is injective

$$e^{tA_p^*} f(x) = T_t^* f(x)$$

for all $t \geq 0$ and $x \in \mathbb{X}$, that is,

$$e^{tA_p^*} f = T_t^* f, \qquad t \geq 0.$$

In particular, $\{e^{tA_p^*}, t \geq 0\}$ being a strongly continuous semigroup, $t \mapsto T_t^* f$ is continuous. This means that $f \in \mathbb{X}^{\odot}$. We have thus proved that

$$cl\mathcal{D}(A^*) \subset \mathbb{X}^{\odot}$$

and that on $cl\mathcal{D}(A^*)$, $e^{tA_p^*}$ coincides with T_t^*.

It remains to show that $\mathbb{X}^{\odot} \subset cl\mathcal{D}(A^*)$. Let, for the sake of this argument, B be the generator of the sun-dual semigroup. Clearly, $\mathcal{D}(B) \subset \mathbb{X}^{\odot}$ is dense in $\mathbb{X}^{\odot}$ and so it suffices to show that $\mathcal{D}(B) \subset \mathcal{D}(A^*)$. To this end, let $f \in \mathcal{D}(B)$, so that the limit $\lim_{t\to 0+} t^{-1}(T_t^* f - f)$ exists in the norm of $\mathbb{X}^*$. In particular, there is a $C = C(f)$ such that

$$\|t^{-1}(T_t^* f - f)\| \leq C(f), \qquad t \in (0, 1].$$

Therefore, for $x \in \mathcal{D}(A)$,

$$|f(Ax)| = |\lim_{t\to 0+} t^{-1} f(T_t x - x)| = |\lim_{t\to 0+} t^{-1}(T_t^* f - f)(x)| \leq C(f)\|x\|.$$

This shows that $f \in \mathcal{D}(A^*)$, and completes the proof. □

1.5.7 Example

Consider the semigroup $\{T_t = e^{tA}, t \geq 0\}$ of Example 1.2.10; we seek a characterization of $\mathbb{X}^{\odot}$ and A_p^* (see the neighborhood of (1.29)).

By the theorem of Steinhaus (see, e.g., [14], Section 5.2.16), members of $\mathbb{X}^*$ have a dual status: they may be treated as functionals on $\mathbb{X}$ and as (classes of) essentially bounded measurable functions on $[0, 1]$. Note that, because of the Dirac measure featuring in the definition of μ, a class in $\mathbb{X}^*$ is composed of essentially bounded functions on $[0, 1]$ that differ from each other on sets of Lebesgue measure zero, but at $\tau = 1$ have the same finite value.

By 1.5.6, $\mathbb{X}^{\odot}$ is the closure of $\mathcal{D}(A^*)$ (in $\mathbb{X}^*$) and, by 1.5.5, $\mathcal{D}(A^*)$ is the range of $R_\lambda^* = [(\lambda - A)^{-1}]^*$. Thus we will take a closer look at this range.

Recall from 1.2.10 (see (1.15) and (1.16) in particular) that, for $y \in \mathbb{X}$,

$$R_\lambda y(\tau) = \begin{cases} a\frac{y(1)+C_\lambda(y)}{\lambda+a(1-e^{-\lambda})} e^{-\lambda\tau} + e^{-\lambda\tau}\int_0^\tau e^{\lambda\sigma} y(\sigma)\,d\sigma, & \tau \in [0, 1), \\ \frac{y(1)+C_\lambda(y)}{\lambda+a(1-e^{-\lambda})}, & \tau = 1, \end{cases}$$

where $C_\lambda(y) = \mathrm{e}^{-\lambda}\int_0^1 \mathrm{e}^{\lambda\sigma} y(\sigma)\,\mathrm{d}\sigma$. Therefore, for any (representative of a) $g \in \mathbb{X}^*$,

$$\begin{aligned}
g(R_\lambda y) &= \int_{[0,1]} g\, R_\lambda y\, \mathrm{d}\mu = \int_0^1 g(\tau) R_\lambda y(\tau)\,\mathrm{d}\tau + g(1) R_\lambda y(1) \\
&= a\frac{y(1) + C_\lambda(y)}{\lambda + a(1-\mathrm{e}^{-\lambda})} D_\lambda(g) + \int_0^1 \mathrm{e}^{-\lambda\tau}\int_0^\tau \mathrm{e}^{\lambda\sigma} y(\sigma)\,\mathrm{d}\sigma\, g(\tau)\,\mathrm{d}\tau \\
&+ \frac{y(1) + C_\lambda(y)}{\lambda + a(1-\mathrm{e}^{-\lambda})} g(1),
\end{aligned}$$

where $D_\lambda : \mathbb{X}^* \to \mathbb{R}$ is given by

$$D_\lambda(g) = \int_0^1 \mathrm{e}^{-\lambda\tau} g(\tau)\,\mathrm{d}\tau, \qquad g \in \mathbb{X}^*.$$

Rearranging and changing the order of integration in the second term yields

$$\begin{aligned}
g(R_\lambda y) &= y(1)\frac{aD_\lambda(g) + g(1)}{\lambda + a(1-\mathrm{e}^{-\lambda})} + C_\lambda(y)\frac{aD_\lambda(g) + g(1)}{\lambda + a(1-\mathrm{e}^{-\lambda})} \\
&+ \int_0^1 \mathrm{e}^{\lambda\sigma}\int_\sigma^1 \mathrm{e}^{-\lambda\tau} g(\tau)\,\mathrm{d}\tau\; y(\sigma)\,\mathrm{d}\sigma. \qquad (1.28)
\end{aligned}$$

By definition of $C_\lambda(y)$, the last two terms add up to $\int_0^1 f(\sigma)y(\sigma)\,\mathrm{d}\sigma$, where

$$f(\sigma) = \frac{aD_\lambda(g) + g(1)}{\lambda + a(1-\mathrm{e}^{-\lambda})}\mathrm{e}^{\lambda(\sigma-1)} + \mathrm{e}^{\lambda\sigma}\int_\sigma^1 \mathrm{e}^{-\lambda\tau} g(\tau)\,\mathrm{d}\tau, \qquad \sigma \in [0,1]. \quad (1.29)$$

The latter function is manifestly continuous on the entire interval $[0, 1]$ and $f(1) = \frac{aD_\lambda(g)+g(1)}{\lambda+a(1-\mathrm{e}^{-\lambda})}$. Hence, (1.28) takes the form

$$g(R_\lambda y) = \int_{[0,1]} f\, y\,\mathrm{d}\mu = f(y).$$

We see that, for any g, the class of $R_\lambda^* g$ contains a function that is continuous on $[0, 1]$, namely, f given by (1.29). Since the essential supremum of $|f| = |R_\lambda^* g|$ coincides with the supremum norm of f, the range of R_λ^* may be identified with a subspace of $C[0, 1]$.

With this identification in mind, we claim that $\mathbb{X}^\odot = C[0, 1]$,

$$\mathcal{D}(A_\mathrm{p}^*) = \{f \in C^1[0,1];\ f'(1) = a(f(0) - f(1))\} \qquad (1.30)$$

and

$$A_\mathrm{p}^* f = f'.$$

For the sake of this argument, let $\mathcal{D}$ be the right-hand side of equality (1.30). If $f \in \mathcal{D}$, then $\lambda f - f'$ belongs to $C[0, 1] \subset \mathbb{X}^*$. Integrating by parts, we see that

$$\int_\sigma^1 \mathrm{e}^{-\lambda\tau}\left(\lambda f(\tau) - f'(\tau)\right) \mathrm{d}\tau = \mathrm{e}^{-\lambda\sigma} f(\sigma) - \mathrm{e}^{-\lambda} f(1), \qquad \sigma \in [0, 1],$$

and in particular $D_\lambda(\lambda f - f') = f(0) - \mathrm{e}^{-\lambda} f(1)$. Thus, looking at the definition of $\mathcal{D}$, we conclude that

$$\frac{a D_\lambda(\lambda f - f') + \lambda f(1) - f'(1)}{\lambda + a(1 - \mathrm{e}^{-\lambda})} = f(1)$$

and that the right-hand side of (1.29) with g replaced by $\lambda f - f'$ returns $f(\sigma)$. In other words, any $f \in \mathcal{D}$ is of the form (1.29) for $g = \lambda f - f'$. It follows that

$$\mathcal{D} \subset \mathcal{D}(A^*)$$

and since $\mathcal{D}$ is dense in $C[0, 1]$, we obtain $\mathbb{X}^\odot = cl\, \mathcal{D}(A^*) = C[0, 1]$, the first part of our claim.

Since $\left(\lambda - A_{\mathrm{p}}^*\right)^{-1}$ is $(\lambda - A^*)^{-1}$ restricted to $\mathbb{X}^\odot$, any member f of $\mathcal{D}(A_{\mathrm{p}}^*)$ is of the form (1.29) with $g \in C[0, 1]$. Such an f manifestly belongs to $C^1[0, 1]$ and a bit of algebra, based on (1.29), shows that $f'(1) = a(f(0) - f(1))$. Thus

$$\mathcal{D}(A_{\mathrm{p}}^*) \subset \mathcal{D}.$$

We have also seen that for $f \in \mathcal{D}$, the right-hand side of (1.29) returns $f(\sigma)$ provided we take $g = \lambda f - f'$. Since this g belongs to $C[0, 1] = \mathbb{X}^\odot$, this reveals that such an f belongs to $\mathcal{D}(A_{\mathrm{p}}^*)$, implying $\mathcal{D} \subset \mathcal{D}(A_{\mathrm{p}}^*)$, and completing the proof of (1.30).

Finally, differentiating (1.29) we see that $f' = \lambda f - g$ or

$$(R_\lambda^* g)' = \lambda R_\lambda^* g - g, \qquad g \in C[0, 1].$$

On the other hand, since for such g, $R_\lambda^* g = \left(\lambda - A_{\mathrm{p}}^*\right)^{-1}$, we have also

$$A_{\mathrm{p}}^* R_\lambda^* g = \lambda R_\lambda^* g - g, \qquad g \in C[0, 1].$$

Thus $A_{\mathrm{p}}^* R_\lambda^* g = (R_\lambda^* g)'$ for all $g \in C[0, 1]$, and this is the same as saying that $A_{\mathrm{p}}^* f = f'$ for all $f \in \mathcal{D}(A_{\mathrm{p}}^*)$.

1.5.8 Exercise

Let $\mathbb{X}$ be a Banach space and let $x_0 \in \mathbb{X}$ and $f_0 \in \mathbb{X}^*$ be fixed. Find the dual operator for $A \in \mathcal{L}(\mathbb{X})$ given by $Ax = f_0(x)x_0$.

1.5.9 Exercise

Let A be a densely defined, not necessarily closed linear operator. Show that A^* is closed.

1.6 Appendix: On Convergence in l^1

A lion's share of sequences of semigroups considered in this book will be composed of Markov operators in $l^1 := l^1(\mathbb{I})$, where $\mathbb{I}$ is a countable set. Hence, in this section we gather some facts concerning convergence in l^1. First we recall Scheffé's Theorem, which we will find useful in many places in this book.

1.6.1 *Convergence in l^1 and Scheffé's Theorem*

We start by noting that for a sequence $(x_n)_{n\geq 1}$ of elements of l^1, where, say, $x_n = \left(\xi_{n,i}\right)_{i\in\mathbb{I}}$, to converge to an $(\xi_i)_{i\in\mathbb{I}}$ it is necessary for numerical sequences $\left(\xi_{n,i}\right)_{n\geq 1}, i \in \mathbb{I}$ to converge to ξ_i. This is easy to see since $|\xi_{n,i} - \xi_i| \leq \|x_n - x\|$. However, if $\mathbb{I}$ is not finite, the converse is not true. For example, taking $\mathbb{I} = \mathbb{N}$ and defining

$$e_n = \left(\delta_{n,i}\right)_{i\geq 1}, \tag{1.31}$$

where $\delta_{n,i}$ equals 1 if $n = i$ and zero otherwise ($\delta_{i,n}$ is the Kronecker delta), we see that $\lim_{n\to\infty} \delta_{n,i} = 0$ for all $i \in \mathbb{N}$ and yet $(e_n)_{n\geq 1}$ does not converge to the zero vector since for all $n \geq 1$, $\|e_n - 0\| = \|e_n\| = 1$. Hence, **convergence in norm** involves more than **coordinate convergence**.

A noteworthy exception is the case where all elements of the sequence $(x_n)_{n\geq 1}$ *and* the limit x are **distributions**; an $x \in l^1$ is said to be a distribution if it is nonnegative and its coordinates add up to 1. (The example presented above makes it clear that the *and* in the previous sentence cannot be omitted.) Then, coordinate-wise convergence is necessary and sufficient for convergence in norm, the statement known as **Scheffé's Theorem**. To see this, let $A_n = \{i \in \mathbb{I}, \xi_i \geq \xi_{n,i}\}$. We have

$$\begin{aligned}\sum_{i\in\mathbb{I}} |\xi_{n,i} - \xi_i| &= \sum_{i\in A_n} |\xi_{n,i} - \xi_i| + \sum_{i\notin A_n} (\xi_{n,i} - \xi_i) \\ &= \sum_{i\in A_n} |\xi_{n,i} - \xi_i| + 1 - 1 + \sum_{i\in A_n} (\xi_i - \xi_{n,i}) \\ &= 2\sum_{i\in A_n} |\xi_{n,i} - \xi_i| = \sum_{i\in\mathbb{I}} \eta_{n,i},\end{aligned}$$

where $\eta_{n,i}$ equals $2|\xi_{n,i} - \xi_i|$ for $i \in A_n$ and zero otherwise. Since $\eta_{n,i} \le 2\xi_i$ for all $n \ge 1$ and $i \in \mathbb{I}$, the claim follows by the Lebesgue Dominated Convergence Theorem.

For completeness, let us also prove the latter theorem which, by the way, may be seen as describing another case where convergence in coordinates implies convergence in norm: it says that if a sequence $(x_n)_{n\ge1}$ of elements of l^1 converges to an $x = (\xi_i)_{i\in\mathbb{I}} \in l^1$ in coordinates and there is a $y = (\eta_i)_{i\in\mathbb{I}} \in l^1$ such that $|x_n| \le y$ (i.e., $|\xi_{n,i}| \le \eta_i$ for all $i \in \mathbb{I}$ and $n \in \mathbb{N}$) then $(x_n)_{n\ge1}$ converges to x also in the norm: $\lim_{n\to\infty} \|x_n - x\| = 0$. For, given $\epsilon > 0$ we may find a finite subset $A \subset \mathbb{N}$ such that $\sum_{i\in\mathbb{I}\setminus A} |\eta_i| < \frac{\epsilon}{4}$. Since, by assumption, $|\xi_i| \le \eta_i$ for all $i \in \mathbb{I}$, it follows that

$$\sum_{i\in\mathbb{I}\setminus A} |\xi_{n,i} - \xi_i| < \sum_{i\in\mathbb{I}\setminus A} 2\eta_i < \frac{\epsilon}{2}.$$

On the other hand, since A is finite, the sum

$$\sum_{i\in A} |\xi_{n,i} - \xi_i|$$

may be made smaller than $\frac{\epsilon}{2}$ by taking n sufficiently large. This means that, for such n, $\|x_n - x\| < \epsilon$, and thus completes the proof.

1.6.2 *Scheffé's Theorem (continued)*

It is convenient to have at our disposal an apparently more general version of Scheffé's Theorem: it says that a sequence $(x_n)_{n\ge1}$ of nonnegative elements of l^1 converges in norm to a nonnegative element $x \ne 0$ as long as it converges in coordinates and $\lim_{n\to\infty} \|x_n\| = \|x\|$. To prove this version, we note that under these assumptions, $\|x_n\| > 0$ for all sufficiently large n, and it makes sense to define $x'_n := \frac{x_n}{\|x_n\|}$ and $x' = \frac{x}{\|x\|}$. These newly formed vectors are distributions and the original statement of Scheffé's Theorem tells us that $\lim_{n\to\infty} x'_n = x'$. It follows that $\lim_{n\to\infty} x_n = \lim_{n\to\infty} \|x_n\| x'_n = \|x\| x' = x$. □

1.6.3 l^1 as a Kantorovič–Banach space

Here is another characteristic of l^1 which distinguishes it from other Banach spaces. Suppose $(x_n)_{n\ge 1}$ is a sequence of elements of l^1 such that $0 \le x_n \le x_{n+1}, n \ge 1$, and $\|x_n\| \le M, n \ge 1$, for some $M > 0$. We will show that these conditions imply existence of the limit $\lim_{n\to\infty} x_n$; this, by definition, means that l^1 is a Kantorovič–Banach space.

To this end, let $x_n = (\xi_{n,i})_{i\in\mathbb{I}}$. By assumption, each of the numerical sequences $(\xi_{n,i})_{n\ge 1}, i \in \mathbb{I}$ is nonnegative, nondecreasing and bounded, and therefore has a limit, say, $\xi_i := \lim_{n\to\infty} \xi_{n,i} \ge 0$. I claim that $x := (\xi_i)_{i\in\mathbb{I}}$ is a member of l^1. For, if $\sum_{i\in\mathbb{I}} \xi_i = \infty$, a finite subset $\mathbb{I}' \subset \mathbb{I}$ may be chosen so that $\sum_{i\in\mathbb{I}'} \xi_i \ge 2M$. Moreover, an n_0 can be chosen so that for all $n \ge n_0$, and $i \in \mathbb{I}'$

$$|\xi_i - \xi_{n,i}| = \xi_i - \xi_{n,i} < Mk^{-1}, \qquad \text{i.e.} \qquad \xi_{n,i} > \xi_i - Mk^{-1}$$

where k is the number of elements of $\mathbb{I}'$. Then, for such n,

$$\|x_n\| \ge \sum_{i\in\mathbb{I}'} \xi_{n,i} > 2M - M = M,$$

contradicting our assumption. Once we know that x is a member of l^1, however, convergence of $(x_n)_{n\ge 1}$ to x is a consequence of the Lebesgue Dominated Convergence Theorem.

The space l^1 is truly exceptional. On the one hand, all separable Banach spaces are (isomorphic to) quotients of l^1 (see, e.g., [20]); on the other it is the only classical, infinite-dimensional Banach space where weak convergence of sequences implies their strong convergence. We prove this fact, due to J. Schur [79], in our next section.

1.6.4 Theorem of Schur: Weakly convergent sequences in l^1 converge strongly

To recall, a sequence $(x_n)_{n\ge 1}$ of elements of a Banach space $\mathbb{X}$ is said to converge to an $x \in \mathbb{X}$ weakly if for any continuous functional f on $\mathbb{X}$ we have $\lim_{n\to\infty} f(x_n) = f(x)$. Clearly, all convergent sequences are weakly convergent, but not vice versa. In l^1, however, all weakly convergent sequences are convergent.

For a proof of this surprising statement, we recall that, by the Banach–Steinhaus Theorem (i.e., the Uniform Boundedness Principle), a weakly convergent sequence is bounded in norm. Moreover, by linearity, it suffices to show that if a sequence $(x_n)_{n\ge 1}$ of elements of l^1 converges weakly to

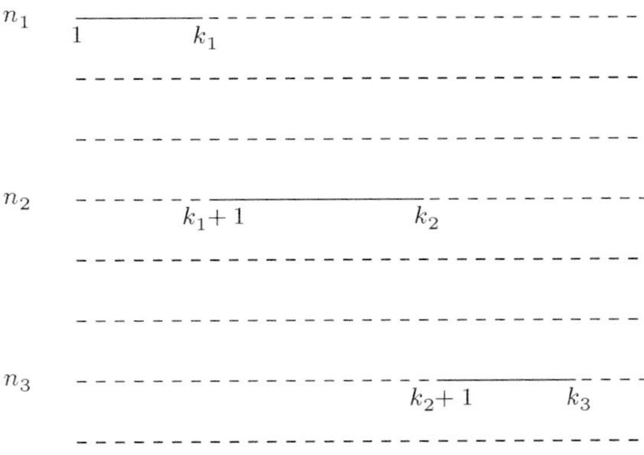

Figure 1.1 Choosing sequences $(n_j)_{j\ge1}$ and $(k_j)_{j\ge1}$.

0, then it converges strongly, as well. Suppose that it is not so. Then, $\|x_n\|$ does not converge to 0, as $n \to \infty$. Any bounded sequence of nonnegative numbers that does not converge to 0 contains a positive subsequence converging to a positive number. Moreover, a subsequence of a weakly convergent sequence converges weakly. Hence, without loss of generality, we may assume that $\lim_{n\to\infty}\|x_n\| = r > 0$, and that $\|x_n\| \neq 0$. Then the sequence $(y_n)_{n\ge1}$, where $y_n := \frac{1}{\|x_n\|}x_n =: \left(\eta_{n,i}\right)_{i\ge1}$, converges weakly to zero, whereas $\|y_n\| = \sum_{i=1}^{\infty}|\eta_{n,i}| = 1$. We will show that such a sequence may not exist.

To simplify the argument, but without loss of generality, in what follows we assume that $\mathbb{I} = \mathbb{N}$. Since any bounded sequence $(\alpha_i)_{i\ge1}$ induces a bounded linear functional on l^1 (by the formula $f\,(\xi_i)_{i\ge1} = \sum_{i=1}^{\infty}\alpha_i\xi_i$), by assumption we have $\lim_{n\to\infty}\sum_{i=1}^{\infty}\alpha_i\eta_{n,i} = 0$. In particular, for any $\ell \ge 1$ we may take $\alpha_i = \delta_{\ell,i}, i \ge 1$. This shows that

$$\lim_{n\to\infty}\eta_{n,\ell} = 0, \quad \ell \ge 1. \tag{1.32}$$

To complete the proof, we inductively define two sequences of integers: $(n_j)_{j\ge1}$ and $(k_j)_{j\ge1}$ (see Figure 1.1). First we put $n_1 = 1$ and choose k_1 so large that $\sum_{i=1}^{k_1}|\eta_{n_1,i}| \ge \frac{3}{5}$. By (1.32), having chosen n_j and k_j we may choose $n_{j+1} > n_j$ large enough to have $\sum_{i=1}^{k_j}|\eta_{n_{j+1},i}| < \frac{1}{5}$ and then, since $\|y_n\| = 1$ for all $n \ge 1$, we may choose a $k_{j+1} > k_j + 1$ so that

$$\sum_{i=k_j+1}^{k_{j+1}}|\eta_{n_{j+1},i}| > \frac{3}{5}. \tag{1.33}$$

Now, define

$$\alpha_i = \operatorname{sgn} \eta_{n_j,i} \quad \text{for} \quad i \in A_j := \{k_{j-1}+1, \dots, k_j\}$$

where $k_0 := 0$, and let f be a continuous linear functional on l^1 related to this bounded sequence. Then, by (1.33),

$$\begin{aligned} f(y_{n_j}) &= \sum_{i \in A_j} \alpha_i \eta_{n_j,i} + \sum_{i \notin A_j} \alpha_i \eta_{n_j,i} = \sum_{i \in A_j} |\eta_{n_j,i}| + \sum_{n \notin A_i} \alpha_i \eta_{n_j,i} \\ &\geq \sum_{i \in A_j} |\eta_{n_j,i}| - \sum_{n \notin A_i} |\eta_{n_j,i}| = 2 \sum_{i \in A_j} |\eta_{n_j,i}| - 1 > \frac{1}{5}, \end{aligned}$$

for all j, contradicting the fact that $\lim_{j\to\infty} f(y_{n_j}) = 0$. This contradiction completes the proof.

An interesting consequence of this theorem will be discussed in 5.1.4.

1.7 Notes

The theory of semigroups of operators is described in many monographs, including [3, 14, 24, 25, 32, 34, 36–39, 47, 51, 53, 55, 58, 66, 67, 73, 83, 90, 91, 95], and each of these books covers slightly different aspects of the theory. Convergence and approximation theory for semigroups can be found in my own [16]. J. van Neerven's monograph [87] is a treasury of results on sun-dual semigroups (see also his subsequent papers); our Section 1.5 contains only basic facts needed to understand later chapters. A different proof of Schur's Theorem may be found in [32], p. 295 and [95], p. 122; see also, for example, [19] and [20].

Concerning usefulness of semigroups in the theory of stochastic processes alluded to at the beginning of this chapter: Over 50 years ago, D.G. Kendall has conjectured, with certain probabilistic applications in mind, that all Markov semigroups in l^1 that can be extended to groups are generated by bounded operators. Subsequently, after a joint effort and partial positive results of several mathematicians, a possible way to a counterexample to this conjecture was outlined and discussed in a couple of influential papers by J.F.C. Kingman. It was only recently that J. Glück, using semigroup-theoretic techniques, has proved an elegant theorem showing that this way is a dead end (see J. Glück's 'On the decoupled Markov group conjecture' arXiv:2004.05995 [math.FA] and the references given there).

2

Generators versus Intensity Matrices

In this chapter, we take up the main subject of the book: illustrating the fact that generators describe Markov chains in a more complete way than intensity matrices do. After defining the latter in Section 2.2 we show in Section 2.3 that in the uniform case the generators may be identified with intensity matrices. Then, in Section 2.4 we discuss the general case and the Hille–Yosida Theorem for Markov semigroups in l^1, thus characterizing generators of Markov chains. This theorem is exemplified with two examples due to Kolmogorov, Kendall and Reuter; besides well illustrating the main idea of the book, these examples are really beautiful applications of the Hille–Yosida Theorem. Moreover, we introduce the generator of a certain birth and death process which will later, in the next chapter, be of particular interest. Section 2.5 is devoted to infinitesimal, local, description of Markov chains, providing an intuitive meaning to the elements of intensity matrices. Information gathered in this section allows then a more detailed description of the Kolmogorov–Kendall–Reuter examples, presented in Sections 2.6 and 2.7. The latter sections discuss the vital information on the Kolmogorov–Kendall–Reuter processes that is contained in their generators but is missing in the corresponding intensity matrices. The chapter is concluded with Section 2.8 where Blackwell's example of a Markov chain is presented in which all states are instantaneous.

2.1 Transition Matrices and Markov Operators in l^1

Intuitively, a Markov chain $X(t), t \geq 0$ where $X(t)$ are random variables with values in $\mathbb{N}$, or, more generally, in a finite or countable state-space $\mathbb{I}$, is a stochastic process 'without memory.' More precisely, for each $t \geq 0$ the information on values of $X(s)$, $s \le t$, that is, 'the past,' influences the future, that is, values of $X(s)$, $s \ge t$ merely through the present, that is, through $X(t)$.

Such processes are conveniently described by transition probabilities

$$p_{i,j}(t,s) = \mathbb{P}(X(t) = j | X(s) = i),$$

where the right-hand side is the conditional probability that at time t the process is in the state j given that at time $s \le t$ it is in i. In this book we will deal merely with time-homogeneous Markov chains, that is, the chains with transition probabilities that do not change in time: for any $h > 0$,

$$p_{i,j}(t+h, s+h) = p_{i,j}(t,s).$$

For such processes, it suffices to consider the transition probabilities $p_{i,j}(t,0)$ denoted simply $p_{i,j}(t)$, $t \ge 0$. For future reference we note that

$$\sum_{j \in \mathbb{I}} p_{i,j}(t) = 1, \qquad t \ge 0. \tag{2.1}$$

This formula simply says that the process is honest: the probability that the process starting at an i will be somewhere in $\mathbb{I}$ at time $t \ge 0$ is 1.

Transition probabilities are customarily gathered in transition matrices

$$P(t) = \left(p_{i,j}(t)\right)_{i,j \in \mathbb{I}}, \qquad t \ge 0,$$

and an analytical treatment of Markov chains is based on the fact that transition matrices satisfy the **Chapman–Kolmogorov equation**,

$$P(s+t) = P(s)P(t), \qquad s,t \ge 0, \tag{2.2}$$

where the right-hand side denotes the matrix product of two matrices. To prove this all-important relation, consider $s,t \ge 0$ and states $i, j \in \mathbb{I}$. Given that $X(0) = i$, the probability that $X(s+t) = j$ is clearly $p_{i,j}(s+t)$. On the other hand, $X(s)$ attains one of the values in $\mathbb{I}$ and so $\mathbb{P}(X(s+t) = j | X(0) = i)$ equals (see Figure 2.1)

$$\sum_{k \in \mathbb{I}} \mathbb{P}(X(s+t) = j | X(0) = i \text{ and } X(s) = k)\mathbb{P}(X(s) = k | X(0) = i)$$

which, by the Markovian nature of the process, and its time-homogeneity, is

$$\sum_{k \in \mathbb{I}} \mathbb{P}(X(s+t) = j | X(s) = k) p_{i,k}(s) = \sum_{k \in \mathbb{N}} p_{i,k}(s) p_{k,j}(t).$$

It follows that $p_{i,j}(s+t)$ is the product of the ith row of the matrix $P(s)$ and of the jth column of the matrix $P(t)$, thus establishing (2.2).

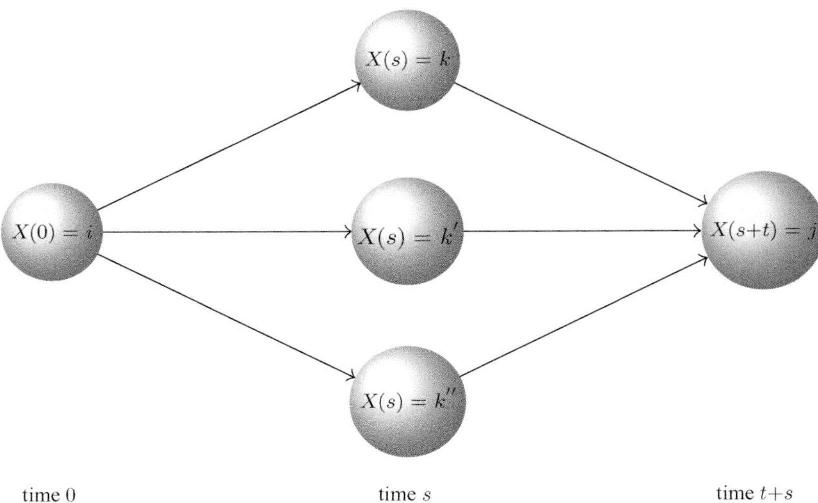

Figure 2.1 An illustration of the Chapman–Kolmogorov equation.

For instance, if $\alpha, \beta \geq 0$ are such that $\alpha + \beta > 0$, then

$$P(t) = \frac{1}{\alpha + \beta} \begin{pmatrix} \beta + \alpha \mathrm{e}^{-(\alpha+\beta)t} & \alpha - \alpha \mathrm{e}^{-(\alpha+\beta)t} \\ \beta - \beta \mathrm{e}^{-(\alpha+\beta)t} & \alpha + \beta \mathrm{e}^{-(\alpha+\beta)t} \end{pmatrix} \tag{2.3}$$

is a semigroup of transition matrices (here, $\#\mathbb{I} = 2$). The matrices of Section 1.1.6 also form such a semigroup (here, $\#\mathbb{I} = 3$). In yet another example, which is of interest in itself, usually referred to as **Speakman's example** (see [82] or [45], [p. 271]), there are two transition semigroups: The first of them is

$$P(t) = \begin{pmatrix} p_1(t) & p_2(t) & p_3(t) \\ p_3(t) & p_1(t) & p_2(t) \\ p_2(t) & p_3(t) & p_1(t) \end{pmatrix}$$

where

$$p_1(t) = \frac{1}{3} + \frac{2}{3}\mathrm{e}^{-\frac{3}{2}t} \cos\left(\frac{\sqrt{3}}{2}t\right),$$

$$p_2(t) = \frac{1}{3} + \frac{2}{3}\mathrm{e}^{-\frac{3}{2}t} \cos\left(\frac{\sqrt{3}}{2}t - \frac{2}{3}\pi\right),$$

$$p_3(t) = \frac{1}{3} + \frac{2}{3}\mathrm{e}^{-\frac{3}{2}t} \cos\left(\frac{\sqrt{3}}{2}t + \frac{2}{3}\pi\right).$$

The second is

$$P^{\diamond}(t) = \begin{pmatrix} p_1^{\diamond}(t) & p_2^{\diamond}(t) & p_2^{\diamond}(t) \\ p_2^{\diamond}(t) & p_1^{\diamond}(t) & p_2^{\diamond}(t) \\ p_2^{\diamond}(t) & p_2^{\diamond}(t) & p_1^{\diamond}(t) \end{pmatrix},$$

where $p_1^{\diamond}(t) = \frac{1}{3} + \frac{2}{3}\mathrm{e}^{-\frac{3}{2}t}$ and $p_2^{\diamond}(t) = \frac{1}{3} - \frac{1}{3}\mathrm{e}^{-\frac{3}{2}t}$.

The main interest in this example comes from the fact that, although these two semigroups are clearly different, we have $P(nc) = P^{\diamond}(nc)$ for all natural n and $c = \frac{4\pi}{\sqrt{3}}$. Indeed, a bit of elementary trigonometry shows that $p_1(nc) = p_1^{\diamond}(nc)$ and $p_2(nc) = p_3(nc) = p_2^{\diamond}(nc)$.

2.1.1 *Transition matrices as Markov operators in l^1*

Suppose that at time $t = 0$, our Markov chain $\{X(t), t \geq 0\}$ is at state $i \in \mathbb{I}$ with probability p_i. The total probability formula shows then that

$$\mathbb{P}(X(t) = j) = \sum_{i \in \mathbb{I}} p_i p_{i,j}(t).$$

In matrix notation, this may be expressed by saying that the distribution at time t is the matrix product of the distribution at time 0, treated as a row-vector $x = (p_i)_{i\in\mathbb{I}}$, and of the transition matrix $P(t)$:

$$y = x \cdot P(t). \tag{2.4}$$

(Here, y is a row-vector, as well.) Yet in other words, the jth coordinate in y is the scalar product of x and of the jth column in $P(t)$.

Let l^1 be the space of absolutely summable sequences $x = (\xi_i)_{i\in\mathbb{I}}$, that is, such $(\xi_i)_{i\in\mathbb{I}}$ that $\sum_{i\in\mathbb{I}} |\xi_i| < \infty$. It is useful to think of formula (2.4) as of a map assigning to any $x \in l^1$, not necessarily being a distribution, the product of x and $P(t)$. Clearly, such a multiplication is possible here even if $\mathbb{I}$ is infinite, because x is absolutely summable, and all $p_{i,j}(t) \geq 0$ do not exceed 1. Also, because of (2.1), for $x = (\xi_i)_{i\in\mathbb{I}} \in l^1$,

$$\|y\| = \sum_{j\in\mathbb{I}} \left| \sum_{i\in\mathbb{I}} \xi_i p_{i,j}(t) \right| \leq \sum_{i\in\mathbb{I}} |\xi_i| \sum_{j\in\mathbb{I}} p_{i,j}(t) = \sum_{i\in\mathbb{N}} |\xi_i| = \|x\|, \tag{2.5}$$

so that y indeed belongs to l^1.

The map described above will in what follows be denoted by $P(t)$ (and thus will not be distinguished from the transition matrix) and, to comply with the customs regarding linear operators, we will write $P(t)x$ instead of $x \cdot P(t)$:

$$y = P(t)x. \tag{2.6}$$

This should not lead to misunderstandings: whereas (2.4) stresses the fact that y is obtained as a result of multiplication of a vector and a matrix, in writing (2.6) we stress that the manipulations we are performing are linear.

Equation (2.4) makes it clear that $P(t)$ maps nonnegative x into nonnegative y. Moreover, for such x, inequalities in (2.5) become equalities. This shows that $P(t)$ is a Markov operator.

2.1.2 *Strong continuity*

As seen in the previous section, transition matrices of a Markov chain may be identified with a family $\{P(t), t \geq 0\}$ of Markov operators in l^1. Since $p_{i,i}(0) = 1$ and $p_{i,j}(0) = 0$ for $j \neq i$, we have

$$P(0) = I, \tag{2.7}$$

and the Chapman–Kolmogorov equation says that these operators form a semigroup.

The so-obtained semigroup, however, need not *a priori* be strongly continuous. We claim that a semigroup of Markov operators $P(t) = (p_{i,j}(t))_{i,j\in\mathbb{I}}$, $t \geq 0$ is strongly continuous iff

$$\lim_{t\to 0+} p_{i,i}(t) = 1, \qquad i \in \mathbb{I}. \tag{2.8}$$

This condition is necessary, since strong continuity implies in particular that $\lim_{t\to 0} P(t)e_i = e_i$, where

$$e_i = \left(\delta_{i,j}\right)_{j\in\mathbb{I}}, \tag{2.9}$$

and the ith coordinate of $P(t)e_i$ is $p_{i,i}(t)$. To prove the converse, we take an arbitrary $x = (\xi_i)_{i\geq 1} \in l^1$ to calculate

$$\begin{aligned}
\|P(t)x - x\| &= \sum_{j\in\mathbb{I}} \left| \sum_{i\in\mathbb{I}} \xi_i p_{i,j}(t) - \xi_j \right| \\
&\leq \sum_{j\in\mathbb{I}} [1 - p_{j,j}(t)]|\xi_j| + \sum_{j\in\mathbb{I}} \sum_{i\in\mathbb{I}, i\neq j} |\xi_i| p_{i,j}(t) \\
&= \sum_{j\in\mathbb{I}} [1 - p_{j,j}(t)]|\xi_j| + \sum_{i\in\mathbb{I}} |\xi_i| \sum_{j\in\mathbb{I}, j\neq i} p_{i,j}(t) \\
&= 2\sum_{j\in\mathbb{I}} [1 - p_{j,j}(t)]|\xi_j|,
\end{aligned} \tag{2.10}$$

with the last equality following by $\sum_{j\in\mathbb{I}, j\neq i} p_{i,j}(t) = 1 - p_{i,i}(t)$. Since (2.8) implies $\lim_{t\to 0} \sum_{j\in\mathbb{I}} [1 - p_{j,j}(t)]|\xi_j| = 0$ by the Dominated Convergence Theorem, we are done.

In the remainder of this chapter we assume that (2.8) is satisfied. In other words, all semigroups in this chapter are strongly continuous. Note that this condition implies continuity of all $t \mapsto p_{i,j}(t)$. *In particular,* $\lim_{t\to 0+} p_{i,j}(t) = 0, i \neq j$.

2.1.3 Exercise

If the second condition in (2.1) is replaced by

$$\sum_{j\in\mathbb{I}} p_{i,j}(t) \le 1, \qquad t \ge 0,$$

the resulting semigroup is not Markov but sub-Markov in that $\|P(t)x\| \le \|x\|$ for all positive x. Check the details of the proof in the previous subsection to see that also in this case (2.8) is necessary and sufficient for the semigroup to be strongly continuous.

2.2 The Matrix of Intensities

2.2.1 *Infinitesimal description*

It is probably clear to the reader that explicit formulae for transition matrices are usually not available, especially for more complicated Markov chains. Neither are they very informative, even if available. Let us look, for example, at the upper left corner of transition matrix (2.3). It tells us that the probability that a chain starting at a certain state (call it 1 to distinguish it from the other state, say, 2) will after time $t \ge 0$ be back there is

$$\frac{\beta + \alpha e^{-(\alpha+\beta)t}}{\alpha+\beta}.$$

Well, should I say 'back there' or 'still there'? In fact, it is both. The process might have stayed at 1 all the time, but could have also jumped back and forth between 1 and 2 to land at state 1 at time t; the probability given above combines all such events together. In particular, without further analysis we cannot tell from this exact formula how long the process stays at 1 before jumping to 2. (As we shall see later, the probability that the process is 'still there' is $e^{-\alpha t}$.)

Such insight may be gained from **infinitesimal description**. We are already familiar with one description of this type – the one furnished by the notion of the generator of a semigroup. Remarkably, Markov chains rarely are described by their generators; more often, Q-matrices, called also Kolmogorov matrices or intensity matrices, are used for this purpose. As we shall see, these two notions are closely related but obviously not identical. In what follows we will have much to say on the interrelations between infinitesimal generators and Q-matrices. Below, step by step we prove existence of Q-matrices for Markov chains.

2.2.2 A theorem on subadditive functions

Let $\phi\colon [0,\infty) \to [0,\infty)$ be subadditive, that is, let $\phi(s+t) \le \phi(s) + \phi(t)$, $s,t \ge 0$, and assume that $\lim_{t\to 0+}\phi(t) = 0$. Then,

$$\lim_{t\to 0+} \frac{\phi(t)}{t} \tag{2.11}$$

exists (but may be infinite) and equals $\sup_{t>0} \frac{\phi(t)}{t}$.

Proof Let $t_n > 0$ be such that $\lim_{n\to\infty} t_n = 0$ and $\liminf_{t\to 0+} \frac{\phi(t)}{t} = \lim_{n\to\infty} \frac{\phi(t_n)}{t_n}$. Next, fix $s > 0$ and for each n choose a natural number k_n and an $h_n \in [0, t_n)$ so that $s = k_n t_n + h_n$. Since ϕ is subadditive,

$$\frac{\phi(s)}{s} \le \frac{k_n\phi(t_n)}{s} + \frac{\phi(h_n)}{s} \le \frac{\phi(t_n)}{t_n} + \frac{\phi(h_n)}{s}.$$

Because of $0 \le h_n < t_n$, we have $\lim_{n\to\infty}\phi(h_n) = 0$. Thus, letting $n \to \infty$, we see that $\frac{\phi(s)}{s} \le \liminf_{t\to 0+} \frac{\phi(t)}{t}$. It follows that $\sup_{t>0} \frac{\phi(t)}{t} \le \liminf_{t\to 0+} \frac{\phi(t)}{t}$. Since by definition $\limsup_{t\to 0+} \frac{\phi(t)}{t} \le \sup_{t>0} \frac{\phi(t)}{t}$, this proves that the limit (2.11) exists and equals $\sup_{t>0} \frac{\phi(t)}{t}$. □

2.2.3 Existence of q_i's

Let $\{P(t), t \ge 0\}$ be a strongly continuous semigroup of sub-Markov operators in l^1. Then, for each $i \in \mathbb{I}$, the limit

$$\boxed{q_i := \lim_{t\to 0+} \frac{1 - p_{i,i}(t)}{t}}$$

exists, is nonnegative but may be infinite.

Proof Since $\lim_{x\to 1} \frac{x-1}{\ln x} = 1$ and, by assumption, $\lim_{t\to 0+} p_{i,i}(t) = 1$, it suffices to show existence of the limit $\lim_{t\to 0+} \frac{-\ln p_{i,i}(t)}{t}$; then both limits will be the same. On the other hand, $p_{i,i}(s+t) = \sum_{j\in\mathbb{I}} p_{i,j}(s)p_{j,i}(t) \ge p_{i,i}(s)p_{i,i}(t)$, proving that $\phi(t) = -\ln p_{i,i}(t)$ is subadditive: $\phi(s+t) \le \phi(s) + \phi(t)$. Since $\lim_{t\to 0+}\phi(t) = 0$ and $\phi(t) \ge 0$, assumptions of 2.2.2 are satisfied, and so we see that

$$q_i := \lim_{t\to 0+} \frac{1 - p_{i,i}(t)}{t} = \lim_{t\to 0+} \frac{-\ln p_{i,i}(t)}{t} = \sup_{t>0} \frac{-\ln p_{i,i}(t)}{t}. \qquad \square$$

2.2.4 Definition

q_i is the intensity of jump from i. If $q_i < \infty$, the state $i \in \mathbb{I}$ is said to be **stable**. Otherwise, it is called **instantaneous**.

2.2.5 *Notation*

It will become useful to define:

$$q_{i,i} = -q_i, \qquad i \in \mathbb{I}.$$

2.2.6 Corollary

Let i be a stable state. Then $p_{i,i}(t) \geq \mathrm{e}^{-q_i t}, t \geq 0$. For, the last line of the proof presented in Section 2.2.3 reveals that

$$-\ln p_{i,i}(t) \leq q_i t, \qquad \text{for all} \quad t > 0.$$

This is equivalent to our statement. (The case $t = 0$ is trivial.)

2.2.7 *Existence of $q_{i,j}$'s*

Let $i \neq j$. The limit

$$\boxed{q_{i,j} := \lim_{t \to 0+} \frac{p_{i,j}(t)}{t} \geq 0}$$

exists and is finite; $q_{i,j}$ is the intensity of jump from i to j.

Proof Let $q_{i,j} := \limsup_{t \to 0+} t^{-1} p_{i,j}(t)$, and let $t_n > 0$ be such that $\lim_{n\to\infty} t_n = 0$ and $\lim_{n\to\infty} t_n^{-1} p_{i,j}(t_n) = q_{i,j}$. (Clearly, $q_{i,j} \geq 0$ but we do not exclude the possibility $q_{i,j} = \infty$ yet.)

Let $t > 0$ and an integer k be fixed, and let $X_0, X_1, \ldots, X_k$ be a discrete-parameter, finite Markov chain with initial state i and transition matrix $P = P(t)$, that is, suppose that for each states $i_1, \ldots, i_k \in \mathbb{I}$,

$$\mathbb{P}(X_0 = i, X_1 = i_1, \ldots, X_k = i_k) = p_{i,i_1}(t) p_{i_2,i_3}(t) \cdots p_{i_{k-1},i_k}(t).$$

We define $\tau = \tau(\omega)$ as the smallest integer $\alpha \in \{1, \ldots, k\}$ such that $X_\alpha = j$; if no such α exists, we agree that $\tau = k + 1$. We claim that

$$p_{i,j}(kt) \geq \sum_{\ell=1}^{k-1} \mathbb{P}(X_\ell = i, \tau > \ell) p_{i,j}(t) p_{j,j}((k - \ell - 1)t). \tag{2.12}$$

Indeed, the ℓth term on the right-hand side is the probability that a path starting at i is at i again at $\ell \geq 1$, visits j for the first time at time $\ell + 1$, and then (possibly leaves and) returns to j after the final $k - \ell - 1$ steps. Clearly, such path leads from i to j, but not all paths leading from i to j have this form. Moreover, the probabilities corresponding to different ℓ in the sum come from pairwise disjoint events: hence the claim.

On the other hand, for each $\ell \in \{1, \ldots, k-1\}$,

$$p_{i,i}(\ell t) = \mathbb{P}(X_\ell = i, \tau > \ell) + \sum_{m=1}^{\ell-1} \mathbb{P}(\tau = m) p_{j,i}((\ell - m)t).$$

Indeed, each path starting and ending at i after ℓ steps, either never goes through j, or there is the first time it visits j and then the path leads from j to i in the remaining time. Since $\sum_{m=1}^{\ell-1} \mathbb{P}(\tau = m) = 1 - \mathbb{P}(\tau \geq \ell) \leq 1$, this relation implies

$$\mathbb{P}(X_\ell = i, \tau > \ell) \geq p_{i,i}(\ell t) - \max_{m=1,\ldots,\ell-1} p_{j,i}(mt). \tag{2.13}$$

Armed with (2.12) and (2.13), we can complete the proof: Given $\epsilon \in (0, 1)$, we find $h = h(\epsilon)$ such that

$$p_{j,i}(s) < \frac{\epsilon}{2}, \qquad p_{i,i}(s) > 1 - \frac{\epsilon}{2} \qquad \text{and} \qquad p_{j,j}(s) > 1 - \epsilon,$$

provided $s \in (0, h)$. Fix such s's. Then k_n's defined as integral parts of st_n^{-1} are integers such that $\lim_{n\to\infty} k_n t_n = s$. Thus, $k_n t_n < h$ provided n is sufficiently large, and each of these k_n's may play the role of k in the analysis leading to (2.12) and (2.13). Therefore, since m and ℓ in (2.13) satisfy $m < \ell < k_n$, using this relation with t replaced by t_n, we obtain

$$\mathbb{P}(X_\ell = i, \tau > \ell) \geq p_{i,i}(\ell t_n) - \frac{\epsilon}{2} > 1 - \epsilon.$$

Also, $p_{j,j}((k_n - 1 - \ell)t_n) > 1 - \epsilon$, for each $\ell \in \{1, \ldots, k_n - 1\}$. Inequality (2.12) now shows that

$$p_{i,j}(k_n t_n) \geq (1 - \epsilon)^2 (k_n - 1) p_{i,j}(t_n).$$

It follows that

$$\frac{p_{i,j}(s)}{s} = \lim_{n\to\infty} \frac{p_{i,j}(k_n t_n)}{k_n t_n} \geq (1-\epsilon)^2 \lim_{n\to\infty} \frac{k_n - 1}{k_n} \frac{p_{i,j}(t_n)}{t_n} = (1-\epsilon)^2 q_{i,j}.$$

As a first consequence of this inequality, $q_{i,j} < \infty$. Furthermore, since $s \in (0, h)$ has been chosen arbitrarily

$$\liminf_{s\to 0+} \frac{p_{i,j}(s)}{s} \geq (1-\epsilon)^2 q_{i,j},$$

and then $\liminf_{t\to 0+} \frac{p_{i,j}(t)}{t} \geq \limsup_{t\to 0+} \frac{p_{i,j}(t)}{t}$, $\epsilon \in (0, 1)$ being arbitrary as well. Hence, $\lim_{t\to 0+} \frac{p_{i,j}(t)}{t}$ exists and is finite. □

2.2.8 Definition

A matrix $(q_{i,j})_{i,j\in\mathbb{I}}$ is termed a **Kolmogorov matrix** (or a **Q-matrix**) if

- $\sum_{j\in\mathbb{I}} q_{i,j} = 0$ for all $i \in \mathbb{I}$,
- $q_{i,j} \ge 0, i \ne j$.

It often happens that the first condition here is replaced by a weaker one: $\sum_{j\in\mathbb{I}} q_{i,j} \le 0$. It will be convenient to refer to such a matrix as an **intensity matrix** (although in the literature, the terms *Kolmogorov matrix*, *Q-matrix*, and *intensity matrix* are often used interchangeably).

2.2.9 Corollary

If all states of a sub-Markov transition matrix are stable, then $(q_{i,j})_{i,j\in\mathbb{I}}$ defined in Sections 2.2.3, 2.2.5, and 2.2.7 is an intensity matrix.

Proof Since $t^{-1}\sum_{j\ne i} p_{i,j}(t) \le t^{-1}(1 - p_{i,i}(t))$, Fatou's lemma implies $\sum_{j\ne i} q_{i,j} \le q_i$. □

Hence, to any semigroup of Markov or sub-Markov operators one can assign an intensity matrix: the q_i appearing in it (with the minus sign) as a diagonal element is the infinitesimal intensity with which the probability mass escapes from a state i to other states, whereas the off-diagonal element $q_{i,j}$ is the infinitesimal intensity with which the probability mass escapes from the state i to the state j. In Section 2.5 we will be able to interpret the elements of $(q_{i,j})_{i,j\in\mathbb{I}}$ in terms of times the Markov chain involved spends at a given state and probabilities of jumps to other states after the chain leaves this state.

The reader should note that our analysis in no way proves that different semigroups must have different intensity matrices. In fact, some intensity matrices, and even Kolmogorov matrices, correspond to many different semigroups. Of course, this is not the case with the generator: there is a one-to-one correspondence between generators and semigroups. Our main question in this chapter becomes thus even more clear: how are the generator and the intensity matrix related? In the next section we treat the case where these two objects may be identified.

2.2.10 Exercise

Probabilistically, why

$$\frac{\beta}{\alpha+\beta} + \frac{\alpha}{\alpha+\beta}\mathrm{e}^{-(\alpha+\beta)t} \ge \mathrm{e}^{-\alpha t}$$

for all $t \ge 0$ and nonnegative α, β such that $\alpha + \beta > 0$? Give a calculus proof of this relation.

2.3 The Uniform Case

2.3.1 *A criterion for continuity in operator norm*

The following are equivalent.

(i) The generator of a strongly continuous semigroup of sub-Markov operators in l^1 is bounded.

(ii) All states $i \in \mathbb{I}$ are stable, and

$$\sup_{i \in \mathbb{I}} q_i < \infty. \tag{2.14}$$

(iii) The limit (2.8) is uniform in $i \in \mathbb{I}$.

Proof

(i) $\Longrightarrow$ (ii) If $P(t) = \mathrm{e}^{tA}$ for a bounded operator A, then

$$\left\| t^{-1}\left(P(t) - I_{l^1}\right)\right\| \le t^{-1}\left(\mathrm{e}^{t\|A\|} - 1\right) \underset{t \to 0}{\longrightarrow} \|A\|.$$

Hence, there exists a constant $C > 0$ such that

$$\left\| t^{-1}\left(P(t)x - x\right)\right\| \le C\|x\|, \qquad x \in l^1, t \in (0, 1].$$

Taking e_i defined in (2.9), and noting that the ith coordinate of $P(t)e_i - e_i$ is $p_{i,i}(t) - 1$ we conclude that

$$t^{-1}\left(1 - p_{i,i}(t)\right) \le t^{-1}\|P(t)e_i - e_i\| \le C, \qquad t \in (0, 1], i \in \mathbb{I}.$$

This implies (ii).

(ii) $\Longrightarrow$ (iii) By Corollary 2.2.6, $1 - p_{i,i}(t) \le 1 - \mathrm{e}^{-q_i t} \le 1 - \mathrm{e}^{-qt}$, where $q = \sup_{i \in \mathbb{I}} q_i$.

(iii) $\Longrightarrow$ (i) By (2.10),

$$\begin{aligned}\|P(t)x - x\| &\le 2\sum_{j \in \mathbb{I}}[1 - p_{j,j}(t)]|\xi_j| \le 2\sup_{i \in \mathbb{I}}(1 - p_{i,i}(t))\sum_{j \in \mathbb{I}}|\xi_j| \\ &= 2\sup_{i \in \mathbb{I}}(1 - p_{i,i}(t))\,\|x\|, \qquad x = (\xi_i)_{i \in \mathbb{I}} \in l^1.\end{aligned}$$

Hence, $\|P(t) - I_{l^1}\| \le 2\sup_{i \in \mathbb{I}}(1 - p_{i,i}(t))$, implying $\lim_{t \to 0+}\|P(t) - I_{l^1}\| = 0$ and thus completing the proof (see Section 1.1.11). $\square$

2.3.2 Bounded linear operators in l^1 as matrices

Each bounded linear operator in l^1 may be identified with a matrix $(a_{i,j})_{i,j\in\mathbb{I}}$ such that $\sup_{i\in\mathbb{I}}\sum_{j\in\mathbb{I}}|a_{i,j}| < \infty$. In this identification,

$$\|A\| = \sup_{i\in\mathbb{I}}\sum_{j\in\mathbb{I}}|a_{i,j}|. \tag{2.15}$$

To see this, let A be a bounded linear operator in l^1 and let

$$\left(a_{i,j}\right)_{j\in\mathbb{I}} := Ae_i, \qquad i\in\mathbb{I}$$

where e_i is defined in (2.9). For the matrix $\left(a_{i,j}\right)_{i,j\in\mathbb{I}}$ we have then

$$A(\xi_i)_{i\in\mathbb{I}} = A\sum_{i\in\mathbb{I}}\xi_i e_i = \sum_{i\in\mathbb{I}}\xi_i Ae_i = \sum_{i\in\mathbb{I}}\xi_i\left(a_{i,j}\right)_{j\in\mathbb{I}} = \left(\sum_{i\in\mathbb{I}}\xi_i a_{i,j}\right)_{j\in\mathbb{I}}$$

(since convergence in norm of l^1 implies convergence in coordinates), which may be written as

$$A(\xi_i)_{i\in\mathbb{I}} = (\xi_i)_{i\in\mathbb{I}}\cdot\left(a_{i,j}\right)_{i,j\in\mathbb{I}}, \tag{2.16}$$

where the right-hand side is the product of a row-vector $(\xi_i)_{i\in\mathbb{I}}$ and the matrix $\left(a_{i,j}\right)_{i,j\in\mathbb{I}}$.

We note that, for each i,

$$\sum_{j\in\mathbb{I}}|a_{i,j}| = \left\|\left(a_{i,j}\right)_{j\in\mathbb{I}}\right\| = \|Ae_i\| \le \|A\|\,\|e_i\| = \|A\|, \tag{2.17}$$

and, on the other hand, for any $x = (\xi_i)_{i\in\mathbb{I}}\in l^1$,

$$\begin{aligned}\|Ax\| &= \left\|\left(\sum_{i\in\mathbb{I}}\xi_i a_{i,j}\right)_{j\in\mathbb{I}}\right\| = \sum_{j\in\mathbb{I}}\left|\sum_{i\in\mathbb{I}}\xi_i a_{i,j}\right| \le \sum_{j\in\mathbb{I}}\sum_{i\in\mathbb{I}}|\xi_i|\,|a_{i,j}| \\ &= \sum_{i\in\mathbb{I}}|\xi_i|\sum_{j\in\mathbb{I}}|a_{i,j}| \le \left(\sup_{i\in\mathbb{I}}\sum_{j\in\mathbb{I}}|a_{i,j}|\right)\sum_{i\in\mathbb{I}}|\xi_i| \\ &= \left(\sup_{i\in\mathbb{I}}\sum_{j\in\mathbb{I}}|a_{i,j}|\right)\|x\|.\end{aligned}$$

Combining this with (2.17) we obtain (2.15).

Conversely, if $(a_{i,j})_{i,j\in\mathbb{I}}$ is a matrix such that $\sup_{i\in\mathbb{I}}\sum_{j\in\mathbb{I}}|a_{i,j}|$ is finite, then formula (2.16) defines a bounded linear operator such that (2.15) holds.

2.3.3 *Generator is a Q-matrix*

As we have seen, condition (2.14) implies norm-continuity of the semigroup and so, see Section 1.1.11 (details may be found, e.g., in [14], Sections 7.4.19 and 7.4.20) existence of the bounded linear operator

$$Q = \lim_{t \to 0+} \frac{P(t) - I_{l^1}}{t}; \tag{2.18}$$

this operator is the generator of the semigroup. By 2.3.2, Q may be identified with a matrix $(q_{i,j})_{i,j \in \mathbb{I}}$. The entries in this matrix coincide with the limits obtained in Sections 2.2.3, 2.2.5, and 2.2.7. In fact, (2.18) shows that in this case these limits are uniform in the following sense (see 2.3.2 again):

$$\lim_{t \to 0+} \sup_{i \in \mathbb{I}} \sum_{j \in \mathbb{I}} \left| \frac{p_{i,j}(t) - \delta_{i,j}}{t} - q_{i,j} \right| = 0. \tag{2.19}$$

Moreover, $(q_{i,j})_{i,j \in \mathbb{I}}$ is a Kolmogorov matrix (and not just an intensity matrix): the first condition in Definition 2.2.8 is a consequence of (2.19) combined with $\sum_{j \in \mathbb{I}} (p_{i,j}(t) - \delta_{i,j}) = 0$.

Hence, in the uniform case, the notions of the generator and of the Q-matrix coincide. As we shall see later, in general the generator is quite a bit more than the Q-matrix.

2.4 The Generator

To be able to see how the Kolmogorov matrices and generators differ, we need to characterize the latter first. Therefore, the main goal of this section is the Hille–Yosida Theorem for Markov semigroups in l^1. The key role in our analysis is played by the linear functional $\Sigma \in (l^1)^*$ defined by

$$\Sigma (\xi_i)_{i \in \mathbb{I}} = \sum_{i \in \mathbb{I}} \xi_i.$$

In terms of this functional, an operator P is a Markov operator if Px is nonnegative and $\Sigma P x = \Sigma x$ provided x is nonnegative.

2.4.1 *A criterion for the resolvent to be a Markov operator*

Let A be an operator in l^1 (think of A as a candidate for the generator), and suppose that for $\lambda > 0$ and nonnegative $y \in l^1$ there exists a unique solution x to the resolvent equation $\lambda x - Ax = y$, and this solution is nonnegative. By linearity it follows that the resolvent equation has a unique solution for every $y \in l^1$. When is the map $y \mapsto \lambda x$ a Markov operator?

For nonnegative $y \in l^1$,

$$\lambda \Sigma x = \Sigma A x + \Sigma y.$$

Hence, the map in question is a Markov operator provided for all nonnegative $x \in \mathcal{D}(A)$, $\Sigma A x = 0$. (This map is a sub-Markov operator provided that $\Sigma A x \leq 0$.)

Conversely, if A is the generator of a Markov semigroup $\{P(t), t \geq 0\}$, then for all nonnegative $x \in \mathcal{D}(A)$,

$$\Sigma A x = \Sigma \lim_{t \to 0+} \frac{P(t)x - x}{t} = \lim_{t \to 0+} \frac{\Sigma P(t)x - \Sigma x}{t} = 0,$$

Σ being continuous. (For sub-Markov semigroups, $\Sigma A x \leq 0$.)

2.4.2 Example

Suppose $A = a(P - I)$ where $a > 0$ and P is a Markov operator. For $\lambda > 0$ and $y \in l^1$, the solution to $\lambda x - A x = y$ is the solution to $(\lambda + a)x - aPx = y$, and thus is given by the Neumann series:

$$x = \frac{1}{\lambda + a} \sum_{n=0}^{\infty} \left(\frac{a}{\lambda + a} \right)^n P^n y.$$

Since for all nonnegative x, $\Sigma a(Px - x) = 0$, the map $y \mapsto \lambda x$ is a Markov operator.

2.4.3 Theorem

(Hille–Yosida Theorem for Markov and sub-Markov semigroups in l^1) An operator A is the generator of a Markov semigroup in l^1 iff

(i) A is densely defined,
(ii) for any $\lambda > 0$ and nonnegative $y \in l^1$, the resolvent equation $\lambda x - A x = y$ has precisely one solution $x \in \mathcal{D}(A)$, and this solution is nonnegative,
(iii) for nonnegative $x \in \mathcal{D}(A)$, $\Sigma A x = 0$.

If (iii) is replaced by the weaker condition, that is, for nonnegative $x \in \mathcal{D}(A)$, $\Sigma A x \leq 0$, the semigroup is composed of sub-Markov operators.

Proof Necessity of (i) is clear. It is easy to check that if $\{P(t), t \geq 0\}$ is a Markov semigroup then $\lambda R_\lambda := \lambda\, (\lambda - A)^{-1} = \lambda \int_0^\infty \mathrm{e}^{-\lambda t} P(t)\, \mathrm{d}t$ are Markov operators. Hence, (ii) and (iii) are consequences of 2.4.1.

Conversely, if (ii) and (iii) hold, then, by 2.4.1, the map $y \mapsto \lambda x$, where x is the solution to the resolvent equation, is a Markov operator. In particular $\lambda - A$ is invertible with a bounded inverse, implying that A is closed. Hence, A satisfies all the conditions of the Hille-Yosida Theorem and thus generates a contraction semigroup in l^1.

We need to show that this semigroup is a Markov semigroup. We already know that λR_λ is a Markov operator. Hence, so are the semigroups generated by the Yosida approximation $A_\lambda = \lambda^2 R_\lambda - \lambda I_{l^1}$. For,

$$\Sigma e^{-\lambda t} e^{t\lambda^2 R_\lambda} x = e^{-\lambda t} \sum_{n=0}^{\infty} \frac{t^n \lambda^n}{n!} \Sigma(\lambda R_\lambda)^n x = e^{-\lambda t} \sum_{n=0}^{\infty} \frac{t^n \lambda^n}{n!} \Sigma x = \Sigma x,$$

provided $x \geq 0$. Since the strong limit of Markov operators is a Markov operator, we are done. The case of sub-Markov operators is analogous. □

In what follows instead of 'generator of a (sub-)Markov semigroup' we will often simply say '**(sub-)Markov generator**.'

We turn to two important examples of Kolmogorov [59], described in the context of Markov semigroups by Kendall and Reuter [57] (see also [56] and Section II.20 in [21]). In these two examples $\mathbb{I}$ is the set of natural numbers, and so l^1 is the space of absolutely summable $x = (\xi_i)_{i \geq 1}$. Similarly to (2.9) we will use the vectors

$$e_n = \left(\delta_{i,n}\right)_{i \geq 1}, \qquad n \geq 1. \tag{2.20}$$

2.4.4 *First Kolmogorov–Kendall–Reuter semigroup*

Given positive numbers a_i, $i \geq 2$ such that

$$\sum_{i=2}^{\infty} a_i^{-1} < \infty, \tag{2.21}$$

we define the following operator, denoted A, in l^1: its domain $\mathcal{D}(A)$ is the set of $x = (\xi_i)_{i \geq 1}$ such that

$$\sum_{i=2}^{\infty} |a_i \xi_i - \xi_1| < \infty, \tag{2.22}$$

and for x in $\mathcal{D}(A)$ we agree that $Ax = (\eta_i)_{i \geq 1}$ where

$$\eta_1 = \sum_{j=2}^{\infty} (a_j \xi_j - \xi_1) \qquad \text{and} \qquad \eta_i = \xi_1 - a_i \xi_i, \quad i \geq 2. \tag{2.23}$$

Note that (2.22) implies $\lim_{i\to\infty} a_i\xi_i = \xi_1$ and thus existence of a constant K such that $|\xi_i| \le K{a_i}^{-1}, i \ge 2$. It follows that $\mathcal{D}(A) \subset l^1$. Also, $e_n \in \mathcal{D}(A), n \ge 2$ and $e_1 \notin \mathcal{D}(A)$, but

$$(1, a_2^{-1}, a_3^{-1}, \dots) \in \mathcal{D}(A);$$

(note assumption (2.21)). It is not hard to deduce from these relations that $\mathcal{D}(A)$ is dense in l^1. Moreover, it is clear by definition of A that $\Sigma\,(\eta_i)_{i\ge1} = \sum_{i\ge1} \eta_i = 0$ for any x whether positive or not.

Hence, to prove that A is a Markov generator it remains to show condition (ii) in 2.4.3. This we do in the following lemma.

2.4.5 Lemma

The resolvent equation $\lambda x - Ax = y$ has precisely one solution for nonnegative y, and this solution is nonnegative.

Proof We write the resolvent equation in coordinates:

$$\lambda\xi_1 - \sum_{j=2}^{\infty}(a_j\xi_j - \xi_1) = \eta_1, \tag{2.24}$$

$$(\lambda + a_i)\xi_i - \xi_1 = \eta_i, \qquad i \ge 2. \tag{2.25}$$

Here $\lambda > 0$ and $y = (\eta_i)_{i\ge1} \in l^1$ are given, and $x = (\xi_i)_{i\ge1} \in \mathcal{D}(A)$ is searched-for. Conditions (2.25) are met if

$$\xi_i = \frac{\xi_1 + \eta_i}{\lambda + a_i}, \qquad i \ge 2, \tag{2.26}$$

and so it is clear that all we need to do to solve (2.24)–(2.25) is to make sure that ξ_1 satisfies (2.24), and that so constructed $(\xi_i)_{i\ge1}$ belongs to $\mathcal{D}(A)$. Plugging (2.26) into (2.24) and noting that

$$a_j\xi_j - \xi_1 = \frac{a_j\eta_j - \lambda\xi_1}{\lambda + a_j}, \tag{2.27}$$

we find that (2.24) is satisfied iff

$$\lambda C(\lambda)\xi_1 = \eta_1 + \sum_{j=2}^{\infty} \frac{a_j\eta_j}{\lambda + a_j}, \tag{2.28}$$

where $C(\lambda) := 1 + \sum_{j=2}^{\infty}(\lambda + a_j)^{-1} \ne 0$. Since $\sum_{j=2}^{\infty}(\lambda + a_j)^{-1} < \sum_{j=2}^{\infty} a_j^{-1} < \infty$ and $\sum_{j=2}^{\infty}\left|\frac{a_j\eta_j}{\lambda+a_j}\right| < \sum_{j=2}^{\infty}|\eta_j| < \infty$, both series featuring here are absolutely convergent, and so ξ_1 is well defined by (2.28). By the same token, (2.27) shows that $\sum_{i=2}^{\infty}|a_i\xi_i - \xi_i| < \infty$, that is, $(\xi_i)_{i\ge1} \in \mathcal{D}(A)$.

We have proved that the resolvent equation has precisely one solution. Since relations (2.26) and (2.28) make it clear that ξ_i's are nonnegative provided so are η_i's, this completes the proof. □

2.4.6 *A criterion for existence of an infinite product*

For the analysis of the second Kolmogorov–Kendall–Reuter semigroup we need the following criterion of classical analysis (see, e.g., Theorem 118 in [80]):

The product $\prod_{n=1}^{\infty} a_n$, where $a_n \in (0,1]$, converges (to a nonzero limit) iff so does the series $\sum_{n=1}^{\infty}(1-a_n)$.

Proof Without loss of generality we assume that $a_n \neq 1$ for all $n \geq 1$. Let $b_n > 0$ be chosen so that $a_n = \mathrm{e}^{-b_n}$. Then existence of the product $\prod_{n=1}^{\infty} a_n$ (recall once again that by convention this implies $\prod_{n=1}^{\infty} a_n > 0$) is equivalent to convergence of the series $\sum_{n=1}^{\infty} b_n$. On the other hand, $\lim_{n\to\infty} \frac{b_n}{1-a_n} = \lim_{n\to\infty} \frac{\ln a_n}{a_n - 1} = \lim_{x\to 1} \frac{\ln x}{x-1} = 1$ implies that this series converges iff so does $\sum_{n=1}^{\infty}(1-a_n)$. □

2.4.7 *Second Kolmogorov–Kendall–Reuter semigroup*

We start with a sequence $(a_n)_{n\geq 1}$ of positive numbers such that $a_1 = 1, a_2 = 0$ and

$$\sum_{n=3}^{\infty} a_n^{-1} < \infty. \tag{2.29}$$

The domain of A is defined as the set of all $x = (\xi_i)_{i\geq 1}$ satisfying

$$\sum_{i=1}^{\infty} |a_{i+1}\xi_{i+1} - a_i\xi_i| < \infty \qquad \text{and} \qquad \lim_{i\to\infty} a_i\xi_i = \xi_1, \tag{2.30}$$

and we define $Ax = (\eta_i)_{i\geq 1}$ where

$$\eta_i = a_{i+1}\xi_{i+1} - a_i\xi_i, \quad i \geq 1. \tag{2.31}$$

(Note that the existence of the limit in the second part of (2.30) is a consequence of the first part since absolutely convergent series converge; it is the fact that the limit is to be equal to ξ_1 that is an additional assumption.) As in the previous example we see that the domain of A is a subset of l^1, and $e_n \in \mathcal{D}(A), n \geq 2$. Also, $e_1 \notin \mathcal{D}(A)$ (because the second condition in (2.30) fails) but

$$(1, 0, a_3^{-1}, a_4^{-1}, \dots) \in \mathcal{D}(A)$$

(note assumption (2.29)). It follows that A is densely defined. Also, by definition, $\Sigma Ax = 0$ for any $x \in \mathcal{D}(A)$ (the second condition in (2.30) is used here).

Hence, to prove that A is a Markov generator we need to check the second condition in 2.4.3. This will be done in the following two lemmas.

2.4.8 Lemma

For a nonnegative y the resolvent equation $\lambda x - Ax = y$ has a nonnegative solution.

Proof We write the resolvent equation in coordinates:

$$\begin{aligned} (\lambda + 1)\xi_1 &= \eta_1, \\ \lambda\xi_2 - a_3\xi_3 &= \eta_2, \\ (\lambda + a_i)\xi_i - a_{i+1}\xi_{i+1} &= \eta_i, \qquad i \geq 3. \end{aligned} \tag{2.32}$$

Then, given $\lambda > 0$ and a nonnegative $(\eta_i)_{i\geq 1} \in l^1$, we define

$$\begin{aligned} \xi_1 &= \frac{\eta_1}{\lambda + 1}, \\ \xi_i &= \pi_i\eta_1 + \frac{1}{\lambda + a_i}\zeta_i, \qquad i \geq 2, \end{aligned} \tag{2.33}$$

where

$$\pi_i = \pi_i(\lambda) = \frac{1}{(\lambda + a_i)(\lambda + 1)} \prod_{j=i+1}^{\infty} \frac{a_j}{\lambda + a_j}, \qquad i \geq 2,$$

$$\zeta_i = \sum_{j=i}^{\infty} \left(\prod_{i<k\leq j} \frac{a_k}{\lambda + a_k} \right) \eta_j, \qquad i \geq 2,$$

and, by convention, the product over the empty set is 1. Using Criterion 2.4.6 and assumption (2.29) we see that the infinite product $\prod_{j=3}^{\infty} \frac{a_j}{\lambda + a_j}$ converges (to a nonzero limit), and thus π_i, $i \geq 2$ are well defined. Moreover, since

$$\prod_{i<k\leq j} \frac{a_k}{\lambda + a_k} \leq 1 \tag{2.34}$$

for all $i \geq 3$ and $j \geq i + 1$, the series that define ζ_i's converge absolutely and so ξ_i's are well defined also.

Since $\zeta_i - \frac{a_{i+1}}{\lambda + a_{i+1}}\zeta_{i+1} = \eta_i$, $i \geq 2$, we have

$$\lambda\xi_2 - a_3\xi_3 = (\lambda\pi_2 - a_3\pi_3)\eta_1 + \zeta_2 - \frac{a_3}{\lambda + a_3}\zeta_3 = \eta_2,$$

and

$$(\lambda + a_i)\xi_i - a_{i+1}\xi_{i+1} = ((\lambda + a_i)\pi_i - a_{i+1}\pi_{i+1})\eta_1 + \zeta_i - \frac{a_{i+1}}{\lambda + a_{i+1}}\zeta_{i+1}$$

$$= \eta_i, \qquad i \geq 3.$$

This proves that $(\xi_i)_{i\geq 1}$ satisfies (2.32), but we still need to make sure that $(\xi_i)_{i\geq 1} \in \mathcal{D}(A)$.

In view of (2.34), $\lim_{i\to\infty} \zeta_i = 0$. Since by assumption $\lim_{i\to\infty} a_i = \infty$,

$$\lim_{i\to\infty} a_i\xi_i = \eta_1 \lim_{i\to\infty} a_i\pi_i = \frac{\eta_1}{\lambda + 1} = \xi_1.$$

Thus, we are left with showing the first condition in (2.30). Moreover, because of (2.32) and the fact that $(\eta_i)_{i\geq 1} \in l^1$, all we need to prove is that $\sum_{i=1}^{\infty} \xi_i$ is finite (ξ_i's are nonnegative since so are η_i's). To this end, we note first that

$$\sum_{i=3}^{\infty} \pi_i \leq \frac{1}{\lambda + 1} \sum_{i=3}^{\infty} \frac{1}{\lambda + a_i} < \infty.$$

Therefore $\sum_{i=3}^{\infty} \xi_i$ is finite iff so is $\sum_{i=3}^{\infty} \frac{1}{\lambda + a_i}\zeta_i$. The sum of the latter series, however, does not exceed

$$\sum_{i=3}^{\infty} \frac{1}{\lambda + a_i} \sum_{j=i}^{\infty} \eta_j \leq \| (\eta_i)_{i\geq 1} \| \sum_{i=3}^{\infty} \frac{1}{\lambda + a_i} < \infty,$$

completing the proof. □

2.4.9 Lemma

The solutions to the resolvent equation are unique.

Proof By linearity, it suffices to show that the equation $\lambda x - Ax = 0$ has only trivial solution $x = 0$. Hence, assume that in (2.32) all η's are zero. It follows that

$$\xi_1 = 0, \quad \xi_3 = \lambda a_3^{-1}\xi_2, \qquad \text{and} \qquad \xi_{i+1} = (\lambda + a_i)a_{i+1}^{-1}\xi_i, \quad i \geq 3.$$

By induction,

$$\xi_i = \frac{\lambda\xi_2}{a_i \prod_{j=3}^{i-1} \frac{a_j}{\lambda + a_j}}, \qquad i \geq 3.$$

If this sequence is to belong to $\mathcal{D}(A)$, the second requirement in (2.30) must be met. Recalling that $\xi_1 = 0$, we see that this holds iff

$$0 = \frac{\lambda\xi_2}{\prod_{j=3}^{\infty} \frac{a_j}{\lambda + a_j}}.$$

It follows that $\xi_2 = 0$ and thus all the ξ_i's are zero. □

We complete this section with a study of a generator which later will be seen to be related to a birth and death process, and will be used to illustrate one of the points of Kato's Theorem (see, in particular, Example 3.2.7 and Section 3.4). We start with the description of the generator of a **pure birth process** with large intensities of jumps (see also Example 3.2.2).

2.4.10 *A pure birth process generator*

Let $(a_n)_{n\geq 1}$ be a sequence of positive numbers such that $\sum_{n=1}^{\infty} a_n^{-1} < \infty$. Consider the operator A in l^1 defined on the domain

$$\mathcal{D}(A) = \left\{ (\xi_i)_{i\geq 1} ; \sum_{i\geq 1} |a_{i-1}\xi_{i-1} - a_i\xi_i| < \infty \right\}$$

by the formula

$$A\,(\xi_i)_{i\geq 1} = (a_{i-1}\xi_{i-1} - a_i\xi_i)_{i\geq 1}\,,$$

where, for simplicity of definition, we agree that $a_0\xi_0$ is 0. As in Section 2.4.4, we note that for $(\xi_i)_{i\geq 1} \in \mathcal{D}(A)$, the limit $\lim_{n\to\infty} a_n\xi_n$ exists, and from this we deduce that $\mathcal{D}(A) \subset l^1$. We claim that A is a sub-Markov generator.

Since all $e_i, i \geq 1$ belong to $\mathcal{D}(A)$, A is densely defined. Moreover, for nonnegative $(\xi_i)_{i\geq 1} \in \mathcal{D}(A)$, $\Sigma A\,(\xi_i)_{i\geq 1} = -\lim_{n\to\infty} a_n\xi_n \leq 0$. Turning to (remaining) point (ii) of the Hille–Yosida Theorem, we note first that for $\lambda > 0$ and $(\eta_i)_{i\geq 1} \in l^1$, the resolvent equation for A:

$$\lambda\,(\xi_i)_{i\geq 1} - (a_{i-1}\xi_{i-1} - a_i\xi_i)_{i\geq 1} = (\eta_i)_{i\geq 1} \tag{2.35}$$

may be written in coordinates as

$$\begin{aligned}\xi_1 &= \frac{1}{\lambda + a_1}\eta_1,\\ \xi_i &= \frac{a_{i-1}}{\lambda + a_i}\xi_{i-1} + \frac{1}{\lambda + a_i}\eta_i.\end{aligned}$$

Solving this recurrence we see that a solution to the resolvent equation must be of the form

$$\xi_i = \frac{\pi_{i-1}}{\lambda + a_i}\sum_{j=1}^{i}\frac{\eta_j}{\pi_{j-1}}, \tag{2.36}$$

where $\pi_j = \prod_{k=1}^{j}\frac{a_k}{\lambda + a_k}$, $j \geq 1$ and $\pi_0 = 1$. Since, as it is easy to check,

$$\pi_{i-1} - \pi_i = \frac{\lambda\pi_{i-1}}{\lambda + a_i}, \qquad i \geq 1,$$

we have $\sum_{i=j}^{\infty} \frac{\pi_{i-1}}{\lambda+a_i} = \lambda^{-1}(\pi_{j-1} - \pi_\infty) < \lambda^{-1}\pi_{j-1}$, where the infinite product $\pi_\infty := \lim_{i\to\infty} \pi_i > 0$ exists by 2.4.6 and assumption on $(a_n)_{n\geq 1}$. It follows that $(\xi_i)_{i\geq 1}$ is a member of l^1:

$$\sum_{i=1}^{\infty} |\xi_i| \leq \sum_{i=1}^{\infty} \frac{\pi_{i-1}}{\lambda + a_i} \sum_{j=1}^{i} \frac{|\eta_j|}{\pi_{j-1}} \leq \sum_{j=1}^{\infty} \frac{|\eta_j|}{\pi_{j-1}} \sum_{i=j}^{\infty} \frac{\pi_{i-1}}{\lambda + a_i} \leq \lambda^{-1} \sum_{j=1}^{\infty} |\eta_j|.$$

Equation (2.35) shows now that $(a_{i-1}\xi_{i-1} - a_i\xi_i)_{i\geq 1}$ belongs to l^1, too, that is, that $(\xi_i)_{i\geq 1}$ belongs to $\mathcal{D}(A)$ and is a true solution to the resolvent equation. Since it is clear, by (2.36), that this solution is nonnegative as long as $(\eta_i)_{i\geq 1}$ is, we are done.

So prepared we turn to the following example of a birth and death chain generator with birth chain part described by $a_n = 2 \cdot 3^n$, $n \geq 1$.

2.4.11 *A birth and death process generator*

Let $\mathcal{D}(A)$ be the set of $(\xi_i)_{i\geq 1}$ such that

$$\sum_{i=1}^{\infty} 3^i |3\xi_i - \xi_{i-1}| < \infty \tag{2.37}$$

with convention $\xi_0 = 0$, and let $A\,(\xi_i)_{i\geq 1} = (\eta_i)_{i\geq 1}$ where

$$\eta_1 = 9\xi_2 - 6\xi_1,$$
$$\eta_i = 3^{i-1}(9\xi_{i+1} - 9\xi_i + 2\xi_{i-1}), \quad i \geq 2.$$

We note that

$$\eta_i = 3^{i-1}(3(3\xi_{i+1} - \xi_i) - 2(3\xi_i - \xi_{i-1})), \quad i \geq 2, \tag{2.38}$$

and so convergence of the series (2.37) implies that $(\eta_i)_{i\geq 1}$ belongs to l^1. Also, for $(\xi_i)_{i\geq 1} \in \mathcal{D}(A)$, the limit $\lim_{n\to\infty} 3^n\xi_n$ exists; this implies that $\mathcal{D}(A)$ is contained in l^1.

Our aim is to show that A is a sub-Markov generator. It is clear that A is densely defined because all e_i's belong to $\mathcal{D}(A)$. Moreover, it can be checked that if $(\xi_i)_{i\geq 1} \in \mathcal{D}(A)$ is nonnegative, then $\Sigma A\,(\xi_i)_{i\geq 1} = -\lim_{n\to\infty} 3^n\xi_n \leq 0$ (see Exercise 2.4.18). Unfortunately, a frontal assault at the resolvent equation for A may result in heavy losses of confidence in linear algebra skills, and we resort to a less obvious way, expounded in the following sections (Sections 2.4.12–2.4.16).

More specifically, we use a combination of perturbation (cf., e.g., [47], p. 39) and approximation arguments. To this end, first of all, we split A into two parts:

$$A = B + D$$

where ('b' for 'birth')

$$B\,(\xi_i)_{i\geq 1} = 2\left(3^{i-1}\xi_{i-1} - 3^i\xi_i\right)_{i\geq 1},$$

and ('d' for 'death') $D\,(\xi_i)_{i\geq 1} = (\eta_i)_{i\geq 1}$ where

$$\begin{aligned}\eta_1 &= 9\xi_2,\\ \eta_i &= 3^{i+1}\xi_{i+1} - 3^i\xi_i, \qquad i \geq 2,\end{aligned}$$

for $(\xi_i)_{i\geq 1} \in \mathcal{D}(D) := \mathcal{D}(B) := \mathcal{D}(A)$. To recall, in the previous section we have proved that B is a sub-Markov generator.

2.4.12 *Solutions to the resolvent equation for* A_r

We start our investigations by proving that for each $r \in [0, 1)$ (but not yet for $r = 1$), the resolvent equation for

$$A_r := B + rD$$

has a unique solution, as long as $\lambda > \frac{18r}{1-r}$.

To this end, we note that, for $x = (\xi_i)_{i\geq 1} \in \mathcal{D}(A)$,

$$\|Dx\| = \sum_{i=2}^{\infty} |3^{i+1}\xi_{i+1} - 3^i\xi_i| + 9|\xi_2| \leq \frac{1}{2}\|Bx\| + 9\|x\|. \tag{2.39}$$

It follows that (recall that $\|\lambda\,(\lambda - B)^{-1}\| \leq 1$)

$$\begin{aligned}\|rD\,(\lambda - B)^{-1}\,y\| &\leq \frac{r}{2}\|B\,(\lambda - B)^{-1}\,y\| + 9r\|\,(\lambda - B)^{-1}\,y\|\\ &= \frac{r}{2}\|\lambda\,(\lambda - B)^{-1}\,y - y\| + 9r\|\,(\lambda - B)^{-1}\,y\|\\ &\leq r\|y\| + \frac{9r}{\lambda}\|y\|\\ &\leq \frac{1+r}{2}\|y\|, \qquad y \in l^1,\end{aligned} \tag{2.40}$$

provided that $\lambda > \frac{18r}{1-r}$.

Now, suppose an $x \in \mathcal{D}(A)$ solves the resolvent equation for A_r:

$$\lambda x - Bx - rDx = y \tag{2.41}$$

for a given $\lambda > \frac{18r}{1-r}$ and a $y \in l^1$. Applying $(\lambda - B)^{-1}$ to both sides of this equation yields

$$x = r\,(\lambda - B)^{-1}\,Dx + (\lambda - B)^{-1}\,y,$$

and so, by induction, for any $n \geq 1$,

$$x = (\lambda - B)^{-1} [rD(\lambda - B)^{-1}]^n Dx + (\lambda - B)^{-1} \sum_{k=0}^{n} [rD(\lambda - B)^{-1}]^k y.$$

Since, by (2.40), the first term here converges to 0, as $n \to \infty$, and $\sum_{k=0}^{\infty} \|rD(\lambda - B)^{-1}\|^k < \infty$, we must have

$$x = (\lambda - B)^{-1} \sum_{k=0}^{\infty} [rD(\lambda - B)^{-1}]^k y.$$

On the other hand, a simple calculation shows that the x defined by the formula above is a true solution to (2.41).

2.4.13 *An approximation*

The formula for the solution to the resolvent equation obtained in the previous section does not allow to check easily that $x \geq 0$ as long as $y \geq 0$. Therefore, to prove that A_r, $r \in [0, 1)$ are sub-Markov generators (see Section 2.4.14) we resort to the following approximation.

For $n \geq 3$ and $r \in [0, 1]$ ($r = 1$ included), let $B_{n,r}$ with $\mathcal{D}(B_{n,r}) = \mathcal{D}(A)$ be given by $B_{n,r}\,(\xi_i)_{i\geq 1} = (\eta_i)_{i\geq 1}$ where

$$\begin{aligned}
\eta_i &= 2(3^{i-1}\xi_{i-1} - 3^i\xi_i), \qquad 1 \leq i \leq n-1,\\
\eta_n &= 2 \cdot 3^{n-1}\xi_{n-1} - (2-r)3^n\xi_n,\\
\eta_i &= (2-r)(3^{i-1}\xi_{i-1} - 3^i\xi_i), \qquad i \geq n+1.
\end{aligned}$$

Also, let the operator $D_{n,r}$ with domain equal to the entire l^1 be given by $D_{n,r}\,(\xi_i)_{i\geq 1} = (\eta_i)_{i\geq 1}$ where

$$\begin{aligned}
\eta_1 &= 9r\xi_2,\\
\eta_i &= r(3^{i+1}\xi_{i+1} - 3^i\xi_i), \qquad 2 \leq i \leq n-1,\\
\eta_n &= -r3^n\xi_n,\\
\eta_i &= 0, \qquad i \geq n+1.
\end{aligned}$$

We claim that, for each $r \in [0, 1]$ and $x = (\xi_i)_{i\geq 1} \in \mathcal{D}(A)$,

$$\lim_{n\to\infty} A_{n,r}x = Bx + rDx, \tag{2.42}$$

where

$$A_{n,r} := B_{n,r} + D_{n,r}.$$

For, (a) the first $n-1$ coordinates of $Bx + rDx$ and $A_{n,r}x$ are the same, (b) the nth coordinates differ by

$$r(3^{n+1}\xi_{n+1} - 3^n\xi_n)$$

and (c) the remaining coordinates differ by

$$r(3^{i+1}\xi_{i+1} - 3^i\xi_i) + r(3^{i-1}\xi_{i-1} - 3^i\xi_i), \qquad i \ge n+1.$$

It follows that

$$\|Bx + rDx - A_{n,r}x\| \le r|3^{n+1}\xi_{n+1} - 3^n\xi_n| + r\sum_{i=n+1}^{\infty}\Big[|3^{i+1}\xi_{i+1} - 3^i\xi_i| + |3^{i-1}\xi_{i-1} - 3^i\xi_i|\Big].$$

This implies (2.42) by the definition of $\mathcal{D}(A)$.

2.4.14 *Operators* A_r, $r \in [0,1)$ *are sub-Markov generators*

Since we know from Section 2.4.11 that, for each n and r, $B_{n,r}$ is a sub-Markov generator, and $D_{n,r}$ is bounded, $A_{n,r}$ is a generator also (by the Phillips Perturbation Theorem). Furthermore, the operator $D_{n,r}$ is related to a bounded Kolmogorov matrix. Thus, arguing as in Example 1.3.3 we find that $D_{n,r}$ is a Markov generator. Hence, by Trotter's Product Formula, $A_{n,r}$ is a sub-Markov operator, as well.

Now, (2.42) says that the extended limit of the operators $A_{n,r}, n \ge 1$ contains the operator A_r. Since, by 2.4.12, for sufficiently large λ the resolvent equation for A_r has a solution, the limit $\lim_{n\to\infty}\left(\lambda - A_{n,r}\right)^{-1}$ exists by the Sova–Kurtz Convergence Theorem. Also, A_r being densely defined, the regularity space for the semigroups $\{e^{tA_{n,r}}, t \ge 0\}$ is the entire l^1. This means that the limit

$$T_r(t)x := \lim_{n\to\infty} e^{tA_{n,r}}x$$

exists for all $x \in l^1$ and $\{T_r(t), t \ge 0\}$ is a strongly continuous semigroup in l^1. Since $\{e^{tA_{n,r}}, t \ge 0\}, n \ge 1$ are sub-Markov semigroups, so is $\{T_r(t), t \ge 0\}$.

Condition (2.42) also shows that the generator, say, G_r, of the latter semigroup extends A_r. Our goal will be reached once we prove that

$$G_r = A_r, \tag{2.43}$$

that is, that G_r is not a proper extension of A_r.

To this end, suppose x belongs to $\mathcal{D}(G_r) \setminus \mathcal{D}(A_r)$, and take $\lambda > \frac{18r}{1-r}$. Then $y := \lambda x - G_r x$ belongs to l^1 and, since the resolvent equation for A_r has a solution (see Section 2.4.12), there is an $x_0 \in \mathcal{D}(A_r)$ such that $\lambda x_0 - A_r x_0 = y$. Thus $\lambda(x - x_0) - G_r(x - x_0) = 0$, because G_r extends A_r. Since $x - x_0 \neq 0$, this contradicts the fact that G_r is a sub-Markov generator. This contradiction establishes (2.43), completing our proof.

2.4.15 *Solutions to the resolvent equation for* A

We are finally ready to show that solutions to the resolvent equation for A exist and are unique for large $\lambda > 0$. To this end, we consider $r \in (0, 1)$. By (2.39),

$$\|A_r x\| = \|Bx + rDx\| \geq \|Bx\| - r\|Dx\| \geq \|Bx\| - \frac{r}{2}\|Bx\| - 9r\|x\|$$

and thus

$$\|Bx\| \leq \frac{2}{2-r}\|A_r x\| + \frac{18r}{2-r}\|x\|, \qquad x \in \mathcal{D}(A).$$

Therefore,

$$\begin{aligned}
\|(1-r)Dx\| &\leq \frac{1-r}{2}\|Bx\| + 9(1-r)\|x\| \\
&\leq \frac{1-r}{2-r}\|A_r x\| + \frac{9r(1-r)}{2-r}\|x\| + 9(1-r)\|x\| \\
&\leq \frac{1-r}{2-r}\|A_r x\| + 12\|x\|, \qquad x \in \mathcal{D}(A).
\end{aligned}$$

Since A_r is a sub-Markov generator, arguing as in (2.40) we obtain

$$\begin{aligned}
\|(1-r)D\left(\lambda - A_r\right)^{-1} y\| &\leq \frac{2-2r}{2-r}\|y\| + \frac{12}{\lambda}\|y\|, \\
&\leq \frac{r'+1}{2}\|y\|, \qquad y \in l^1,
\end{aligned}$$

provided that $\lambda > \frac{24}{1-r'}$, where $r' := \frac{2-2r}{2-r}$ is strictly smaller than 1 and so is $\frac{r'+1}{2}$.

This allows repeating the argument from the latter part of Section (2.4.12): writing the resolvent equation for A as

$$\lambda x - A_r x - (1-r)Dx = y,$$

we conclude that $x = (\lambda - A_r)^{-1} \sum_{k=0}^{\infty}[(1-r)D\left(\lambda - A_r\right)^{-1}]^k y$ is the unique solution to this equation.

2.4.16 *A is a sub-Markov generator*

When combined with the result just obtained, relation (2.42) with $r = 1$ shows, as in Section 2.4.14, that the sub-Markov semigroups $\{e^{tA_{n,1}}, t \ge 0\}$ converge, as $n \to \infty$, to a strongly continuous sub-Markov semigroup. By the same relation, the generator, say, G, of the limit semigroup extends A. Arguing as in the latter part of 2.4.14 we see, however, that G cannot be a proper extension of A because for large $\lambda > 0$ the resolvent equation for A has a solution: we must have $A = G$. This completes the proof that A is a sub-Markov generator.

2.4.17 Exercise

In the definition of the domain of the second Kolmogorov–Kendall–Reuter generator (see (2.30)), change $\lim_{i\to\infty} a_i\xi_i = \xi_1$ to $\lim_{i\to\infty} a_i\xi_i = \alpha\xi_1$, where $\alpha \in (0, 1)$ is a given parameter. By modifying the proofs of Lemmas 2.4.8 and 2.4.9, check that this new operator is a generator of a sub-Markov semigroup.

2.4.18 Exercise

Check that if a nonnegative $(\xi_i)_{i\ge 1}$ is a member of the domain of the operator of Section 2.4.11, then $\Sigma A\,(\xi_i)_{i\ge 1} = -\lim_{n\to\infty} 3^n\xi_n \le 0$.

2.5 Intensities, Generators, and Infinitesimal Description

Up to this point, the presentation of the examples of the previous section may seem to be somewhat 'raw.' Without some knowledge on the way the related Markov chains behave, all calculations seem a bit abstract, and the definition of the generator – mysterious. In this section we will get some insight into the way the generators and local behavior of the process are connected. Armed with this insight, and with some additional tools, we will come back to a detailed description of the Kolmogorov–Kendall–Reuter examples in Sections 2.6 and 2.7. We will come back to the birth and death process generator in Chapter 3.

Our first result, contained in Section 2.5.1, says roughly that, if $e_n \in \mathcal{D}(A)$, then the nth row of the intensity matrix may be recovered from the generator; Section 2.5.5 extends this result to the sub-Markovian case. Moreover, in Sections 2.5.7 and 2.5.8 we will see how the local (in time) behavior of the process can be reconstructed from the intensity matrix: a chain starting at i stays there for an exponential time with parameter q_i and then jumps to a state

$j \neq i$ with probability $\frac{q_{i,j}}{q_i}$. In the meantime, in Section 2.5.6, we will see that the condition $e_i \in \mathcal{D}(A)$, $i \in \mathbb{I}$ is equivalent to the all-important Kolmogorov differential equations.

2.5.1 $q_{i,j}$'s and the generator

Suppose $\{P(t), t \geq 0\}$ is a Markov semigroup with generator A, and let $i \in \mathbb{I}$ be fixed. The following are equivalent.

(1) $q_i < \infty$ and $\sum_{j \in \mathbb{I}} q_{i,j} = 0$,
(2) $e_i \in \mathcal{D}(A)$.

If any of these conditions hold,

$$\boxed{Ae_i = \left(q_{i,j}\right)_{j \in \mathbb{I}}.} \tag{2.44}$$

Proof (2) $\Longrightarrow$ (1) The ith coordinate of the vector

$$v(t) := t^{-1}(P(t)e_i - e_i)$$

is $t^{-1}(p_{i,i}(t) - 1)$. Since $\lim_{t \to 0+} v(t) = Ae_i$ and strong convergence in l^1 implies coordinate-wise convergence, $-q_i = \lim_{t \to 0+} t^{-1}(p_{i,i}(t) - 1)$ is the ith coordinate of Ae_i, and in particular, it is finite.

Similarly, $q_{i,j}$ is the jth coordinate of Ae_i (this, by the way, proves the final claim). Moreover, $\Sigma(v(t)) = 0$, for each $t \geq 0$, since $P(t)$ is a Markov operator. Therefore, $\sum_{j \in \mathbb{I}} q_{i,j} = \Sigma Ae_i = \lim_{t \to 0+} \Sigma v(t) = 0$, Σ being continuous.

To prove the converse, consider first the case where $q_i = 0$. Then, since $P(t)e_i$ is the ith 'row' in the matrix $P(t)$,

$$\begin{aligned}\|t^{-1}(P(t)e_i - e_i)\| &= t^{-1}\sum_{j \neq i} p_{i,j}(t) + t^{-1}[1 - p_{i,i}(t)] \\ &= 2t^{-1}(1 - p_{i,i}(t)).\end{aligned}$$

As $t \to 0+$, this converges to $2q_i = 0$. It follows that $e_i \in \mathcal{D}(A)$ and $Ae_i = 0$.

For $q_i > 0$, let $u(t)$ be the vector with ith coordinate equal 0, and jth coordinate equal to $\frac{1}{t} p_{i,j}(t)$, $j \neq i$. The claim reduces to showing that $u(t)$ converges, as $t \to 0$, to $u := (q_{i,j}(1-\delta_{i,j}))_{j \in \mathbb{I}}$ (u is obtained from $(q_{i,j})_{j \in \mathbb{I}}$ by replacing its ith coordinate by 0). Clearly, it suffices to show that $\frac{t}{1-p_{i,i}(t)}u(t)$ converges to $\frac{1}{q_i}u$. This, however, follows by Schéffe's Theorem 1.6.1 since all the vectors involved here are distributions (we use the second part of assumption (1)), and they converge coordinate-wise. □

2.5.2 *Absorbing states*

The result obtained as a by-product of the analysis of the case $q_i = 0$ is worth noting. We proved that in this case $Ae_i = 0$. It follows (use, e.g., Corollary 7.4.26 in [14]) that $P(t)e_i = e_i$, that is, $p_{i,i}(t) = 1$ for all $t \geq 0$. This means that i is an absorbing state: the process reaching this state stays there for ever.

2.5.3 Corollary

An intensity matrix is a Kolmogorov matrix iff $e_i \in \mathcal{D}(A)$ for all $i \in \mathbb{I}$.

2.5.4 *Relation between the generator and the intensity matrix*

Suppose conditions of Section 2.5.1 are satisfied for all $i \in \mathbb{I}$. Then all finite combinations x of basic vectors are members of $\mathcal{D}(A)$ and (2.44) becomes

$$Ax = x \cdot Q,$$

where on the right-hand side we have the product of a row-vector with a matrix. However, as we shall see in Sections 3.3 and 3.5, this equality need not hold for all $x \in \mathcal{D}(A)$ (e.g., see equations (3.35), (3.36) and (3.39)).

2.5.5 $q_{i,j}$*'s and the generator (continued)*

The case where $\{P(t), t \geq 0\}$ is a semigroup of sub-Markov operators may be reduced to that considered in 2.5.1. More specifically: Suppose $\{P(t), t \geq 0\}$ is a sub-Markov semigroup with generator A, and let $i \in \mathbb{I}$ be fixed. Then, defining 'dishonesty function' $d_i(t) := 1 - \sum_{j \in I} p_{i,j}(t)$, we have

(A) $d_i'(0) := \lim_{t \to 0+} \frac{d_i(t)}{t}$ exists and is finite,
(B) the following are equivalent:
  (1) $q_i < \infty$ and $\sum_{j \in \mathbb{I}} q_{i,j} + d'(0) = 0$,
  (2) $e_i \in \mathcal{D}(A)$.
  If any of these conditions hold,

$$Ae_i = \left(q_{i,j}\right)_{j \in \mathbb{I}}. \tag{2.45}$$

Sketch of proof Let a ('a' for 'additional') be such that $a \notin \mathbb{I}$, and consider an extended state-space $\widetilde{\mathbb{I}} = \mathbb{I} \cup \{a\}$. Then

$$\widetilde{p_{i,j}}(t) = \begin{cases} p_{i,j}(t), & i, j \in \mathbb{I}, \\ d_i(t), & i \in \mathbb{I}, j = a, \\ 1, & i = j = a, \\ 0, & i = a, j \in \mathbb{I} \end{cases}$$

are transition probabilities in $\widetilde{\mathbb{I}}$ such that

$$\sum_{j \in \widetilde{\mathbb{I}}} \widetilde{p_{i,j}}(t) = 1 \qquad \text{and} \qquad \lim_{t \to 0+} \widetilde{p_{i,i}}(t) = 1, \qquad i \in \widetilde{\mathbb{I}}.$$

Hence, (A) is a direct consequence of 2.2.7 and (B) be is a consequence of 2.5.1. Details of this reasoning may be found in Section 4.2.1. □

2.5.6 *Kolmogorov backward equations*

Let $\{P(t), t \geq 0\}$ be a semigroup of sub-Markov operators with generator A, and suppose $e_i \in \mathcal{D}(A)$ for all $i \in \mathbb{I}$. Then, by 1.1.15, the vector-valued functions $t \mapsto P(t)e_i$ are differentiable on the entire half-line with (see (2.45))

$$\frac{\mathrm{d}}{\mathrm{d}t}(P(t)e_i) = P(t)Ae_i = P(t)\left(q_{i,j}\right)_{j \in \mathbb{I}}, \qquad t \geq 0.$$

Since convergence in the sense of l^1 norm implies convergence in coordinates, the real-valued functions $t \mapsto p_{i,j}(t), i, j \in \mathbb{I}$ are differentiable at all $t \geq 0$ also, and we obtain

$$p'_{i,j}(t) = \sum_{k \in \mathbb{I}} q_{i,k} p_{k,j}(t), \qquad i, j \in \mathbb{I}. \tag{2.46}$$

These are the celebrated **Kolmogorov backward equations**.

Conversely, suppose all states are stable, and the Kolmogorov backward equations are satisfied. I claim that then all e_i's are members of $\mathcal{D}(A)$. To see this, for $\lambda > 0$, let $\left(r_{i,j}(\lambda)\right)_{i,j \in \mathbb{I}}$ be the matrix representing $(\lambda - A)^{-1}$. Since the resolvent is the Laplace transform of the semigroup and convergence in the sense of l^1 norm implies convergence in coordinates, we have

$$r_{i,j}(\lambda) = \int_0^\infty \mathrm{e}^{-\lambda t} p_{i,j}(t)\,\mathrm{d}t, \qquad i, j \in \mathbb{I}.$$

The functions featuring on the right-hand side of the Kolmogorov equations (2.46) are bounded and continuous, and the series converges uniformly in $t \geq 0$. Also, on the left-hand side we have the derivative of a differentiable, bounded function with continuous, bounded derivative. Thus, calculating the Laplace transform yields

$$\lambda r_{i,j}(\lambda) - \delta_{i,j} = \sum_{k \in \mathbb{I}} q_{i,k} r_{k,j}(\lambda). \tag{2.47}$$

Now, fix i and think of $\left(q_{i,j}\right)_{j \in \mathbb{I}} \in l^1$. The quantity on the right-hand side of (2.47) is the jth coordinate of $(\lambda - A)^{-1} \left(q_{i,j}\right)_{j \in \mathbb{I}}$. Hence, this equation says that

$$e_i = \lambda (\lambda - A)^{-1} e_i - (\lambda - A)^{-1} \left(q_{i,j}\right)_{j \in \mathbb{I}}.$$

This implies that $e_i \in \mathcal{D}(A)$.

2.5.7 *Exponential time spent at state i*

Suppose sample paths of a Markov chain corresponding to a Markov semigroup $\{P(t), t \geq 0\}$ are right-continuous[1] and that i is a stable state. Then, the time spent at i is exponential with parameter q_i. More specifically, if $X(0) = i$, then the time to the first moment when $X(t) \neq i$ is a random variable with specified exponential distribution.

Proof For $n \geq 1$, let τ_n be the smallest of times of the form $t = k2^{-n}$ such that $X(t) \neq i$. It is clear that τ is a random variable, since

$$\{\tau_n = k2^{-n}\} = \{X(\ell 2^{-n}) = i, \ell = 1, \ldots, k-1 \text{ and } X(k2^{-n}) \neq i\}.$$

Since the set of times to choose from for $n+1$ is larger than that for n, $\tau_{n+1} \leq \tau_n$, and right continuity of paths implies that $\lim_{n \to \infty} \tau_n = \tau$ (monotonically), where τ is the smallest of $t > 0$ such that $X(t) \neq i$. In particular, τ is a random variable.

Using $\{\tau \geq t\} = \bigcap_{n \geq 1} \{\tau_n \geq t\}$ (with $\{\tau_{n+1} \geq t\} \subset \{\tau_n \geq t\}$),

$$\begin{aligned} \mathbb{P}(\tau \geq t) &= \lim_{n \to \infty} \mathbb{P}(\tau_n \geq t) = \lim_{n \to \infty} \mathbb{P}\left(X(k2^{-n}) = i, k = 1, \ldots, [2^n t]\right) \\ &= \lim_{n \to \infty} \left(p_{i,i}(2^{-n})\right)^{[2^n t]} = \lim_{n \to \infty} \exp\left(\frac{[2^n t]}{2^n} \frac{\ln p_{i,i}(2^{-n})}{2^{-n}}\right). \end{aligned}$$

Since $\lim_{n \to \infty} \frac{[2^n t]}{2^n} = t$ and, by $\lim_{x \to 1} \frac{x-1}{\ln x} = 1$, $\lim_{h \to 0+} \frac{\ln p_{i,i}(h)}{h} = \lim_{h \to 0+} \frac{p_{i,i}(h) - 1}{h} = -q_i$,

$$\mathbb{P}(\tau \geq t) = \mathrm{e}^{-t q_i},$$

completing the proof. □

[1] For a construction of a Markov chain with paths that are right-continuous and have left-hand limits see, for example, [72], pp. 88–90.

2.5.8 *Probabilities of jumps from state i*

Suppose sample paths of a Markov chain corresponding to a Markov semigroup $\{P(t), t \geq 0\}$ are right-continuous, e_i belongs to $\mathcal{D}(A)$, and $q_i > 0$. (See Corollary 2.5.2 for the case $q_i = 0$.) Then, the process starting at i stays there for an exponential time with parameter q_i and after this time elapses the process jumps to $j \neq i$ with probability

$$\frac{q_{i,j}}{q_i}.$$

Proof Let τ_n be defined as in the previous subsection. Since

$$\{X(\tau_n) = j\} = \bigcup_{k \geq 1} \{\tau_n = k2^{-n}, X(k2^{-n}) = j\},$$

$X(\tau_n)$ is a random variable (note importance of the fact that τ_n has countably many values). Right continuity of paths implies that $X(\tau) = \lim_{n\to\infty} X(\tau_n)$. Thus, $X(\tau)$ is a random variable, and we are to show that

$$\mathbb{P}(X(\tau) = j) = \frac{q_{i,j}}{q_i}.$$

We have

$$\begin{aligned} \mathbb{P}(X(\tau_n) = j) &= \sum_{k=1}^{\infty} \mathbb{P}(\tau_n = k2^{-n}, X(k2^{-n}) = j) \\ &= \sum_{k=1}^{\infty} \mathbb{P}\big(X(m2^{-n}) = i, m = 1, \dots, k-1, X(k2^{-n}) = j\big). \end{aligned}$$

The kth summand in the last series is $\big(p_{i,i}(2^{-n})\big)^{k-1} p_{i,j}(2^{-n})$ implying

$$\mathbb{P}(X(\tau_n) = j) = \frac{p_{i,j}(2^{-n})}{1 - p_{i,i}(2^{-n})} = \frac{p_{i,j}(2^{-n})}{2^{-n}} \frac{2^{-n}}{1 - p_{i,i}(2^{-n})}.$$

Thus $\lim_{n\to\infty} \mathbb{P}(X(\tau_n) = j) = \frac{q_{i,j}}{q_i}$, and we are left with showing that $\mathbb{P}(X(\tau) = j) = \lim_{n\to\infty} \mathbb{P}(X(\tau_n) = j)$. However, by right continuity of paths, $X(\tau(\omega)) = j$ iff $X(\tau_n(\omega)) = j$ for almost all $n \geq 1$, and so the claim follows for example by the Lebesgue Dominated Convergence Theorem (write $\mathbb{P}(X(\tau) = j) = \mathbb{E}\, 1_{\{X(\tau)=j\}}$). □

2.5.9 Remark

By the second condition in (1) of Section 2.5.1, our proposition shows that the probability that the process jumps from i to one of the states $j \neq i$ equals 1. If the semigroup is composed of sub-Markov operators, $e_i \in \mathcal{D}(A)$ does not imply the condition just mentioned, and then with nonzero probability the process may jump 'nowhere,' that is, simply disappear. (This is what *dishonest* Markov chains do by their evil nature and wicked character.)

It may of course happen also that condition (1) is not satisfied even if $q_i < \infty$ and the semigroup is composed of Markov operators (see, e.g., Section 2.7.1). Interestingly, the proof of Proposition 2.5.8 still works in this case. The difference is that then the proposition does not tell the whole story of a particle leaving the state i; the particle may do something else than jumping to one of the states $j \neq i$.

2.6 Back to the First Kolmogorov–Kendall–Reuter Example

2.6.1 *In the first Kolmogorov–Kendall–Reuter semigroup $i = 1$ is an instantaneous state*

In the first example of Kolmogorov, Kendall and Reuter, $e_1 \notin \mathcal{D}(A)$, so that one of the conditions listed in (1) of Section 2.5.1 fails. As we will see, $q_1 = \infty$ so that the state $i = 1$ is instantaneous.

Here is a proof. Suppose $q_1 < \infty$. Then, there is a $t_0 > 0$ such that $1 - p_{1,1}(t) \le 2q_1 t$, for $0 \le t < t_0$. For $t \ge t_0$, on the other hand, $1 - p_{1,1}(t) \le t_0^{-1} t$. Thus, there is an integer n_0 such that

$$1 - p_{1,1}(t) \le n_0 t, \qquad t \ge 0,$$

and so

$$\lambda^2 \int_0^\infty \mathrm{e}^{-\lambda t} \left(1 - p_{1,1}(t)\right) \mathrm{d}t \le n_0, \qquad \lambda > 0. \tag{2.48}$$

However, (2.28) reveals in particular that the Laplace transform of $p_{1,1}$ is the inverse of $\lambda \left(1 + \sum_{j=2}^\infty (\lambda + a_j)^{-1}\right)$. It follows that the left-hand side of (2.48) equals

$$\frac{\lambda \sum_{j=2}^\infty (\lambda + a_j)^{-1}}{1 + \sum_{j=2}^\infty (\lambda + a_j)^{-1}} \ge \frac{\lambda \sum_{j=2}^{n_0+2} (\lambda + a_j)^{-1}}{1 + \sum_{j=2}^\infty (\lambda + a_j)^{-1}}.$$

Therefore, the limit, as $\lambda \to \infty$, of the left-hand side in (2.48) is no smaller than $n_0 + 1$ (since $\lim_{\lambda \to \infty} \sum_{j=2}^\infty (\lambda + a_j)^{-1} = 0$ by assumption (2.21)), a contradiction proving our claim that $q_1 = \infty$.

2.6.2 Approximation of the first Kolmogorov–Kendall–Reuter semigroup

Some further insight into the first Kolmogorov–Kendall–Reuter semigroup may be gained by the following analysis. For $n \ge 1$, let A_n be the bounded operator in l^1 represented by a matrix with all entries outside the upper-left corner of size $(n+1) \times (n+1)$ equal zero, and the remaining ones given by

$$\begin{pmatrix} -n & 1 & 1 & \cdots & 1 & 1 \\ a_2 & -a_2 & 0 & \cdots & 0 & 0 \\ a_3 & 0 & -a_3 & 0 & \cdots & 0 \\ \vdots & \vdots & \ddots & \ddots & \ddots & 0 \\ a_{n+1} & 0 & 0 & \cdots & 0 & -a_{n+1} \end{pmatrix}.$$

We will see that the semigroups generated by A_n's converge to the first Kolmogorov–Kendall–Reuter semigroup.

To this end, one may repeat calculations of Lemma 2.4.5 with minor changes to see that A_n are generators of Markov semigroups, and in doing this to find an explicit form of the resolvent of A_n. This allows finding the limit $\lim_{n\to\infty} (\lambda - A_n)^{-1}$ and checking that it coincides with the resolvent of the first Kolmogorov–Kendall–Reuter semigroup calculated in Lemma 2.4.5, so that the claim follows by the Trotter–Kato Convergence Theorem.

The same result may, however, be obtained with almost no algebra. For, A_n is a bounded linear operator, and so for $\lambda > \|A_n\|$ its resolvent equation has a unique solution. Moreover, since the rows of the A_n matrix add up to zero, $\Sigma A_n x = 0$ for all $x \in l^1$. It follows (see 2.4.1) that the resolvent of A_n is a Markov operator. Thus, the Yosida approximation is composed of Markov operators (for sufficiently large λ) and, *a fortiori*, so must be $\{e^{tA_n}, t \ge 0\}$. (For yet another proof, see Exercise 2.7.7.)

Finally, for $x = (\xi_i)_{i\ge1} \in \mathcal{D}(A)$, $A_n x = (\eta_i)_{i\ge1}$, where

$$\eta_1 = \sum_{j=2}^{n+1} (a_j\xi_j - \xi_1),$$

$$\eta_i = \xi_1 - a_i\xi_i, \qquad i = 2, \dots, n+1,$$

$$\eta_i = 0, \qquad\qquad i \ge n+2.$$

Since for $x \in \mathcal{D}(A)$, condition (2.22) is met, it is clear that as $n \to \infty$, $A_n x$ converge to Ax given by formula (2.23) (we have $\|A_n x - Ax\| = 2\sum_{j=n+2}^{\infty} |a_j\xi_j - \xi_1|$). Hence, the claim follows by the Sova–Kurtz Theorem.

2.6.3 *Some probabilistic intuition*

A Markov chain described by A_n of Section 2.6.2 behaves as follows: All states $i \geq n+2$ are absorbing. Starting at a state $2 \leq i \leq n+1$ the process spends an exponential time with parameter a_i there, and then jumps immediately to $i = 1$. Here, the exponential distribution parameter is n, and after the visit at $i = 1$ is over, the process jumps to one of the states $i = 2, \ldots, n+1$, all states being equally probable.

As $n \to \infty$, the set of absorbing states becomes smaller and smaller but the most important, though gradual, change occurs at $i = 1$: the expected time of the visit converges to zero, and we know that in the limit the state becomes instantaneous. Since all positions after a jump from $i = 1$ are equally probable for each n, it is natural to expect that in the limit all states should also 'have equal rights.' In fact, as Kendall and Reuter show [57], $q_{1,i} = 1$ for all $i \geq 2$; this also agrees with the intuitions gained in the previous section.

A question of interest arises therefore of how these equal rights are claimed [74]. This can be figuratively explained as follows (a naive approach that after a jump each state is chosen with equal probability will not work: since there are infinitely many states to jump to, this 'equal probability' would need to be zero).

Imagine that there is an alarm clock assigned to each state $i \geq 2$ (a clock-maker paradise!), that each of them goes off after an exponential time with parameter 1, and that all these clocks are independent of each other. The process starting at the state 1 waits for the first of the clocks to signal (see the next section), and if it is the ith clock that goes off, the process jumps to the state i. Since the infimum of infinitely many exponential times, all with the same parameter equal 1, is a.s. zero, the jump occurs 'immediately.' During the process's visit to ith state, which is of course exponential with parameter a_i, all clocks are switched off. They start to compete again once the process returns to 1, and so the procedure continues.

Remarkably, 'equal rights' are expressed in the fact that all clocks go off after exponential times with the same parameter. In particular, one should not think that expected durations of visits to the states $i \geq 2$ are the same. On the contrary, the states with smaller a_i will be visited for longer times if as often as other states.

2.6.4 *Some probabilistic intuition, continued*

A curious reader may wonder where the idea of infinitely many clocks of the previous section comes from. Here is a hint.

Suppose we are given n independent, exponentially distributed random variables $\tau_2, \ldots, \tau_{n+1}$, each with parameter 1. I claim that a process related to A_n of Section 2.6.2, starting at the state 1, may be described as follows: the process waits at i for the random time

$$\tau = \min_{i=2,\ldots,n+1} \tau_i$$

and then jumps to an $i \in \{2, \ldots, n+1\}$ iff

$$\tau = \tau_i. \tag{2.49}$$

For, τ so defined is exponentially distributed with parameter n, being the minimum of n independent, exponentially distributed random variables, each with parameter 1. Moreover, the probability of event (2.49) is n^{-1}. Since this description agrees with that of Section 2.5.8 we are done.

An advantage of this construction is that, whereas, as we have seen, in the limit as $n \to \infty$, it is impossible to describe the process in terms of probabilities of jumps after sojourn in the state 1, the description in terms of random variables τ_n is still possible: in the limit, instead of finitely many clocks (random variables τ_n), we have infinitely many of them.

Possibly beguiled by (2.49), the reader should however refrain himself from thinking that for the first Kolmogorov–Kendall–Reuter process starting at 1 there is 'a first clock to go off.' The phrase 'the first of the clocks,' used in the previous section, was merely a convenient figure of speech, poorly designed to convey an underlying intuition. As a matter of fact, as we will see in the next section, for any $i \geq 2$, before the ith clock goes off and the state i is visited for the first time, infinitely many other clocks go off and corresponding states are visited, and they are visited many times!

2.6.5 *Kendall–Reuter's construction*

This complex, curious, convoluted, confusing, exciting, extraordinary, fascinating, intriguing, intricate, puzzling and perhaps somewhat perplexing situation may be modeled as follows (following, of course, Kendall and Reuter [57]). Suppose we are given independent, exponentially distributed random variables

$$X_{i,n}, Y_{i,n} \qquad n \in \mathbb{N}, i = 2, 3, \ldots, \tag{2.50}$$

such that

$$\mathbb{E}\, X_{i,n} = 1 \qquad \text{and} \qquad \mathbb{E}\, Y_{i,n} = a_n^{-1}.$$

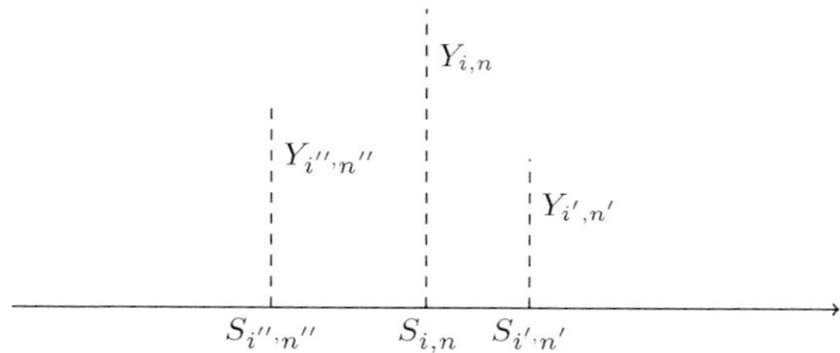

Figure 2.2 Time spent at state 1 (the axis) and visits to other states (the peaks).

Leaving the $Y_{i,n}$'s temporarily aside, let

$$S_{i,0} = 0 \text{ and } S_{i,n} = X_{i,1} + X_{i,2} + \cdots + X_{i,n}, \qquad n \geq 1, i = 2, 3, \ldots.$$

Disregarding a set of probability zero, we may safely assume that $S_{i,n} < S_{i,n+1}$ for all i and n. Thus, with each elementary event ω we have an increasing sequence of numbers $S_{i,0}(\omega) = 0 < S_{i,1}(\omega) < S_{i,2}(\omega) < \ldots$. Think of $S_{i,n}$ as the time when the ith clock goes off for the nth time, and put all these times, for all states, at one time axis, modeling the time spent at the state 1. We imagine that at time $S_{i,n}, n \geq 1$ ($n = 0$ is excluded) the process leaves the state 1 to visit the state i for the nth time, and when its visit there is over, it comes back to 1. While the process is away from the state 1, all clocks are switched off (see Figure 2.2): time is not running at 1.

Suppose now that such a process has already been defined, fix a state i and a moment n: we would like to describe the total time $T_{i,n}$ the process starting at 1 has wandered here and there and lingered at 1 before the ith clock has gone off for the nth time. To this end, think of another state, say, $j \neq i$, and of the related sequence $\left(S_{j,n}\right)_{n\geq 0}$. It is clear that there is precisely one $m \in \{0, 1, 2, \ldots\}$, depending on i, j, n, and ω, such that

$$S_{j,m} < S_{i,n} \leq S_{j,m+1}.$$

In fact, disregarding yet another set of probability zero, we may be sure that

$$S_{j,m} < S_{i,n} < S_{j,m+1}.$$

This simply says that before the ith clock signaled for the nth time, clock number j signaled precisely m times. Therefore, since a visit to a state $j \geq 2$ is of exponential distribution with parameter a_j, $T_{i,n}$ is the sum of the following components:

(a) $S_{i,n}$ itself; this is the time spent at the state 1,

(b) $Y_{i,1} + \cdots + Y_{i,n-1}$; these are $n-1$ visits to the state i,
(c) $\sum_{j\neq i}(Y_{j,1} + \cdots + Y_{j,m})$; the sum $Y_{j,1} + \cdots + Y_{j,m}$ describes a random number of exponential visits to the state j, each exponentially distributed with parameter a_j.

In other words,

$$T_{i,n} = S_{i,n} + (Y_{i,1} + \cdots + Y_{i,n-1}) + \sum_{j\neq i}(Y_{j,1} + \cdots + Y_{j,m}) \qquad (2.51)$$

with m depending on i, j, n and ω.

It is a fundamental fact that with probability 1, $T_{i,n}$'s are finite (see our Lemma 2.6.6). It follows that we can turn things right side up and define our process with the help of $T_{i,n}$'s: since at $S_{i,n}$ the ith clock has gone off for the nth time, it is natural to define the process in the interval $[T_{i,n}, T_{i,n} + Y_{i,n})$, where $Y_{i,n}$ is the variable introduced in (2.50), as staying in the state i; apart from the union of these time-intervals, the process is at the state 1.

For this definition to be consistent, however, we need to make sure that $T_{i,n} + Y_{i,n} \leq T_{i',n'}$ as long as $S_{i,n} < S_{i',n'}$. To this end, given i, n and ω, think of all pairs (i'', n'') such that $S_{i'',n''}(\omega) < S_{i,n}(\omega)$ (the case $i'' = i$ is not excluded), and attach the random variable $Y_{i'',n''}$ to each such pair. Then $T_{i,n}$ is the sum of $S_{i,n}$ and all these variables. Now, if $S_{i,n} < S_{i',n'}$, then the sum $T_{i',n'}$ contains

(a) $S_{i',n'}$, which is larger than $S_{i,n}$,
(b) $Y_{i,n}$,
(c) all the variables composing $T_{i,n}$ (excluding $S_{i,n}$ which, however, is handled by (a)), and
(d) possibly many other (nonnegative) variables.

This shows our claim.

For the proof that the process defined above is really related to the first Kolmogorov–Kendall–Reuter semigroup, the reader should consult [57]. But the main point in our analysis is that, as we claimed before, and as formula (2.51) now clearly reveals, in the process considered, before the first visit to the state i there are usually many, many visits to infinitely many other states. For any i and any infinite subsequence of $X_{j,1}, j \neq i$, conditional on any positive value of $X_{i,1}$, the probability that all variables from the subsequence are larger than $X_{1,i}$ is zero.

2.6.6 Lemma

For all i and n, $\mathbb{P}(T_{i,n} < \infty) = 1$.

Proof Since $S_{i,n}$ and $Y_{i,1} + \cdots + Y_{i,n}$ are both almost surely finite by definition, it suffices to show that $R_{i,n} := \sum_{j\neq i}(Y_{j,1} + \cdots + Y_{j,m})$ is finite. To this end, we claim that for $\lambda > 0$,

$$\mathbb{E}\,\mathrm{e}^{-\lambda R_{i,n}} = \mathbb{E}\,\mathrm{e}^{-\lambda S_{i,n}\sum_{j\neq i}\frac{1}{\lambda+a_j}}. \tag{2.52}$$

If m is given, then for any j, we have that $\mathbb{E}\,\mathrm{e}^{-\lambda(Y_{j,1}+\cdots+Y_{j,m})}$, being the Laplace transform of the sum of m independent identically distributed random variables, equals $\left(\frac{a_j}{\lambda+a_j}\right)^m$. Next, if $S_{i,n}$ is given, then m is Poisson distributed with parameter $S_{i,n}$ (see Section 7.5.5 in [14]) for any $j \neq i$. Thus,

$$\mathbb{E}\left[\mathrm{e}^{-\lambda(Y_{j,1}+\cdots+Y_{j,m})}|S_{i,n}\right] = \mathrm{e}^{S_{i,n}(\frac{a_j}{\lambda+a_j}-1)} = \mathrm{e}^{-S_{i,n}\frac{\lambda}{\lambda+a_j}}.$$

By independence,

$$\mathbb{E}\left[\mathrm{e}^{-\lambda\sum_{j\neq i}(Y_{j,1}+\cdots+Y_{j,m})}|S_{i,n}\right] = \mathrm{e}^{-S_{i,n}\sum_{j\neq i}\frac{\lambda}{\lambda+a_j}},$$

and this implies (2.52).

Finally, $\lim_{\lambda\to 0+} S_{i,n}(\omega)\sum_{j\neq i}\frac{1}{\lambda+a_j} = S_{i,n}(\omega)\sum_{j\neq i}a_j^{-1}$ for all ω, and the last series converges by assumption. Therefore, by the Lebesgue Dominated Convergence Theorem, the right-hand side of (2.52) converges to 1, because $S_{i,n}$ is finite. On the other hand, $\mathrm{e}^{-\lambda R_{i,n}(\omega)}$ converges to 1 iff $R_{i,n}(\omega) < \infty$, and is zero otherwise. Hence, the left-hand side converges to $\mathbb{P}(R_{i,n}(\omega) < \infty)$. □

2.7 Back to the Second Kolmogorov–Kendall–Reuter Example

2.7.1 *The curious state $i = 1$ in the second Kolmogorov–Kendall–Reuter example*

There are (at least) two points of interest in the second Kolmogorov–Kendall–Reuter example. To begin with, we look at the following intensity matrix:

$$Q = \begin{pmatrix} -1 & 0 & 0 & 0 & 0 & 0 & \cdots \\ 0 & 0 & 0 & 0 & 0 & 0 & \cdots \\ 0 & a_3 & -a_3 & 0 & 0 & 0 & \cdots \\ 0 & 0 & a_4 & -a_4 & 0 & 0 & \cdots \\ \vdots & \vdots & \vdots & \vdots & \ddots & \ddots & \ddots \end{pmatrix}. \tag{2.53}$$

If we would like to define the related operator (say, A_0, to distinguish it from A of (2.31)) as in (2.16),

$$A_0\,(\xi_i)_{i\geq 1} = (\xi_i)_{i\geq 1} \cdot \left(q_{i,j}\right)_{i,j\geq 1},$$

then

- the definition of A_0 would coincide with that of A in (2.31), and
- the natural domain would be composed of $(\xi_i)_{i\geq 1}$ satisfying the first condition in (2.30).

The second condition in the latter equation thus makes it clear that A is a restriction of A_0, and a natural question arises: What is the role of this condition? The first answer is that A_0 is too large to be a generator, and so it must be restricted to a smaller domain. But, what does this condition *mean* probabilistically, that is, what does it command the related process *to do*?

The second curious point is the first line in the intensity matrix (2.53). To repeat the comment from Section 2.5.9: the description of the fate of a particle leaving the state $i = 1$ given in Proposition 2.5.8 is still valid, but does not tell the whole story. Here, it means that after spending an exponential time at $i = 1$ the particle jumps to any $i \geq 2$ with probability equal to ... zero. On the other hand, the second Kolmogorov–Kendall–Reuter semigroup is composed of Markov operators, and the particle cannot simply disappear. So, what does it really do?

As we shall see now, there is one answer to both of these questions.

2.7.2 *An intuition*

It will be convenient to think of the second Kolmogorov–Kendall–Reuter process as similar to climbing an infinite ladder with the top at $i = 2$: The process starting at $i \geq 3$, that is, at the $(i-1)$st rung from the top, goes one rung up after an exponential time with parameter a_i. After reaching the top, it stays there for ever. We will argue that the second condition in (2.30) shows how after leaving the state 1 it should go to the bottom of the ladder, that is, to infinity, from where it is to climb up to the top at $i = 2$. Interestingly, the climb is done in finite time.

2.7.3 *Approximation of the second Kolmogorov–Kendall–Reuter semigroup*

To substantiate our claim, we consider the following sequence of approximating Markov semigroups and their Markov processes. For $n \geq 3$, let A_n be the (bounded) Q-matrix whose upper left corner is

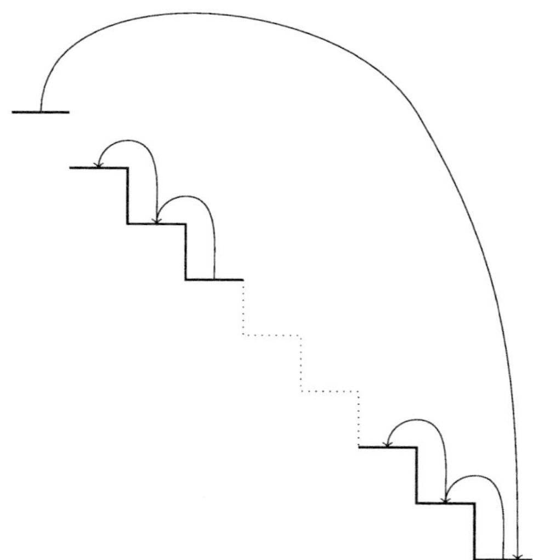

Figure 2.3 A finite chain approximating the second Kolmogorov–Kendall–Reuter example.

$$\begin{pmatrix} -1 & 0 & 0 & 0 & 0 & \cdots & 1 \\ 0 & 0 & 0 & 0 & 0 & \cdots & 0 \\ 0 & a_3 & -a_3 & 0 & 0 & \cdots & \vdots \\ 0 & 0 & a_4 & -a_4 & 0 & \cdots & \vdots \\ \vdots & \vdots & \vdots & \vdots & \ddots & \ddots & \vdots \\ 0 & 0 & 0 & 0 & 0 & a_n & -a_n \end{pmatrix}, \tag{2.54}$$

and the remaining entries are zero. The proof that A_n's generate Markov semigroups is the same as in Section 2.6.2.

In the process related to A_n, 'the ladder' (or 'the staircase' if we want to use Figure 2.3) is finite and it is clear what happens with the process leaving $i = 1$: it jumps to $i = n$ where it starts to climb up to $i = 2$, and finishes the climb in a finite time. So, if we will show that the semigroups generated by A_n's converge to the second Kolmogorov–Kendall–Reuter semigroup, it will become intuitively clear that in the limit process, the particle leaving the state $i = 1$ should jump to an additional state, 'the infinity,' or 'the bottom of the infinite ladder.'

To this end, suppose $x = (\xi_i)_{i\geq 1}$ is a member of $\mathcal{D}(A)$, so that (2.30) holds. By definition, the first $n-1$ coordinates of $A_n x$ and Ax are the same, and

$$\|A_n x - Ax\| = |a_{n+1}\xi_{n+1} - \xi_1| + \sum_{i=n+1}^{\infty} |a_{i+1}\xi_{i+1} - a_i\xi_i|.$$

This, by (2.30) converges to 0, as $n \to \infty$, completing the proof (by the Sova–Kurtz Theorem). Remarkably, both conditions listed in (2.30) are needed for this result, revealing that appropriate characterization of the domain of A is a rather delicate matter.

2.7.4 Remark

Assumption (2.29) was undoubtedly crucial in the proof that the second Kolmogorov–Kendall–Reuter operator (defined by (2.30) and (2.31)) generates a Markov semigroup. The analysis performed above allows interpreting this assumption in probabilistic terms. To this end, first let τ_n's be exponential random variables with parameters a_n, respectively. Since

$$\mathbb{E}\sum_{n=3}^{\infty} \tau_n = \sum_{n=3}^{\infty} \mathbb{E}\,\tau_n = \sum_{n=3}^{\infty} a_n^{-1} < \infty,$$

the sum $\sum_{n=3}^{\infty} \tau_n$ is a.s. finite. Intuitively, this means that the time needed for the second Kolmogorov–Kendall–Reuter process to climb even from the very bottom of the ladder to its top is a.s. finite. This makes the process honest – if (2.29) is not met, the climb started at the bottom may not finalize in a finite time. For the matter of the fact, the climb would never reach any of the rungs. Hence, with nonzero probability the process after jumping from $i = 1$ 'to infinity' would never come back, that is, would simply disappear.

For completeness, before we close this section, we need to clarify the relation between Q defined in (2.53) and the second Kolmogorov–Kendall–Reuter semigroup.

2.7.5 *Q of* (2.53) *is the intensity matrix of the second Kolmogorov–Kendall–Reuter semigroup*

To be sure, the analysis presented in the previous section does by no means prove that Q of (2.53) is the intensity matrix of the second Kolmogorov–Kendall–Reuter semigroup, call it Q_0. It merely suggests this fact.

For a formal proof we recall that all e_i, $i \ge 2$ belong to $\mathcal{D}(A)$. Moreover, a short calculation shows that Ae_i is precisely the ith row of the Q-matrix of (2.53). Thus, by 2.5.1, Q and Q_0 may differ only in the the first row.

Now, (2.33) reveals that the first row of the matrix related to the operator $(\lambda - A)^{-1}$ is

$$\left(\frac{1}{\lambda+1}, \pi_2, \pi_3, \pi_4, \dots\right). \tag{2.55}$$

Since $(\lambda - A)^{-1}$ is the Laplace transform of the semigroup,

$$\frac{1}{\lambda+1} = \int_0^\infty \mathrm{e}^{-\lambda t} p_{1,1}(t)\,\mathrm{d}t \qquad \text{and} \qquad \pi_i = \int_0^\infty \mathrm{e}^{-\lambda t} p_{1,i}(t)\,\mathrm{d}t,\ i \ge 2.$$

It follows that $p_{1,1}(t) = \mathrm{e}^{-t}$ and so $q_{1,1} = -1$. By the way, since – as the other rows of the intensity matrix make it clear – a process starting at $i \ge 2$ will never return to $i = 1$, $p_{1,1}(t) = \mathrm{e}^{-t}$ is the probability that a process starting at $i = 1$ is still there (and not 'back there') after time t, proving that the time spent in $i = 1$ is indeed exponential with parameter 1 (as suggested in 2.7.1).

We are left with showing that $q_{1,i} = 0$, $i \ge 2$. (See also Exercise 2.7.9.) To this end, first we rewrite the definition of π_i as follows:

$$\pi_2 = \lambda^{-1}\widetilde{\pi}_2, \qquad \pi_i = \lambda^{-1}(\widetilde{\pi}_i - \widetilde{\pi}_{i-1}), i \ge 3, \tag{2.56}$$

where $\widetilde{\pi}_i = \frac{1}{\lambda+1}\prod_{j=i+1}^\infty \frac{a_j}{\lambda+a_j}$. (Since $\lim_{i\to\infty}\widetilde{\pi}_i = \frac{1}{\lambda+1}$, this representation shows that the sum of coordinates in (2.55) is λ^{-1}, as it should.) Next, let T_i, $i = 1, 3, 4, 5, \dots$ be independent, exponentially distributed random variables such that $\mathbb{E}\, T_i = a_i^{-1}$. We claim that

$$\lambda^{-1}\widetilde{\pi}_i = \int_0^\infty \mathrm{e}^{-\lambda t}\mathbb{P}(T_1 + \sum_{j=i+1}^\infty T_j \le t)\,\mathrm{d}t, \qquad i \ge 2. \tag{2.57}$$

If this is the case, since two different functions, one of them continuous, the other right-continuous, cannot have the same Laplace transform, we have

$$p_{1,2}(t) = \mathbb{P}(T_1 + \sum_{j=3}^\infty T_j \le t)$$

and

$$\begin{aligned} p_{1,i}(t) &= \mathbb{P}(T_1 + \sum_{j=i+1}^\infty T_j \le t) - \mathbb{P}(T_1 + \sum_{j=i}^\infty T_j \le t) \\ &\le \mathbb{P}(T_1 + \sum_{j=i+1}^\infty T_j \le t), \qquad i \ge 3. \end{aligned}$$

(The formula for $p_{1,i}(t)$, $i \ge 3$ says that $p_{1,i}(t)$ is the probability that there was enough time to climb up to the ith rung, but not enough time to jump from that rung to $i-1$.) However, $\mathbb{P}(T_1+\sum_{j=i+1}^{\infty}T_j\le t)\le\mathbb{P}(T_1+T_{i+1}\le t)=\int_0^t g_i(s)\,\mathrm{d}s$, where g_i is the probability density function of T_1+T_{i+1}. Since $g_i(s)=a_{i+1}\int_0^s \mathrm{e}^{-(s-u)}\mathrm{e}^{-a_{i+1}u}\,\mathrm{d}u\le a_{i+1}s$, we obtain $p_{1,i}(t)\le a_{i+1}\frac{t^2}{2}$, and this forces $q_{1,i}=\lim_{t\to0+}\frac{p_{1,i}(t)}{t}=0$.

It remains to show (2.57). Since, for fixed i and $t\ge 0$,

$$\{T_1+\sum_{j=i+1}^{n}T_j\le t\},n\ge i+1$$

is a decreasing sequence of events with intersection $\{T_1+\sum_{j=i+1}^{\infty}T_j\le t\}$, by the Lebesgue Dominated Convergence Theorem,

$$\lim_{n\to\infty}\int_0^\infty \mathrm{e}^{-\lambda t}\mathbb{P}(T_1+\sum_{j=i+1}^{n}T_j\le t)\,\mathrm{d}t=\int_0^\infty \mathrm{e}^{-\lambda t}\mathbb{P}(T_1+\sum_{j=i+1}^{\infty}T_j\le t)\,\mathrm{d}t. \tag{2.58}$$

On the other hand, $\int_0^\infty \mathrm{e}^{-\lambda t}\mathbb{P}(T_1+\sum_{j=i+1}^{n}T_j\le t)\,\mathrm{d}t$ equals

$$\int_0^\infty \mathrm{e}^{-\lambda t}\int_0^t g_{i,n}(s)\,\mathrm{d}s\,\mathrm{d}t=\lambda^{-1}\int_0^\infty \mathrm{e}^{-\lambda t}g_{i,n}(t)\,\mathrm{d}t,$$

where $g_{i,n}$ is the probability density function of the random variable $T_1+\sum_{j=i+1}^{n}T_j$. Since T_j's are independent and exponentially distributed with $\mathbb{E}\,T_j=a_j^{-1}$, the right-hand side in (2.58) is

$$\lim_{n\to\infty}\frac{1}{\lambda}\frac{1}{\lambda+1}\prod_{j=i+1}^{n}\frac{a_j}{\lambda+a_j}.$$

This completes the proof.

2.7.6 Remark

Formulae for $p_{1,i}$ reveal that ∞ works as a 'transport hub': a process going through this point is distributed across the entire state-space according to the law $(0,p_{1,1}(t),p_{1,2}(t),\dots)$. In other words, we are dealing with the first example of an **entrance law** here.

2.7.7 Exercise

Show that the A_n's of Sections 2.6.2 and 2.7.1 are generators of Markov semigroups by representing them in the form $A_n=c_n\left(B_n-I_{l^1}\right)$ where $c_n>0$ and B_n is a Markov operator.

2.7.8 Exercise

Consider the following modification of the second Kolmogorov–Kendall–Reuter operator: in (2.31) let $\eta_2 = p\xi_1 + a_3\xi_3$, where $p \in (0, 1)$. By finding appropriate approximation argue that in this case the process after spending at $i = 1$ an exponential time with parameter 1 jumps to the absorbing state $i = 2$ (with probability p) or to the 'bottom of the ladder' with probability $1 - p$. Check that for this operator the second condition in (2.30) must be replaced by $\lim_{i\to\infty} a_i\xi_i = (1 - p)\xi_1$.

2.7.9 Exercise

An appropriately extended argument of Section 2.2.7 shows ([21], p. 9) that if a state i is stable then the functions $t \mapsto p_{i,j}(t)$, $j \in \mathbb{I}$ are differentiable (thus in particular $q_{i,j} = p'_{i,j}(0)$), and their derivatives are bounded and continuous. Recall also that for a bounded continuous function $f\colon \mathbb{R}^+ \to \mathbb{R}$, $f(0) = \lim_{\lambda\to\infty} \lambda \int_0^\infty \mathrm{e}^{-\lambda t} f(t)\,\mathrm{d}t$. Use these facts to check that, since in the second Kolmogorov–Kendall–Reuter example the state $i = 1$ is stable, $q_{1,i} = \lim_{\lambda\to\infty} \lambda^2\pi_i$, $i \geq 2$, and conclude that $q_{1,i} = 0$, $i \geq 2$.

2.8 Blackwell's Example

In the first Kolmogorov–Kendall–Reuter example, the state $i = 1$ is instantaneous. As it transpires there are, honest!, Markov chains with countably many states all of which are instantaneous. Such examples were presented first by Dobrushin [28] and Feller and McKean [44]. This section is devoted to the discussion of an example of Blackwell (see the original paper [13] or, e.g., p. 297 in [45] or p. 65 in [64]).

2.8.1 *Building blocks*

Let $\mathbb{I}$ be the set of functions $i\colon \mathbb{N} \to \{0, 1\}$ admitting value 1 finitely many times. Since for any $n \geq 1$ there is only a finite number of functions $i\colon \mathbb{N} \to \{0, 1\}$ admitting value 0 from the nth coordinate onward, $\mathbb{I}$ is countable, as a countable union of finite sets. We will find it convenient to write elements $x = (\xi_i)_{i\in\mathbb{I}}$ of $l^1 := l^1(\mathbb{I})$, as $x = \sum_{i\in\mathbb{I}} \xi_i e_i$ (see (2.9)). In fact, the series on the right converges to x in the norm of l^1.

Given positive numbers $\alpha_n, \beta_n, n \geq 1$ we construct a sequence of Markov semigroups in l^1. To this end, we need some notation. For $n \geq 1$ let F_n be the map $F_n : \mathbb{I} \to \mathbb{I}$ changing the nth coordinate of an i from 0 to 1 and vice versa. Also, for a finite set $E = \{n_1, \dots, n_k\} \subset \mathbb{N}$, let

$$F_E := F_{n_1} \circ \dots \circ F_{n_k}$$

be the map that changes an $i \in \mathbb{I}$ at its coordinates $n_1, \dots, n_k$. Finally, let $G_n : \mathbb{I} \to \{0, 1\}$ assign to an i its nth coordinate.

For $n \geq 1$, let B_n be the bounded linear operator determined by its values on e_i as follows:

$$B_n e_i = \begin{cases} -\beta_n e_i + \beta_n e_{F_n(i)}, & \text{if } G_n(i) = 0, \\ -\alpha_n e_i + \alpha_n e_{F_n(i)}, & \text{if } G_n(i) = 1. \end{cases} \tag{2.59}$$

B_n is the generator of a Markov chain on $\mathbb{I}$ in which the nth coordinate of an $i \in \mathbb{I}$ jumps between 0 and 1, the intensity of the 'forward' jump being β_n and that of the 'backward' jump being α_n. In other words, B_n is the generator of the Markov semigroup in l^1 determined by (cf. (2.3))

$$\mathrm{e}^{tB_n} e_i = \begin{cases} p_n(t) e_i + (1 - p_n(t)) e_{F_n(i)}, & \text{if } G_n(i) = 0, \\ q_n(t) e_i + (1 - q_n(t)) e_{F_n(i)}, & \text{if } G_n(i) = 1, \end{cases}$$

where

$$p_n(t) = \frac{\alpha_n}{\alpha_n + \beta_n} + \frac{\beta_n}{\alpha_n + \beta_n} \mathrm{e}^{-(\alpha_n + \beta_n)t},$$
$$q_n(t) = \frac{\beta_n}{\alpha_n + \beta_n} + \frac{\alpha_n}{\alpha_n + \beta_n} \mathrm{e}^{-(\alpha_n + \beta_n)t}.$$

2.8.2 *Approximating semigroups*

The operators B_n commute. It follows that

$$T_n(t) = \prod_{k=1}^{n} \mathrm{e}^{tB_k}$$

defines a strongly continuous semigroup $\{T_n(t), t \geq 0\}$ with generator $A_n = \sum_{k=1}^{n} B_k$. This semigroup describes n combined independent Markov chains, each changing one of the first n coordinates of i as described above, and is determined by

$$T_n(t) e_i = \sum_{E \subset \{1,\dots,n\}} p_{E,n}(t, i) e_{F_E(i)}, \tag{2.60}$$

with summation over 2^n subsets E of $\{1, \dots, n\}$ and

$$p_{E,n}(t, i) = \prod_{k=1}^{n} r_k(t, i, E)$$

where

$$r_k(t, i, E) = \begin{cases} p_k(t), & \text{if } G_k(i) = 0 \text{ and } k \notin E, \\ 1 - p_k(t), & \text{if } G_k(i) = 0 \text{ and } k \in E, \\ q_k(t), & \text{if } G_k(i) = 1 \text{ and } k \notin E, \\ 1 - q_k(t), & \text{if } G_k(s) = 1 \text{ and } k \in E. \end{cases} \tag{2.61}$$

This formula simply says that, if the initial state of the Markov chain related to A_n is given, its state at time $t \geq 0$ may be characterized by listing the coordinates which are different from the original ones – this is the role of the set E. Since each coordinate evolves independently from the other ones, the probability of such a change is the product of probabilities of change or no change on the first, second, third, and remaining coordinates.

2.8.3 *Preparation for the main theorem*

Assume now that

$$\sum_{n=1}^{\infty} \frac{\beta_n}{\alpha_n + \beta_n} < \infty, \tag{2.62}$$

and fix i and a *finite* subset E of $\mathbb{N}$. Then, for n large enough, $E \subset \{1, \dots, n\}$. Also, since i has finitely many coordinates different from 0, for sufficiently large n and $k > n$, $r_k(t, i, E)$ describes the probability of no change from initial 0 in the kth coordinate of i (after perhaps many changes back and forth in the meantime), that is, $r_k(t, i, E) = p_k(t)$. By Criterion 2.4.6, the limit

$$p_E(t, i) := \lim_{n\to\infty} p_{E,n}(t, i) \tag{2.63}$$

exists, because the finite number of terms in the product does not influence convergence, whereas

$$\sum_{n=1}^{\infty}(1 - p_n(t)) = \sum_{n=1}^{\infty} \frac{\beta_n}{\alpha_n + \beta_n}(1 - \mathrm{e}^{-(\alpha_n+\beta_n)t}) \leq \sum_{n=1}^{\infty} \frac{\beta_n}{\alpha_n + \beta_n} < \infty,$$

by assumption.

2.8.4 Lemma

As $n \to \infty$, the right-hand side of (2.60) converges to

$$T(t)e_i := \sum_E p_E(t,i)e_{F_E(i)},$$

where the sum is over all finite subsets E of $\mathbb{N}$.

Proof Clearly, $\sum_{E\subset\{1,\dots,n\}} p_{E,n}(t,i) = 1$ for $i \in \mathbb{I}$, $t \ge 0$ and $n \ge 1$, and the summands are nonnegative. Therefore, by (2.63) and Scheffé's Theorem 1.6.1, all we need to show is that

$$\sum_E p_E(t,i) = 1, \qquad \text{for} \qquad t \ge 0, i \in \mathbb{I}. \tag{2.64}$$

The proof of this relation will be more transparent if we assume $i = (0,0,\dots)$; this we do without loss of generality. Also, for simplicity of notation, we suppress dependence on t and i, and write p_n and p_E instead of $p_n(t)$ and $p_E(t,i)$, respectively. Thus

$$p_E = \prod_{n\in E}(1-p_n)\prod_{n\notin E} p_n.$$

For a (finite) $E \subset \mathbb{N}$, let

$$E_1 = \begin{cases} E\setminus\{1\}, & \text{if } 1 \in E,\\ E\cup\{1\}, & \text{if } 1 \notin E.\end{cases}$$

Then

$$p_E + p_{E_1} = \prod_{n\in E, n\neq 1}(1-p_n)\prod_{n\notin E, n\neq 1} p_n.$$

Therefore, letting $L = \sum_{\text{finite } E\subset\{1,2,\dots\}} p_E$, we have

$$L = \sum_{\text{finite } E\subset\{2,3,\dots\}}\prod_{n\in E}(1-p_n)\prod_{n\notin E, n\neq 1} p_n.$$

Repeating this argument k times, we see that

$$L = \sum_E \prod_{n\in E}(1-p_n)\prod_{n\in(\mathbb{N}\setminus\{1,\dots,k\})\setminus E} p_n,$$

where $E \subset \{k+1, k+2, \dots\}$ are finite sets. Since the sum includes the component corresponding to $E = \emptyset$, $L \ge \prod_{n=k+1}^{\infty} p_n$. On the other hand, the last product may be chosen as close to 1 as we wish by taking k large enough. This completes the proof. □

2.8.5 Theorem

The semigroups $\{T_n(t), t \geq 0\}$ converge to the semigroup $\{T(t), t \geq 0\}$;

$$T(t)x = \sum_{i \in \mathbb{I}} \xi_i T(t) e_i$$

where $T(t)e_i$ is defined in Lemma 2.8.4, and $x = \sum_{i \in \mathbb{I}} \xi_i e_i$. The limit semigroup is strongly continuous and is composed of Markov operators.

Proof The statement concerning convergence is clear by Lemma 2.8.4. Also, $\{T(t), t \geq 0\}$ is a semigroup, being a limit of semigroups, and as direct consequence of (2.64), it is composed of Markov operators. Hence, we are left with showing that $\{T(t), t \geq 0\}$ is strongly continuous, and for this it suffices to show that for all $i \in \mathbb{I}$, $p_{i,i}(t) = p_\emptyset(t, i)$ converges to 1 as $t \to 0$ (see Section 2.1.2). Let $i \in \mathbb{I}$, and let m be so large that all coordinates of i with indices larger than m are zeros. Then

$$p_\emptyset(t, i) = \prod_{k=1}^{n} r_k(t, i, \emptyset) \prod_{k=n+1}^{\infty} p_k(t) \geq \prod_{k=1}^{n} r_k(t, i, \emptyset) \prod_{k=n+1}^{\infty} \frac{\alpha_k}{\alpha_k + \beta_k} \tag{2.65}$$

for $n \geq m$, where (see (2.61))

$$r_k(t, i, \emptyset) = \begin{cases} p_k(t), & \text{if } G_k(i) = 0, \\ q_k(t), & \text{if } G_k(i) = 1. \end{cases}$$

We have $\lim_{t \to 0} p_k(t) = \lim_{t \to 0} q_k(t) = 1$ for all $k \geq 1$. Therefore,

$$\liminf_{t \to 0} p_\emptyset(t, i) \geq \prod_{k=n+1}^{\infty} \frac{\alpha_k}{\alpha_k + \beta_k},$$

and the latter product may be chosen as close to 1 as desired. This shows that $\lim_{t \to 0} p_\emptyset(t, i) = 1$. □

2.8.6 Corollary

Suppose

$$\sum_{n=1}^{\infty} \beta_n = \infty; \tag{2.66}$$

this assumption does not contradict (2.62). Then in the limit Markov semigroup, the Blackwell's semigroup, all states $i \in \mathbb{I}$ are instantaneous.

Proof Fix $i \in \mathbb{I}$. Let m be so large that all coordinates of i with indices $n > m$ are zero. Then, as in (2.65),

$$p_{i,i}(t) = p_{\emptyset}(t, i) = \prod_{k=1}^{n_1} r_k(t, i, \emptyset) \prod_{k=n_1+1}^{\infty} p_k(t) \leq \prod_{k=n_1+1}^{n_2} p_k(t),$$

for all $n_2 > n_1 > m$. It follows that

$$q_i = \lim_{t \to 0+} t^{-1} \left(1 - p_{i,i}(t)\right) \geq \lim_{t \to 0+} t^{-1} \left(1 - \prod_{k=n_1+1}^{n_2} p_k(t)\right)$$
$$= - \lim_{t \to 0+} \sum_{k=n_1+1}^{n_2} p_k'(t) \prod_{\substack{j \neq k \\ n_1 < j \leq n_2}} p_j(t) = \sum_{k=n_1+1}^{n_2} \beta_k,$$

where in the previous-to-last equality we used de l'Hospital's rule. Letting $n_2 \to \infty$, we obtain $q_i = \infty$. □

2.8.7 *What is Blackwell's process like?*

Starting in a state i, Blackwell's process communicates with infinitely many neighbors of i differing from i in one coordinate. Again we may think of infinitely many alarm clocks going off after exponential times, but in this case each exponential distribution may have a different parameter: the clock related to the nth coordinate has expected time to going off equal β_n^{-1} or α_n^{-1} depending on whether the nth coordinate of i is 0 or 1. And again (this time because of assumption (2.66)) the minimum of these exponential times is 0, so that the process at i immediately hears one of the clocks go off and duly changes appropriate coordinate of i. At the new state i' it has a new collection of clocks representing new neighbors (including one calling it to come back to the old state i), and before any time is spent at i' one of them goes off. Hence, the poor guy must change another coordinate (or the same one again), to hear yet another clock go off ... . A clockmaker's paradise may be an obedient's man hell, and a spectacle to a mathematician.

2.9 Notes

Exhaustive treatments of the theory of Markov chains may be found in Chung's [21] and Freedman's [45]. A very readable but more brief account is Norris's [72]; the short Markov chain chapter in Liggett's [64] is commendable as much as the whole book. There are of course other excellent texts with long and interesting chapters on Markov chains, like Feller's [40] and Grimmett and Stirzaker's [48]. One should also note Anderson's thorough exposition of continuous Markov chains given in [2]

Existence of q_i's and $q_{i,j}$'s (see 2.2.3 and 2.2.7) is due to Doob [30]. Both Kolmogorov–Kendall–Reuter examples were first devised by Kolmogorov, and then analysed in the functional-analytic setting by Kendall and Reuter [57]. Along Blackwell's example, which (surprise, surprise!) is due to Blackwell, they constitute remarkable counterexamples to the so-called Stigler's law of eponymy: 'No law, theorem, or discovery is named after its originator' ([48], p. 19).

If you find it hard to imagine what the Blackwell chain is 'really' like, be encouraged: you are not the only one. In Freedman's own words, the entire [46] 'is a monograph explaining one way to think about chains with instantaneous states' ([46], p. v).

3

Boundary Theory: Core Results

Before continuing, let us try to summarize.[1] Except for the uniform case in which the notions of the Q-matrix and of the generator coincide, Q-matrices, despite their popularity, seem to be less useful than generators. First of all, as the first Kolmogorov–Kendall–Reuter example reveals there may be instantaneous states in $\mathbb{I}$ and then the Q-matrix is improper in that some of its entries are infinite. In fact, as Blackwell's example makes amply clear, the main diagonal of (an improper) Q-matrix may be composed solely of $-\infty$'s. Also, from the second Kolmogorov–Kendall–Reuter example we learn that a Q-matrix does not tell the whole story of the related process. It is in the generator or, more precisely, in *the domain of the generator* that the missing information is skilfully hidden. (See also Section 3.4.)

With this background, the reader is not surprised to hear that with one (proper) Q-matrix there may be related (infinitely!) many Markov and sub-Markov semigroups. In fact, Exercise 2.4.17 nearly furnishes such a family of semigroups for the Q-matrix of the second Kolmogorov–Kendall–Reuter example. I have to say 'nearly' because these semigroups are not composed of Markov operators but rather of sub-Markov operators. Moreover, there, the Q-matrix is not proper.

The first main theorem of this chapter, due to T. Kato [51, 54], says that among all those semigroups there is one that is in a sense minimal (Section 3.1): we split Kato's Theorem into a series of chunks of information, but the main statement is Proposition 3.1.5. In Section 3.2 we give criteria for the minimal semigroup of Kato's Theorem to be composed of Markov operators, and study in detail six specific examples. In Section 3.3, we illustrate a different aspect of Kato's Theorem: taking an explosive pure birth process as a

[1] Some readers may perhaps prefer to summerize, though. Or, be lazing on a sunny afternoon, with or without The Kinks, but definitely with the Haley Reinhart cover featuring Scott Bradlee on piano.

case study, we construct explicitly two semigroups dominating the minimal semigroup: the first of them governs the process which after explosion returns to the state 1, the second – the process which returns to the state 2. Section 3.4 is devoted to a discussion of the position of the domain of the minimal semigroup generator between the domains of two other natural operators. In Section 3.5, using the intuitions gained in Section 3.3, we show how semigroups dominating the minimal one may be defined for a general explosive matrix. The main abstract result is Section 3.5.5 containing the formula for the generator of a postexplosion process, but the rest of this section suggests that, since explosion can come about in many ways, this formula should be further refined.

These results form an introduction to boundary theory for Markov chains: in Section 3.6 we study the lattice of certain functionals on l^1 which in the previous sections were revealed to be instrumental in defining postexplosion processes. In particular, the lattice is divided into two classes: the class of passive and that of active functionals, which are duly characterized. In Section 3.7, minimal elements of the active part of the lattice are then shown to describe different ways explosion may come about; these functionals can be seen as additional points of the state-space of the process, and form the discrete exit boundary (see Chapter 4 and Section 4.2 in particular).

The second main result of this chapter is formula (3.92), providing quite a general form of the generator of a postexplosion process (if merely discrete exit boundary needs to be accounted for). Section 3.8 gathers additional information on extremal and minimal functionals.

3.1 Kato's Theorem

3.1.1 *Some notation*

For simplicity of notation in this and the following chapter (with the exception of Sections 4.3 and 4.4) we assume that $\mathbb{I}$ has been ordered so that we may think that $\mathbb{I} = \mathbb{N}$; in the case $\#\mathbb{I} < \infty$ there is only one semigroup related to a Q-matrix, that is, the exponent of this matrix, and Kato's Theorem is trivial. We recall that $(\xi_i)_{i\geq 1} \in l^1$ is said to be a **distribution** (or a **density**) if it is nonnegative, that is, $\xi_i \geq 0, i \geq 1$, and $\Sigma\,(\xi_i)_{i\geq 1} = 1$. A linear, not necessarily bounded operator A in l^1 is said to be nonnegative if it maps $\mathcal{D}(A) \cap (l^1)^+$ into $(l^1)^+$, where $(l^1)^+$ is the nonnegative cone, that is, the set of nonnegative $(\xi_i)_{i\geq 1} \in l^1$. For x and y in l^1 we write $x \leq y$ or $y \geq x$ if $y - x \in (l^1)^+$. For two operators, A and B, in l^1 we write $A \leq B$ or $B \geq A$ iff $B - A$ is nonnegative. An operator A (defined on the whole of l^1) is said to be Markov if it leaves the set of densities invariant (this definition is equivalent to that given in Section 1.2.6); it is said to be sub-Markov iff it is nonnegative and $\Sigma Ax \le \Sigma x$ for any density x.

3.1.2 Definition

Let $Q = (q_{i,j})_{i,j\ge 1}$ be an intensity matrix (recall that this means that $\sum_{j\ne i} q_{i,j}$ may be smaller than q_i). We define the domain of an operator A_0 to be the linear span of e_i, $i \ge 0$, and put $A_0 e_i = (q_{i,j})_{j\ge 1}$. Furthermore, the operator D ('D' for 'diagonal') with domain

$$\mathcal{D}(D) = \{(\xi_i)_{i\ge 1} \in l^1 \,|\, (q_i\xi_i)_{i\ge 1} \in l^1\}$$

is defined by $D\,(\xi_i)_{i\ge 1} = (-q_i\xi_i)_{i\ge 1}$; note that $-D$ is nonnegative.

3.1.3 *Approximating semigroups*

The operator O ('O' for 'off diagonal') given by $O\,(\xi_i)_{i\ge 1} = \left(\sum_{j\ge 1, j\ne i} \xi_j q_{j,i}\right)_{i\ge 1}$ is well defined on $\mathcal{D}(D)$ and

$$\|Ox\| \le \|Dx\|, \qquad x \in \mathcal{D}(D). \tag{3.1}$$

Also,

$$\|Ox\| = \|Dx\|, \qquad x \in \mathcal{D}(D) \tag{3.2}$$

provided $x \ge 0$ and Q is a Kolmogorov matrix. Moreover, for any $0 \le r < 1$, the operator $D + rO$ with domain $\mathcal{D}(D)$ is the generator of a strongly continuous semigroup of sub-Markov operators in l^1, say, $\{S_r(t), t \ge 0\}$.

Proof For $x \in \mathcal{D}(D)$, $\sum_{i=1}^{\infty} \left|\sum_{j\ge 1, j\ne i} \xi_j q_{j,i}\right|$ does not exceed

$$\sum_{i=1}^{\infty} \sum_{j\ge 1, j\ne i} |\xi_j q_{j,i}| = \sum_{j=1}^{\infty} |\xi_j| \sum_{i\ge 1, i\ne j} q_{j,i} \le \sum_{j=1}^{\infty} q_j |\xi_j| = \|Dx\|, \tag{3.3}$$

with equality iff $(\xi_i)_{i\ge 1}$ is nonnegative and $\sum_{i\ge 1, i\ne j} q_{j,i} = q_j$ for all $j \in \mathbb{N}$. This proves the first claim.

As to the rest, we note first that for $\lambda > 0$ we have $(\lambda - D)^{-1}\,(\xi_i)_{i\ge 1} = \left(\frac{1}{\lambda + q_i}\xi_i\right)_{i\ge 1}$ and that

$$B_\lambda := O(\lambda - D)^{-1} \tag{3.4}$$

is well defined. Moreover, by (3.1),

$$\|B_\lambda x\| \le \|D(\lambda - D)^{-1}x\| = \sum_{j\ge 1} \frac{q_j}{\lambda + q_j}|\xi_j| \le \sum_{j\ge 1} |\xi_j| = \|x\|.$$

Hence, B_λ is a contraction and for any $0 \le r < 1$ the series $\sum_{n=0}^\infty r^n B_\lambda^n$ (which equals $(I - rB_\lambda)^{-1}$) converges in the operator norm. Let

$$R_{\lambda,r} = (\lambda - D)^{-1} \sum_{n=0}^\infty r^n B_\lambda^n.$$

By definition $R_{\lambda,r}x$ belongs to $\mathcal{D}(D)$ and $(\lambda - D)R_{\lambda,r}x = \sum_{n=0}^\infty r^n B_\lambda^n x$, and $rOR_{\lambda,r}x = rB_\lambda \sum_{n=0}^\infty r^n B_\lambda^n x = \sum_{n=1}^\infty r^n B_\lambda^n x$. Hence, $(\lambda - D - rO)R_{\lambda,r}x = x, x \in l^1$. Similarly, $R_{\lambda,r}(\lambda - D - rO)x = x, x \in \mathcal{D}(D)$. This shows that $R_{\lambda,r} = (\lambda - D - rO)^{-1}$. Moreover, if $x \ge 0$ then $y = R_{\lambda,r}x$ is nonnegative too; indeed, $(\lambda - D)^{-1} \ge 0$ and $O \ge 0$ and so $B_\lambda \ge 0$. Since $D + rO$ is obviously densely defined and $\Sigma(D + rO)\,(\xi_i)_{i\ge 1} = (r-1)\sum_{i=1}^\infty q_i\xi_i \le 0$ provided $(\xi_i)_{i\ge 1}$ is a nonnegative member of $\mathcal{D}(D)$, we are done by the Hille–Yosida Theorem 2.4.3. □

3.1.4 *Convergence of the approximation*

As $r \uparrow 1$, the semigroups $\{S_r(t), t \ge 0\}$ converge strongly to a strongly continuous semigroup $\{S(t), t \ge 0\}$ of sub-Markov operators generated by an extension of $D + O$ (hence, an extension of A_0 as well).

Proof We have $R_{\lambda,r} \le R_{\lambda,r'}$ for $r \le r'$, and $\|R_{\lambda,r}\| \le \lambda^{-1}$. Hence, by 1.6.3, there exists the strong limit $R_\lambda = \lim_{r\uparrow 1} R_{\lambda,r}$. We would like to know that the closure of the range of R_λ is the entire l^1. Our first step in this direction is the explicit formula for R_λ given in (3.5).

Clearly, for any $N \in \mathbb{N}$,

$$\sum_{n=0}^N (\lambda - D)^{-1} r^n B_\lambda^n \le R_{\lambda,r} \le R_\lambda.$$

Hence, letting $r \uparrow 1$, we obtain $\sum_{n=0}^N (\lambda - D)^{-1} B_\lambda^n \le R_\lambda$. Using 1.6.3 again, we see that the series $\sum_{n=0}^\infty (\lambda - D)^{-1} B_\lambda^n$ converges and $\sum_{n=0}^\infty (\lambda - D)^{-1} B_\lambda^n \le R_\lambda$. On the other hand, $R_{\lambda,r} \le \sum_{n=0}^\infty (\lambda - D)^{-1} B_\lambda^n$ and so, letting $r \uparrow 1$, $R_\lambda \le \sum_{n=0}^\infty (\lambda - D)^{-1} B_\lambda^n$, proving that

$$R_\lambda = \sum_{n=0}^\infty (\lambda - D)^{-1} B_\lambda^n. \tag{3.5}$$

(Note that we *do not* claim that $R_\lambda = (\lambda - D)^{-1} \sum_{n=0}^{\infty} B_\lambda^n$; in fact, the series $\sum_{n=0}^{\infty} B_\lambda^n$ in general diverges. Operators B_λ^n converge to zero as $n \to \infty$ iff the minimal semigroup is composed of Markov operators; see 3.2.1.)

Next, for $x \in \mathcal{D}(D)$ and $N \in \mathbb{N}$,

$$\sum_{n=0}^{N+1} (\lambda - D)^{-1} B_\lambda^n (\lambda - D)x = x + \sum_{n=1}^{N+1} (\lambda - D)^{-1} B_\lambda^{n-1} Ox$$
$$= x + \sum_{n=0}^{N} (\lambda - D)^{-1} B_\lambda^n Ox. \tag{3.6}$$

Letting $N \to \infty$, we obtain $R_\lambda(\lambda - D)x = x + R_\lambda Ox$, that is, $R_\lambda(\lambda - D - O)x = x$. In particular, the range of R_λ contains $\mathcal{D}(D)$ and so $cl(Range R_\lambda) = l^1$. Therefore, by the Trotter–Kato Theorem, the semigroups $\{S_r(t), t \geq 0\}$ converge as $r \uparrow 1$ to a strongly continuous semigroup:

$$\boxed{S(t) = \lim_{r \to 1-} S_r(t).}$$

The limit semigroup is composed of sub-Markov operators, the operators $S_r(t), t \geq 0, 0 \leq r < 1$, being sub-Markov. Finally, for $x \in \mathcal{D}(D)$, $\lim_{r\uparrow 1} Dx + rOx = Dx + Ox$, proving that the extended limit of $D + rO$, which is the generator of the limit semigroup, is an extension of $D + O$. □

3.1.5 *Properties of the limit semigroup*

Let $\{S(t), t \geq 0\}$ be the semigroup defined in 3.1.4 and suppose that the generator A of a strongly continuous semigroup $\{P(t), t \geq 0\}$ is an extension of the operator A_0. Then A is also an extension of $D + O$ and, if $P(t) \geq 0, t \geq 0$, then $S(t) \leq P(t), t \geq 0$. *We say that* $\{S(t), t \geq 0\}$ *is the minimal semigroup related to* Q.

Proof Suppose that $x = \sum_{n=1}^{\infty} \xi_n e_n$ belongs to $\mathcal{D}(D)$. By definition of $\mathcal{D}(D)$, so do $x_N := \sum_{n=1}^{N} \xi_n e_n, N \geq 1$, and since $Ae_n = A_0 e_n = (D + O)e_n$, we have $Ax_N = (D + O)x_N$. Moreover, $\lim_{n\to\infty} Dx_N = Dx$ and so, by $\|O(x_N - x)\| \leq \|D(x_N - x)\|$, $\lim_{N\to\infty} Ox_N = Ox$. Therefore $\lim_{N\to\infty} Ax_N$ exists and equals $(D + O)x$ and, obviously, $\lim_{N\to\infty} x_N = x$. Since A is closed, being the generator of a semigroup, x belongs to $\mathcal{D}(A)$ and $Ax = (D + O)x$, proving the first claim.

Next, we note that $(\lambda - A)^{-1}$ exists for sufficiently large $\lambda > 0$. For $y \in \mathcal{D}(D)$, we may write

$$(\lambda - D - rO)y - (\lambda - A)y = Ay - Dy - rOy = (1 - r)Oy.$$

Taking $y = (\lambda - D - rO)^{-1}x$, $x \in l^1$, and applying $(\lambda - A)^{-1}$ to the leftmost and rightmost sides of the above equality,

$$(1 - r)(\lambda - A)^{-1}O(\lambda - D - rO)^{-1}x = (\lambda - A)^{-1}x - (\lambda - D - rO)^{-1}x.$$

Since all operators on the left-hand side are nonnegative, for large $\lambda > 0$ we have $(\lambda - D - rO)^{-1} \le (\lambda - A)^{-1}$. As we have seen in the previous subsection, $(\lambda - D - rO)^{-1}$ converges to $(\lambda - G)^{-1}$ where

$$G \text{ is the generator of } \{S(t), t \ge 0\}.$$

Hence, $(\lambda - G)^{-1} \le (\lambda - A)^{-1}$ and then $[(\lambda - G)^{-1}]^n \le [(\lambda - A)^{-1}]^n$ for all $n \ge 1$ (see Exercise 3.1.9). Making use of the Yosida approximation as follows:

$$S(t) = \lim_{\lambda \to \infty} e^{-\lambda t} e^{\lambda^2(\lambda - G)^{-1}t} \le \lim_{\lambda \to \infty} e^{-\lambda t} e^{\lambda^2(\lambda - A)^{-1}t} = P(t),$$

we complete the proof. □

3.1.6 Remark

If the minimal semigroup $\{S(t), t \ge 0\}$ is composed of Markov operators, there is no other semigroup of Markov or sub-Markov operators generated by an extension of A_0. For such a semigroup, say, $\{P(t), t \ge 0\}$, would need to dominate the minimal one, and since it would be different from $\{S(t), t \ge 0\}$ there would exist an $x \ge 0$ and a $t > 0$ such that $P(t)x \ne S(t)x$, that is, at least one coordinate of $P(t)x$ would be strictly larger than the corresponding coordinate of $S(t)x$. As a result, we would have $\Sigma P(t)x > \Sigma S(t)x = \Sigma x$, which is impossible for a Markov or sub-Markov operator. Hence, in such cases it is right to speak of *the* Markov generator related to the intensity matrix. In general, however, there are many Markov and sub-Markov generators that extend A_0.

3.1.7 $D + O$ *and* G

The results just established are more mysterious than they appear to be, and so is the relation between $D + O$ and the generator G of the minimal semigroup. In distinction to many semigroup-theoretical constructions, when the minimal semigroup is not a Markov semigroup, G, although it extends $D + O$, *is not* the closure of $D + O$. This is to say that $\mathcal{D}(D + O) = \mathcal{D}(D)$ is not a core for G: there are x's in $\mathcal{D}(G)$ such that for no sequence $(x_n)_{n \ge 1}$ of elements of $\mathcal{D}(D)$ can we have simultaneously $\lim_{n \to \infty} x_n = x$ and $\lim_{n \to \infty} Gx_n = Gx$. (See Sections 1.1.13 and 1.1.14 for necessary definitions; in particular, $D + O$ is closable because its graph is contained in the graph of the closed operator G.)

Before seeing this, observe that for nonnegative $x \in \mathcal{D}(D)$,

$$\Sigma G x = 0, \tag{3.7}$$

as long as Q is a Kolmogorov matrix. This is what (3.2) says once we note that (a) for nonnegative x, $\|Ox\| = \Sigma Ox$ and $\|Dx\| = -\Sigma Dx$, and (b) for $x \in \mathcal{D}(D)$, $Gx = Dx + Ox$. Moreover, $\mathcal{D}(D)$ is a lattice: if $x = (\xi_i)_{i\ge1}$ belongs to $\mathcal{D}(D)$ then so does $|x| := (|\xi_i|)_{i\ge1}$. Thus, any $x \in \mathcal{D}(D)$ may be written as a difference of two nonnegative elements: $x^+ = \frac{1}{2}(x+|x|)$ and $x^- = \frac{1}{2}(|x|-x)$. It follows that, for any $x \in \mathcal{D}(D)$, $\Sigma Gx = \Sigma(D+O)x^+ - \Sigma(D+O)x^- = 0$. This means that (3.7) extends to all $x \in \mathcal{D}(D)$. Since we will use this fact in the later sections as well (see 3.5.4), we record it here as follows:

$$\Sigma G x = 0, \qquad x \in \mathcal{D}(D). \tag{3.8}$$

Coming back to the main subject of this section, (3.8) reveals that, given $x \in \mathcal{D}(G)$, it is impossible to have $\lim_{n\to\infty} Gx_n = Gx$ for a sequence $(x_n)_{n\ge1}$ of elements of $\mathcal{D}(D)$, unless $\Sigma Gx = 0$. In other words,

$$\mathcal{D}(\overline{D+O}) \subset \{x \in \mathcal{D}(G); \Sigma Gx = 0\} \subset \mathcal{D}(G), \tag{3.9}$$

and, as we shall see in the next section, if the minimal semigroup is not a Markov semigroup, the second inclusion in (3.9) is proper. See Section 3.4 for examples of such a proper inclusion. In fact, it is precisely existence of $x \in \mathcal{D}(G)$ with $\Sigma Gx \neq 0$ that allows construction of post-explosion processes (see Section 3.5.5).

3.1.8 Exercise

Check directly that the operator D of Definition 3.1.2 generates the semigroup given by $T(t)\,(\xi_i)_{i\ge1} = \left(\mathrm{e}^{-q_i t}\xi_i\right)_{i\ge1}$.

3.1.9 Exercise

Let A and B be positive operators. Show that $A \ge B$ implies $A^n \ge B^n$.

3.2 The Question of Explosiveness

In this section we discuss the question of when the minimal semigroup is composed of Markov operators, that is, of when a Kolmogorov matrix is nonexplosive. We start with criteria for nonexplosiveness and next study six concrete Kolmogorov matrices. The second, the third, and the fourth of these matrices are shown to be nonexplosive, and the fifth and the sixth to be explosive. In the first two examples we use point (b) of Section 3.2.1 as a test for explosiveness. In the next three examples, point (f) is used; this criterion is quite handy and therefore often found in the literature. In the sixth example, we take a yet different path.

3.2.1 *Criteria for the minimal semigroup to be Markov*

Let $\{S(t), t \geq 0\}$ be the minimal semigroup related to a Kolmogorov matrix Q. The following are equivalent:

(a) $\{S(t), t \geq 0\}$ is a semigroup of Markov operators;
(b) for any $\lambda > 0$, $\lim_{n\to\infty} B_\lambda^n = 0$ strongly;
(c) for any $\lambda > 0$, $Range(\lambda - D - O)$ is dense in l^1;
(d) for any $\lambda > 0$, $Range(\lambda - A_0)$ is dense in l^1;
(e) for any $\lambda > 0$, $Range(I - B_\lambda)$ is dense in l^1;
(f) if for some $\lambda > 0$ and $a = (\alpha_i)_{i\geq 1} \in l^\infty$ we have $Qa = \lambda a$ (where Qa is the product of the matrix Q and the column-vector a), then $a = 0$;
(g) $\overline{D + O} = G$;
(h) for $x \in \mathcal{D}(G)$, $\Sigma G x = 0$.

If one and hence all of these conditions hold, the matrix Q is said to be **nonexplosive**.

Proof Our plan is to show

1. (a) $\Leftrightarrow$ (b);
2. (b) $\Rightarrow$ (c)$\Rightarrow$ (d) $\Rightarrow$ (e) $\Rightarrow$ (b);
3. (d) $\Leftrightarrow$ (f);
4. (b) $\Rightarrow$ (g)$\Rightarrow$ (h) $\Rightarrow$ (a).

Clearly, condition (a) implies that λR_λ is a Markov operator for all $\lambda > 0$, and the reverse implication can be proved using the Yosida approximation. On the other hand, for $n \geq 1$,

$$I + O\sum_{k=0}^{n}(\lambda - D)^{-1}B_\lambda^k = (\lambda - D)\sum_{k=0}^{n}(\lambda - D)^{-1}B_\lambda^k + B_\lambda^{n+1}. \tag{3.10}$$

Therefore, for $x \geq 0$,

$$\begin{aligned}\|x\| &+ \|O\sum_{k=0}^{n}(\lambda - D)^{-1}B_\lambda^k x\| \\ &= \|\lambda\sum_{k=0}^{n}(\lambda - D)^{-1}B_\lambda^k x\| + \|D\sum_{k=0}^{n}(\lambda - D)^{-1}B_\lambda^k x\| + \|B_\lambda^{n+1}x\|,\end{aligned}$$

since $-D \geq 0$. By (3.2) this gives, $\|x\| = \|\lambda \sum_{k=0}^{n}(\lambda - D)^{-1}B_\lambda^k x\| + \|B_\lambda^{n+1}x\|$. Letting $n \to \infty$ we see that $\|x\| = \|\lambda R_\lambda x\|$ if and only if $\lim_{n\to\infty} \|B_\lambda^n x\| = 0$ (see (3.5)) for all $x \geq 0$. Of course, the latter condition is equivalent to $\lim_{n\to\infty} \|B_\lambda^n x\| = 0$, $x \in l^1$. This shows **(a)** $\Leftrightarrow$ **(b)**.

(b) $\Rightarrow$ **(c)** To prove this, we rewrite (3.10) as

$$(\lambda - D - O)\sum_{k=0}^{n}(\lambda - D)^{-1}B_\lambda^k x = x - B_\lambda^{n+1}x, \qquad x \in l^1; \tag{3.11}$$

this relation shows that if (b) holds, any x may be approximated by elements of $Range(\lambda - D - O)$, as desired.

(c) $\Rightarrow$ **(d)** Recall that for any $x \in \mathcal{D}(D)$ there exist $x_n \in \mathcal{D}(A_0)$ such that $\lim_{n\to\infty} x_n = x$ and $\lim_{n\to\infty} A_0 x_n = \lim_{n\to\infty}(D + O)x_n = (D + O)x$ (see the beginning of the proof of 3.1.5); hence the range of $\lambda - D - O$ is contained in the closure of the range of $\lambda - A_0$.

(d) $\Rightarrow$ **(e)** This becomes clear once we write

$$I - B_\lambda = (\lambda - D)(\lambda - D)^{-1} - O(\lambda - D)^{-1} = (\lambda - D - O)(\lambda - D)^{-1}$$

and note that all elements x of $\mathcal{D}(D)$ are of the form $x = (\lambda - D)^{-1}y$ for some $y \in l^1$. Indeed, this shows that the range of $\lambda - D - O$ is equal to the range of $I - B_\lambda$ and we know that $\lambda - D - O$ is an extension of $\lambda - A_0$.

(e) $\Rightarrow$ **(b)** Calculating as in (3.3) one may show that B_λ is sub-Markov. Hence, $\|B_\lambda^n x\| \leq \|B_\lambda^k x\|$ for $x \geq 0$ and $k = 0, 1, \ldots, n$. Therefore, for such x, $\|B_\lambda^n x\| \leq \|C_n x\|$ where $C_n = C_n(\lambda) = \frac{1}{n+1}\sum_{k=0}^{n} B_\lambda^k$. Writing $x = x^+ - x^-$ as in Section 3.1.7, we see that

$$\|B_\lambda^n x\| \leq \|B_\lambda^n x^+\| + \|B_\lambda^n x^-\| \leq \|C_n x^+\| + \|C_n x^-\|.$$

Hence, it suffices to show that C_n converges strongly to 0, as $n \to \infty$. If $x = y - B_\lambda y$ for some $y \in l^1$, we have $C_n x = \frac{1}{n+1}\sum_{k=0}^{n} B_\lambda^k (I - B_\lambda)y = \frac{1}{n+1}(y - B_\lambda^{n+1}y)$. Therefore, for $x \in Range(I - B_\lambda)$, $\lim_{n\to\infty} C_n x = 0$. If (e) holds, the same is true for all $x \in l^1$ since $\|C_n\| \leq 1$.

To show **(d)** $\Leftrightarrow$ **(f)**, we note that (d) holds iff, for any functional F on l^1, the relation $F(\lambda x - A_0 x) = 0$ for all $x \in \mathcal{D}(A_0)$ implies $F = 0$. By definition of $\mathcal{D}(A_0)$, $F(\lambda x - A_0 x) = 0$ for all $x \in \mathcal{D}(A_0)$ iff $F(\lambda e_i - A_0 e_i) = 0$ for all $i \geq 1$. On the other hand, any F may be identified with an $a = \left(\alpha_j\right)_{j\geq 1} \in l^\infty$ and we have $F(\lambda e_i - A_0 e_i) = \lambda\alpha_i - \sum_{j=1}^{\infty} q_{i,j}\alpha_j$. The system $\lambda\alpha_i = \sum_{j=1}^{\infty} q_{i,j}\alpha_j$, $i \geq 1$ may be written as $\lambda a = Qa$.

(b) ⇒ (g) Since G extends $D+O$ and is closed, $D+O$ is closable. For any $x \in l^1$ and $\lambda > 0$, the vectors

$$x_n := \sum_{k=0}^{n} (\lambda - D)^{-1} B_\lambda^k x$$

belong to $\mathcal{D}(D)$ and we have that $\lim_{n\to\infty} x_n = R_\lambda x = (\lambda - G)^{-1} x$ (see (3.5)). Looking back at (3.11) we see that if (b) holds, the limit $\lim_{n\to\infty}(D+O)x_n$ exists and equals $\lambda R_\lambda x - x$. It follows that $R_\lambda x$ belongs to $\mathcal{D}(\overline{D+O})$ and $\overline{D+O} R_\lambda x = \lambda R_\lambda x - x$. Since we know that $G R_\lambda x = \lambda R_\lambda x - x$, this leads to the conclusion that $\overline{D+O}$ extends G. However, G extends $D+O$ and, being closed, extends $\overline{D+O}$, as well. This shows (g).

Implication **(g) ⇒ (h)** is clear from (3.9).

Finally, if (h) holds, condition (iii) in the Hille–Yosida Theorem 2.4.3 is satisfied, and so G is a Markov generator, proving (a). □

3.2.2 Example

Let $(a_n)_{n\geq 1}$ be a sequence of positive numbers. A Markov chain with intensity matrix $Q = (q_{i,j})_{i,j\geq 1}$ where

$$q_{i,j} = \begin{cases} -a_i, & j = i, \\ a_i, & j = i+1, \\ 0, & \text{otherwise}, \end{cases} \tag{3.12}$$

is said to be a **pure birth process** with rates $a_n, n \geq 1$. (In particular, the Poisson process is a pure birth process with a constant rate.) We imagine a population where, provided that the population size is $i \geq 1$, a new member is born after an exponential time with parameter a_i; no deaths are possible.

For such a Q, we have $(\lambda - D)^{-1}(\xi_i)_{i\geq 1} = \left(\frac{\xi_i}{\lambda + a_i}\right)_{i\geq 1}$ and in particular $(\lambda - D)^{-1} e_i = \frac{1}{\lambda + a_i} e_i$. Also $O e_i = a_i e_{i+1}$. Hence, $B_\lambda e_i = \frac{a_i}{\lambda + a_i} e_{i+1}$ and so

$$B_\lambda^n e_i = \left(\prod_{k=i}^{i+n-1} \frac{a_k}{\lambda + a_k} \right) e_{n+i} \tag{3.13}$$

and $\|B_\lambda^n e_i\| = \prod_{k=i}^{i+n-1} \frac{a_k}{\lambda + a_k}$. Since $e_i, i \geq 1$, are linearly dense in l^1 and $\|B_\lambda\| \leq 1$, B_λ^n converges strongly to 0 as $n \to \infty$ iff $\prod_{n=1}^{\infty} \frac{a_n}{\lambda + a_n} = 0$. By 2.4.6, this last condition is equivalent to divergence of $\sum_{n=1}^{\infty} \frac{1}{\lambda + a_n}$ for all $\lambda > 0$. In particular, $\sum_{n=1}^{\infty} a_n^{-1} < \infty$ implies convergence of all these series and thus explosiveness of Q. See Exercise 3.2.10 for a different proof of this result.

3.2.3 Example

Let Q be the intensity matrix (2.53) of the second Kolmogorov–Kendall–Reuter chain with -1 in the first row replaced by 0. The related chain may be thought of as an example of a **pure death-process**: if we imagine that the chain describes the size of a population, then the number of individuals in this population cannot increase. Rather, given that the population size is $i \ge 3$, after an exponential time with parameter a_i one individual dies. The states $i = 1$ and $i = 2$ are absorbing.

We will show that this Q is nonexplosive and that the generator G of the minimal semigroup is given by

$$G\,(\xi_i)_{i\ge 1} = (a_{i+1}\xi_{i+1} - a_i\xi_i)_{i\ge 1}\,, \tag{3.14}$$

where, to recall, $a_2 = 0$ and we agree, for simplicity of notations, that also $a_1 = 0$, on the domain

$$\mathcal{D}(G) = \{(\xi_i)_{i\ge 1} \in l^1;\ (a_{i+1}\xi_{i+1} - a_i\xi_i)_{i\ge 1} \in l^1 \text{ and } \lim_{i\to\infty} a_i\xi_i = 0\}. \tag{3.15}$$

For the first task, we note that $(\lambda - D)^{-1}\,(\xi_i)_{i\ge 1} = \left(\frac{\xi_i}{\lambda + a_i}\right)_{i\ge 1}$, and that $Oe_1 = Oe_2 = 0$ and $Oe_i = a_i e_{i-1}, i \ge 3$. Therefore,

$$B_\lambda e_1 = B_\lambda e_2 = 0 \qquad \text{and} \qquad B_\lambda e_i = \frac{a_i}{\lambda + a_i} e_{i-1}, \quad i \ge 3.$$

It follows that $B_\lambda^n e_1 = B_\lambda^n e_2 = 0, n \ge 1$ and $B_\lambda^n e_i = 0$ for $i \ge 3$ provided that $n \ge i - 2$. Since e_i's are linearly dense, condition (b) of 3.2.1 is met: Q is nonexplosive.

For the sake of the proof of the second part, let G' be the operator defined by (3.14) and (3.15). We note that for $x = (\xi_i)_{i\ge 1} \in \mathcal{D}(G')$,

$$\Sigma G'x = \lim_{i\to\infty} \sum_{j=1}^{i-1} (a_{j+1}\xi_{j+1} - a_j\xi_j) = \lim_{i\to\infty} a_i\xi_i = 0.$$

This can be used to show that G' is closed. For, suppose that $x_n \in \mathcal{D}(G')$ are such that the limits $x := \lim_{n\to\infty} x_n$ and $y := \lim_{n\to\infty} G'x_n$ exist, and let $x = (\xi_i)_{i\ge 1}$. Since convergence in the norm of l^1 implies convergence of coordinates, we must have $y = (a_{i+1}\xi_{i+1} - a_i\xi_i)_{i\ge 1}$, implying that the first condition characterizing members of $\mathcal{D}(G')$ is satisfied by $(\xi_i)_{i\ge 1}$. Also, as we have just seen, $\Sigma G'x_n = 0$, and thus

$$\lim_{i\to\infty} a_i\xi_i = \lim_{i\to\infty} \sum_{j=1}^{i-1} (a_{j+1}\xi_{j+1} - a_j\xi_j) = \Sigma y = \Sigma \lim_{n\to\infty} G'x_n = 0,$$

because Σ is a continuous functional. It follows that the second condition is satisfied also, and so x belongs to $\mathcal{D}(G')$ and $G'x = y$. We have proved that G' is closed.

Moreover, G' extends $D + O$, the latter operator being defined by the same formula on the domain where $\sum_{i=1}^{\infty} |a_i \xi_i| < \infty$. Hence, $G' \supset \overline{D + O}$. On the other hand, since Q is nonexplosive, $G = \overline{D + O}$ (see point (g) in 3.2.1).

Since we set out to prove that $G = G'$, we are left with showing that for any $x = (\xi_i)_{i \geq 1} \in \mathcal{D}(G')$, there are $x_n \in \mathcal{D}(D + O)$ such that $\lim_{n\to\infty} x_n = x$ and $\lim_{n\to\infty}(D + O)x_n = G'x$. To this end, let $x_n = (\xi_1, \xi_2, \ldots, \xi_n, 0, 0, \ldots)$. Clearly, $\lim_{n\to\infty} x_n = x$. Also, $(D + O)x_n$ and $G'x$ differ by $a_{n+1}\xi_{n+1}$ on the nth coordinate and by $a_{i+1}\xi_{i+1} - a_i\xi_i$ on the ith coordinate where $i \geq n + 1$. Hence, $\|G'x - (D + O)x_n\| = |a_{n+1}\xi_{n+1}| + \sum_{i=n+1}^{\infty} |a_{i+1}\xi_{i+1} - a_i\xi_i|$ converges to 0, as $n \to \infty$.

Notably, we have checked that G is a generator without checking the range condition, that is, the second condition in the Hille–Yosida Theorem. This was possible, because Kato's Theorem is a kind of generation theorem.

3.2.4 Example

(See [77, 85].) Let $d, r > 0$. Consider the Kolmogorov matrix $Q = \left(q_{i,j}\right)_{i,j\geq 1}$ given by

$$q_{i,j} = \begin{cases} (i-1)r, & j = i-1, i \geq 2, \\ -(i-1)r - (i+1)d, & j = i, i \geq 1, \\ (i+1)d, & j = i+1, i \geq 1, \end{cases}$$

and 0 otherwise. To show that Q is nonexplosive, we check that condition (f) of Section 3.2.1 is satisfied. The equation $Q\,(\alpha_i)_{i\geq 1} = \lambda\,(\alpha_i)_{i\geq 1}$ considered there may be rewritten as

$$(i-1)r\alpha_{i-1} - [(i-1)r + (i+1)d]\alpha_i + (i+1)d\alpha_{i+1} = \lambda\alpha_i, \qquad i \geq 1,$$

where we put $\alpha_0 = 0$. Then $\alpha_{i+1} = \left[1 + \frac{(i-1)r+\lambda}{(i+1)d}\right]\alpha_i - \frac{i-1}{i+1}\frac{r}{d}\alpha_{i-1}$, or

$$\alpha_{i+1} - \alpha_i = \frac{(i-1)r + \lambda}{(i+1)d}\alpha_i - \frac{i-1}{i+1}\frac{r}{d}\alpha_{i-1}. \tag{3.16}$$

Note that, if $\alpha_i > 0$ for some $i \geq 1$, then $\alpha_{i+1} - \alpha_i > \frac{i-1}{i+1}\frac{r}{d}\left(\alpha_i - \alpha_{i-1}\right)$. Hence, by induction argument, if $\alpha_1 > 0$ then $\alpha_{i+1} - \alpha_i > 0$ and $\alpha_i > 0$ for all $i \geq 1$. Therefore, by (3.16) again, $\alpha_{i+1} - \alpha_i \geq \frac{\lambda}{(i+1)d}\alpha_i$ or $\alpha_{i+1} \geq \left[1 + \frac{\lambda}{(i+1)d}\right]\alpha_i$ resulting in $\alpha_n \geq \alpha_1 \prod_{i=2}^{n}\left(1 + \frac{\lambda}{id}\right)$, $n \geq 1$.

Hence, if $\alpha_1 > 0$, the limit $\lim_{n\to\infty}\alpha_n$ exists and is no smaller than $\prod_{i=2}^{\infty}\left(1+\frac{\lambda}{id}\right)\alpha_1 = \left(\prod_{i=2}^{\infty}\frac{id}{id+\lambda}\right)^{-1} = \infty$ (see Criterion 2.4.6). Since this contradicts $(\alpha_i)_{i\ge 1}\in l^\infty$, we must have $\alpha_1 \le 0$. But, we may not have $\alpha_1 < 0$ for then $(\beta_i)_{i\ge 1} = -(\alpha_i)_{i\ge 1}\in l^\infty$ would satisfy $Q(\beta_i)_{i\ge 1} = \lambda(\beta_i)_{i\ge 1}$ while having its first coordinate positive, which we know is impossible. Thus, $\alpha_1 = 0$ and an induction argument based on (3.16) shows that $\alpha_i = 0$ for all $i \ge 1$.

3.2.5 Example

Given $a, b > 0$, consider the following Kolmogorov matrix:

$$Q = \begin{pmatrix} 0 & 0 & 0 & 0 & 0 & 0 & \cdots \\ a+b & -d & b & 0 & 0 & 0 & \cdots \\ a & a+2b & -2d & 2b & 0 & 0 & \cdots \\ a & a & a+3b & -3d & 3b & 0 & \cdots \\ a & a & a & a+4b & -4d & 4b & \cdots \\ \vdots & \vdots & \vdots & \ddots & \ddots & \ddots & \ddots \end{pmatrix} \tag{3.17}$$

where $d = a+2b$. This matrix describes a particular case of a chain used by R. Durrett and S. Kruglyak in [33] to model life history of so-called microsatellites, that is, sequences of DNA, which involve periodic repetition of motifs of length of two to six DNA base pairs: the state of the chain is the number of such repeats. (For reasons explained in [18], we have, however, slightly modified the first row of Q.) The part of Q built with b is a reflection of possibility of polymerase slippage during replication which results in shortening or extending the number of repeats in the microsatellite by 1; in a microsatellite with i repeats such slippage may occur at $i-1$ sites. The part of Q built with a corresponds to breakdown events that can occur between any repeats, cutting the microsatellite into two parts; one randomly chosen fragment is then considered the continuation of the microsatellite.

We will check that Q is nonexplosive. To see this, given $\lambda > 0$, consider a solution $(\alpha_i)_{i\ge 1}$ to the system $\lambda(\alpha_i)_{i\ge 1} = Q(\alpha_i)_{i\ge 1}$, which in coordinates reads:

$$\lambda\alpha_i = a\sum_{j=1}^{i-1}\alpha_j + (i-1)b\alpha_{i-1} - (i-1)d\alpha_i + (i-1)b\alpha_{i+1}, \qquad i \ge 1, \tag{3.18}$$

where, by convention, $\sum_{j=1}^{0} = 0$. In particular, $\alpha_1 = 0$. We will use induction argument to show that, if $\alpha_2 \ge 0$, then

$$\alpha_i \ge \left(1+\frac{\mu}{i-1}\right)\alpha_{i-1}, \qquad i \ge 3, \tag{3.19}$$

where $\mu = \lambda/b$. For $i = 3$ this is clear, because the second equation in (3.18) says that $b\alpha_3 = (d + \lambda)\alpha_2$. Also, if (3.19) is true for all indices up to some i, then we have also $\alpha_j \le \alpha_i$ for $j = 1, \ldots, i$ and thus the ith equation in (3.18) says that $\lambda\alpha_i \le (a + b)(i - 1)\alpha_i - (i - 1)d\alpha_i + (i - 1)b\alpha_{i+1}$ implying (3.19) with i replaced by $i + 1$, and completing the proof.

Relation (3.19) in turn shows that

$$\alpha_i \ge \frac{\alpha_2}{\prod_{j=2}^{i-2}(1 - \frac{\mu}{i-1})}, \qquad i \ge 2.$$

Now, by 2.4.6, $\lim_{i\to\infty} \prod_{j=2}^{i-2}(1 - \frac{\mu}{i-1}) = 0$ because $\sum_{i=2}^{\infty} \frac{\mu}{i-1} = \infty$. Thus, $(\alpha_i)_{i\ge1}$ is unbounded if $\alpha_2 > 0$, and arguing as in the previous example, we see that $\alpha_2 < 0$ is also impossible. Thus $\alpha_2 = 0$, and the induction argument based on (3.18) shows that $(\alpha_i)_{i\ge1} = 0$.

3.2.6 Example

We will show that Q given by[2]

$$\begin{pmatrix}
0 & 0 & 0 & 0 & 0 & 0 & 0 & 0 & \cdots \\
1 & -(1+1^4) & 1^4 & 0 & 0 & 0 & 0 & 0 & \cdots \\
2 & 1 & -(3+2^4) & 2^4 & 0 & 0 & 0 & 0 & \cdots \\
3 & 2 & 1 & -(6+3^4) & 3^4 & 0 & 0 & 0 & \cdots \\
\vdots & \vdots & \vdots & \vdots & \ddots & \ddots & \vdots & \vdots & \\
i & (i-1) & (i-2) & \cdots & 1 & -\left(\frac{i(i+1)}{2} + i^4\right) & i^4 & 0 & \cdots \\
\vdots & \vdots & \vdots & \vdots & \vdots & \vdots & \vdots & \vdots & \vdots
\end{pmatrix}$$

is explosive: there is an $a = (\alpha_i)_{i\ge1} \in l^\infty$ such that $Qa = a$. Indeed, the latter equation is satisfied iff $\alpha_1 = 0$ and

$$\sum_{j=1}^{i-1}(i - j)\alpha_j - \left(\frac{i(i-1)}{2} + (i-1)^4\right)\alpha_i + (i-1)^4\alpha_{i+1} = \alpha_i, \qquad i \ge 2;$$

we will show that choosing $\alpha_2 > 0$ and defining the rest of the sequence by the recursion

$$(i-1)^4\alpha_{i+1} = \left(1 + \frac{i(i-1)}{2} + (i-1)^4\right)\alpha_i - \sum_{j=1}^{i-1}(i - j)\alpha_j, \qquad i \ge 2, \tag{3.20}$$

leads to a bounded and nondecreasing solution to this system.

[2] I am grateful to R. Rudnicki for his help in constructing this example.

If $\alpha_1 \le \alpha_2 \le \cdots \le \alpha_i$, then $\left(\text{since } \sum_{j=1}^{i-1}(i-j) = \sum_{j=1}^{i-1} j = \frac{i(i-1)}{2}\right)$

$$(i-1)^4\alpha_{i+1} \ge (1 + (i-1)^4)\alpha_i,$$

proving that $\alpha_{i+1} > \alpha_i$. It follows that $(\alpha_i)_{i\ge1}$ is nondecreasing and in particular positive. Now (3.20) shows that

$$(i-1)^4\alpha_{i+1} \le \left(1 + \frac{i(i-1)}{2} + (i-1)^4\right)\alpha_i \le \left(2(i-1)^2 + (i-1)^4\right)\alpha_i$$

and thus $\alpha_{i+1} \le \left(1 + \frac{2}{(i-1)^2}\right)\alpha_i$, $i \ge 2$, implying

$$\lim_{i\to\infty} \alpha_i \le \alpha_2 \prod_{i=2}^{\infty} \beta_i,$$

where $\beta_i = 1 + \frac{2}{(i-1)^2}$. Since $\beta_i^{-1} \in (0,1)$ and

$$\sum_{i=2}^{\infty}(1 - \beta_i^{-1}) = \sum_{i=2}^{\infty} \frac{\frac{2}{(i-1)^2}}{1 + \frac{2}{(i-1)^2}} \le \sum_{i=2}^{\infty} \frac{2}{(i-1)^2} < \infty,$$

by Criterion 2.4.6, $\prod_{i=2}^{\infty} \beta_i^{-1}$ exists and is nonzero. Hence, $\prod_{i=2}^{\infty} \beta_i$ is finite, and we are done.

3.2.7 Example

Let Q be the intensity matrix with the first row equal to $(-6, 6, 0, \dots)$ and the ith row of the form

$$(0, \dots, 3^i, -3 \cdot 3^i, 2 \cdot 3^i, 0, \dots), \quad i \ge 2.$$

The related chain is an example of a **birth and death process**: we imagine that the chain describes the number of individuals in a population where both deaths and births are possible. If the population size is $i \ge 2$, the time to the next birth/death event is exponential with parameter $3 \cdot 3^i$: when this time is over, one member of the population dies, with probability $\frac{1}{3}$, or a new member is born, with probability $2/3$. If there is only one member in the population, death is impossible, and a birth occurs after an exponential time with parameter 6.

In this example, the operator $D + O$ is given by $(D + O)\,(\xi_i)_{i\ge1} = (\eta_i)_{i\ge1}$, where

$$\eta_1 = 9\xi_2 - 6\xi_1,$$
$$\eta_i = 3^{i-1}(9\xi_{i+1} - 9\xi_i + 2\xi_{i-1}), \quad i \ge 2,$$

on the domain $\mathcal{D}(D)$ equal to the set of all $(\xi_i)_{i\ge1}$ such that $\sum_{i=1}^{\infty} 3^i|\xi_i| < \infty$. As the reader might recall, in Sections 2.4.12–2.4.16 we have proved that the operator A formally given by the same formula, but defined on the larger domain where $\sum_{i=1}^{\infty} 3^i |3\xi_i - \xi_{i-1}| < \infty$, is a sub-Markov generator. By 3.1.5, the semigroup generated by A dominates the minimal semigroup:[3]

$$S(t) \le \mathrm{e}^{tA}.$$

But (see Section 2.4.11 and Exercise 2.4.18) there are nonnegative x's in $\mathcal{D}(A)$ such that $\Sigma Ax < 0$ and this implies that $\Sigma\lambda\,(\lambda - A)^{-1}\,x < \Sigma x$. This in turn implies that we cannot have $\Sigma \mathrm{e}^{tA}x = \Sigma x, t \ge 0$. Since e^{tA} dominates $S(t)$, for such x neither can we have $\Sigma S(t)x = \Sigma x, t \ge 0$. This shows that $\{S(t), t \ge 0\}$ is not a Markov semigroup; the matrix Q is explosive.

3.2.8 Exercise

For a sequence $(a_n)_{n\ge 1}$ of positive numbers such that $\lim_{n\to\infty} a_n = \infty$, let Q be the Kolmogorov matrix which

- in the first row has zeros,
- in the ith row: has a_i in the first and $(i+1)$st columns, and $-2a_i$ in the ith column.

The related Markov chain after appropriate exponential time spent at a state $i \ge 2$, jumps either to $i+1$ or to 1, both jump probabilities being equal. Show that Q is nonexplosive even if $\sum_{i=1}^{\infty} a_i^{-1} < \infty$.

3.2.9 Exercise

Arguing as in 2.4.8 and 2.4.9 show that the resolvent of the generator G of (3.14) and (3.15) is given by $(\lambda - G)^{-1}(\eta_i)_{i\ge 1} = (\xi_i)_{i\ge 1}$ where $\xi_1 = \lambda^{-1}\eta_1$ and $\xi_i = \frac{1}{\lambda + a_i}\sum_{j=i}^{\infty}\left(\prod_{i<k\le j}\frac{a_k}{\lambda + a_k}\right)\eta_j, i \ge 2$. Note also that (a) this formula agrees with $(\lambda - G)^{-1} = \sum_{n=0}^{\infty}(\lambda - D)^{-1}B_\lambda^n$ (see (3.5)), and (b) in agreement with our intuition, the factor $\frac{1}{\lambda + a_i}\prod_{i<k\le j}\frac{a_k}{\lambda + a_k}$ is the Laplace transform of the probability that a population of size j has decreased to the size of i (cf. Section 2.7.5).

3.2.10 Exercise

Using the fact that the operator A of Section 2.4.10 is a sub-Markov generator, and arguing as in the latter part of Example 3.2.7, prove that a pure birth Markov chain with $\sum_{n=1}^{\infty} a_n^{-1} < \infty$ is explosive.

[3] In Section 3.4.4, we will see that $\{\mathrm{e}^{tA}, t \ge 0\}$ in fact coincides with $\{S(t), t \ge 0\}$.

Figure 3.1 Zeno's paradox by Radek Bobrowski.

3.3 Pure Birth Process Example

Probabilistically, the reason why there are in general many semigroups related to a given Q-matrix may be explained as follows. Let us recall from Section 2.5 that if $X(t), t \geq 0$ is a Markov chain, with right-continuous paths, related to Q, then given that $X(t) = i$, the chain waits in this state for an exponential time with parameter $q_i = -q_{i,i}$ and then jumps to one of the other states, the probability of jumping to $j \neq i$ being $q_{i,j}/q_i$ (if $q_i = 0$ the process stays at i for ever). It is important to note that in general such a procedure defines the process only *locally in time*. In other words, the process may be left undefined after a certain random time τ, called **explosion**. This process, undefined after explosion, is termed minimal, and the minimal semigroup of Kato's Theorem describes this process.

The situation is quite analogous to the description of Achilles (A) chasing a Tortoise (T) in Zeno's paradox. A runs after T, who escapes with speed $v_T = qv_A$, where $q \in (0, 1)$, and v_A is the speed of Achilles. Before, however, A reaches T, he must pass through the point where T had been initially. If the initial distance was d, then in the meantime, T has moved an extra distance dq away. So, the procedure must be repeated with new, though smaller, initial distance between the competitors, and so on, *ad infinitum*, forevermore. If we believe that this procedure defines the chase for all times $t \geq 0$, we need also to admit that A will never catch T. This is of course not the case: at time $\tau = \frac{d}{v_A(1-q)}$ the chase will be over.

This situation is also well illustrated by the pure birth process of Section 3.2.2. If the process starts at 1, then after exponential time T_1 with parameter

a_1 it will be at 2, and after exponential time T_2 with parameter a_2 it will be at 3, and so on. Let

$$\tau = \sum_{n=1}^{\infty} T_n. \tag{3.21}$$

Is τ finite or infinite? If $\sum_{n=1}^{\infty} a_n^{-1} = \infty$, then $\mathbb{P}\{\tau = \infty\} = 1$, and in the other case, $\mathbb{P}\{\tau < \infty\} = 1$. Indeed, if the series converges, we may not have $\mathbb{P}\{\tau = \infty\} > 0$, as this would imply $\mathbb{E}\,\tau = \infty$, while we have $\mathbb{E}\,\tau = \sum_{n=1}^{\infty} \mathbb{E}\,T_n = \sum_{n=1}^{\infty} a_n^{-1} < \infty$. Conversely, if the series diverges, then so do the series $\sum_{n=1}^{\infty}(\lambda + a_n)^{-1}$, and, as we have seen in 3.2.2, we have, for any $\lambda > 0$, $\prod_{n=1}^{\infty} \frac{a_n}{\lambda + a_n} = 0$. Hence,

$$\mathbb{E}\,\mathrm{e}^{-\lambda\tau} = \prod_{n=1}^{\infty} \mathbb{E}\,\mathrm{e}^{-\lambda T_n} = \prod_{n=1}^{\infty} \frac{a_n}{\lambda + a_n} = 0,$$

showing that $\mathbb{P}\{\tau = \infty\} = 1$.

This means that, provided $\sum_{n=1}^{\infty} a_n^{-1} < \infty$, after the (random) time τ, the process is left undefined. In other words, at *any* time $t > 0$ some paths of the process may no longer be defined (namely, the paths $X(t, \omega)$ such that $\tau(\omega) < t$), and we observe only some of them – hence the probability that the process is somewhere in $\mathbb{N}$ may be (and is) strictly less than 1. The transition probabilities of the process described above form the minimal semigroup defined in 3.1.4.

Now, we may introduce an additional rule for the behavior of the process after τ; for example, we may require that at τ it jumps back to 1 and does the same for all subsequent explosions. Or, we could require that at τ the process starts at a random point of $\mathbb{N}$ with *a priori* prescribed distribution. All choices of distributions lead then to different processes and different semigroups – all of them, however, have transition semigroups dominating the minimal transition semigroup. Similarly, in Zeno's paradox, Achilles's strategy to run as swiftly as he can to the place where he saw the Tortoise last, may be supplemented with a rule telling him for example, to stay back, after catching the Tortoise, wait until the Tortoise is at the distance d away, and then start the chase again. Such a race would last forevermore, indeed.

The main goal of this section is constructing two different semigroups related to the explosive Kolmogorov matrix of the pure birth process; in the first of them, after the explosion the process returns to $i = 1$ (see Sections 3.3.3–3.3.6), and in the second it returns to $i = 2$ (Sections 3.3.7–3.3.8). We will be able to describe the generators of the related semigroups quite explicitly, and thus illustrate the fact that $D + O$ of Kato's Theorem may have different

Figure 3.2 Postexplosion process by Daniel Lipiński.

extensions being generators of Markov semigroups. In fact, we will be able to *derive* the form of these generators from an intuitive approximation. The intuitions gathered here will then be used in Section 3.5, where an infinity of semigroups dominating the minimal one will be explicitly constructed.

In what follows we always assume that

$$\sum_{n=1}^{\infty} a_n^{-1} < \infty. \tag{3.22}$$

3.3.1 Remark

Condition (f) of 3.2.1 has a nice probabilistic interpretation related to τ of (3.21): the vector $a = (\alpha_n)_{n\geq 1}$, where

$$\alpha_n = \mathbb{E}\{e^{-\lambda\tau}|X(0) = n\}$$

solves the equation $Qa = \lambda a$. Moreover, it is maximal in the sense that if $Q\left(\alpha_n'\right)_{n\geq 1} = \lambda\left(\alpha_n'\right)_{n\geq 1}$ for some $\left(\alpha_n'\right)_{n\geq 1} \in l^\infty$ with $\| \left(\alpha_n'\right)_{n\geq 1} \|_{l^\infty} \leq 1$, then $\alpha_n' \leq \alpha_n$ – see, for example, [72], p. 91. Certainly $a \neq 0$ iff $\tau \neq \infty$.

In the pure birth chain's case, equation $Qa = \lambda a$ reads in coordinates:

$$(\lambda + a_i)\alpha_i = a_i\alpha_{i+1},$$

and it is easy to see that all its solutions are of the form

$$\alpha_i = \left(\prod_{j=1}^{i-1} \frac{\lambda + a_j}{a_j}\right) \alpha_1 = \left(\prod_{j=1}^{i-1} \frac{a_j}{\lambda + a_j}\right)^{-1} \alpha_1, \qquad i \geq 2, \tag{3.23}$$

where α_1 is arbitrary. These solutions are bounded since the infinite product $\prod_{j=1}^{\infty} \frac{a_j}{\lambda+a_j}$ exists and differs from 0. On the other hand, for the process starting at 1, since T_n's are independent, and the series (3.21) converges,

$$\mathbb{E}\, e^{-\lambda\tau} = \prod_{j=1}^{\infty} \int_0^{\infty} a_j e^{-(\lambda+a_j)t}\, dt = \prod_{j=1}^{\infty} \frac{a_j}{\lambda + a_j}.$$

Similarly, for the process starting at i,

$$\mathbb{E}\, e^{-\lambda\tau} = \prod_{j=i}^{\infty} \int_0^{\infty} a_j e^{-(\lambda+a_j)t}\, dt = \prod_{j=i}^{\infty} \frac{a_j}{\lambda + a_j}.$$

Thus, the maximal solution is of the form (3.23) with $\alpha_1 = \prod_{j=1}^{\infty} \frac{a_j}{\lambda+a_j}$.

3.3.2 *The generator of the minimal semigroup*

In the pure birth process example, the generator of the minimal semigroup of Kato's Theorem may be described explicitly. To recall, the resolvent of this generator is given by formula (3.5), which we expand here as follows:

$$R_\lambda = \sum_{n=0}^{\infty} (\lambda - D)^{-1} \left[O\,(\lambda - D)^{-1}\right]^n.$$

Now, (3.13) implies (recall notation (3.4)) that, for any $(\eta_i)_{i\geq 1} \in l^1$, the ith coordinate of $R_\lambda\,(\eta_i)_{i\geq 1}$ equals

$$\begin{aligned}\xi_i &= \frac{1}{\lambda + a_i}\left(\eta_i + \frac{a_{i-1}}{\lambda + a_{i-1}}\eta_{i-1} + \cdots + \pi_{i-1}\eta_1\right) \\ &= \frac{\pi_{i-1}}{\lambda + a_i} \sum_{j=1}^{i} \frac{\eta_j}{\pi_{j-1}},\end{aligned} \tag{3.24}$$

where $\pi_j = \prod_{k=1}^{j} \frac{a_k}{\lambda+a_k}$, $j \geq 1$ and $\pi_0 = 1$.

We note in passing that $\frac{1}{\lambda+a_i}$ is the Laplace transform of $t \mapsto e^{-a_i t}$ and $e^{-a_i t}$ is the probability that the pure birth process starting at i has not left this state before time t; similarly, $\frac{1}{\lambda+a_i}\frac{a_{i-1}}{\lambda+a_{i-1}}$ is the Laplace transform of

$$t \mapsto a_{i-1} \int_0^t e^{-a_i(t-s)} e^{-a_{i-1}s}\, ds,$$

and the last integral is the probability that in the time interval $[0, t]$, the process starting at $i-1$ has jumped from this state to i, but has not left i yet, and so on. Hence, (3.24) is the Laplace transform of the probability that the minimal pure birth chain starting at a random position j with probability η_j will be at i at time t.

On the other hand, formula (3.24) is identical to (2.36). This means that $R_\lambda = (\lambda - A)^{-1}$ where A is the operator defined at the beginning of Section 2.4.10, and this is possible only if G coincides with that operator. In other words,

$$\mathcal{D}(G) = \{(\xi_i)_{i\ge 1}; \sum_{i\ge 1} |a_{i-1}\xi_{i-1} - a_i\xi_i| < \infty\} \tag{3.25}$$

and

$$G\,(\xi_i)_{i\ge 1} = (a_{i-1}\xi_{i-1} - a_i\xi_i)_{i\ge 1}\,,$$

where, as before, we agree that $a_0\xi_0$ is 0.

In the rest of this section we take a closer look at two modifications of G that generate Markov semigroups in l^1. G describes the process which is left undefined after explosion; in our first example below we will deal with the process which after explosion returns immediately to $i = 1$.

3.3.3 *Return to* $i = 1$ *after explosion: Intuition*

For $n \ge 2$, let A_n be the bounded linear operator represented by the Kolmogorov matrix whose upper left corner is

$$\begin{pmatrix} -a_1 & a_1 & 0 & 0 & \cdots & 0 & 0 \\ 0 & -a_2 & a_2 & 0 & \cdots & 0 & 0 \\ \vdots & \vdots & \ddots & \ddots & \cdots & a_{n-2} & 0 \\ 0 & 0 & 0 & 0 & \cdots & -a_{n-1} & a_{n-1} \\ a_n & 0 & 0 & 0 & \cdots & 0 & -a_n \end{pmatrix}, \tag{3.26}$$

and has remaining entries equal to 0. The related (right-continuous-paths) Markov chain starting at $i = 1$ goes gradually 'down the ladder' to jump from its lowest rung $i = n$ back to $i = 1$; all the states $i \ge n + 1$ are absorbing. In the limit, as $n \to \infty$, the number of rungs in the ladder increases, but because of (3.22) the limit process should be able to reach the very bottom in finite time (which is the time of explosion) and then immediately come back to $i = 1$. In what follows we will formally prove convergence of the related semigroups and show that the generator of the limit Markov semigroup extends $D + O$

(cf. Section 3.1.5), and is a modification of G, the generator of the minimal semigroup.

3.3.4 *Return to* $i = 1$ *after explosion: Approximating resolvents*

As in Section 2.6.2, we come to the conclusion that A_n is a Markov generator. In coordinates, the resolvent equation

$$\lambda\,(\xi_i)_{i\geq 1} - A_n\,(\xi_i)_{i\geq 1} = (\eta_i)_{i\geq 1}$$

reads

$$\begin{aligned} (\lambda + a_1)\xi_1 - a_n\xi_n &= \eta_1, \\ (\lambda + a_i)\xi_i - a_{i-1}\xi_{i-1} &= \eta_i, \qquad i = 2,\dots,n, \\ \lambda\xi_i &= \eta_i, \qquad i \geq n+1, \end{aligned} \tag{3.27}$$

and, as some amount of linear algebra shows (see Exercise 3.3.10), its unique solution is

$$\begin{aligned} \xi_1 &= \frac{1}{\lambda + a_1}\left[\eta_1 + \frac{\pi_n}{1-\pi_n}\sum_{j=1}^{n}\frac{\eta_j}{\pi_{j-1}}\right], \\ \xi_i &= \lambda^{-1}\eta_i, \qquad i \geq n+1, \end{aligned} \tag{3.28}$$

where, as before, $\pi_i = \prod_{j=1}^{i}\frac{a_j}{\lambda + a_j}$, with the remaining ξ_i's given recursively:

$$\xi_i = \frac{a_{i-1}}{\lambda + a_i}\xi_{i-1} + \frac{1}{\lambda + a_i}\eta_i, \qquad i = 2,\dots,n. \tag{3.29}$$

3.3.5 *Return to* $i = 1$ *after explosion: The limit resolvent*

As n tends to infinity, ξ_1 of (3.28) (which is in fact $\xi_1(n)$) converges to

$$\xi_1 = \frac{1}{\lambda + a_1}\left[\eta_1 + \frac{\pi_\infty}{1-\pi_\infty}\sum_{j=1}^{\infty}\frac{\eta_j}{\pi_{j-1}}\right], \tag{3.30}$$

where $\pi_\infty = \lim_{n\to\infty}\pi_n = \prod_{j=1}^{\infty}\frac{a_j}{\lambda + a_j}$. (Because $\pi_i \to \pi_\infty \neq 0$, the series featuring here converges since so does $\sum_{i=1}^{\infty}|\eta_i|$.) An induction argument then shows that the remaining coordinates converge to ξ_i's given recursively as follows:

$$\xi_i = \frac{a_{i-1}}{\lambda + a_i}\xi_{i-1} + \frac{1}{\lambda + a_i}\eta_i, \qquad i \geq 2 \tag{3.31}$$

(note that this formula differs from (3.29) in the range of i's to which it applies), so that

$$\xi_i = (\lambda + a_1)\frac{\pi_{i-1}}{\lambda + a_i}\xi_1 + \frac{\pi_{i-1}}{\lambda + a_i}\sum_{j=2}^{i}\frac{\eta_j}{\pi_{j-1}},$$
$$= \frac{\pi_{i-1}}{\lambda + a_i}\left[\sum_{j=1}^{i}\frac{\eta_j}{\pi_{j-1}} + \frac{\pi_\infty}{1-\pi_\infty}\sum_{j=1}^{\infty}\frac{\eta_j}{\pi_{j-1}}\right], \qquad i \geq 2. \tag{3.32}$$

What we have just established is but a mere coordinate-wise convergence. By Scheffé's Theorem 1.6.1, to prove strong convergence we need to check that the limiting $(\xi_i)_{i\geq 1}$ when multiplied by λ is a density, provided $(\eta_j)_{j\geq 1}$ is a density. (It is clear that ξ_i's are nonnegative if so are the η_i's.) For this we note the following relation which may be checked by induction argument (see also Exercise 3.3.11),

$$\lambda\sum_{i=1}^{n}\frac{\pi_{i-1}}{\lambda + a_i} = 1 - \pi_n, \tag{3.33}$$

and its two immediate consequences: $\lambda\sum_{i=j}^{n}\frac{\pi_{i-1}}{\lambda+a_i} = \pi_{j-1} - \pi_n$ and

$$\lambda\sum_{i=j}^{\infty}\frac{\pi_{i-1}}{\lambda + a_i} = \pi_{j-1} - \pi_\infty, \qquad j \geq 1. \tag{3.34}$$

Therefore, noting that (3.30) extends (3.32) to $i = 1$, summing over $i \geq 1$ and changing the order of summation in the resulting double sums, we see that

$$\lambda\sum_{i=1}^{\infty}\xi_i = \sum_{j=1}^{\infty}\frac{\eta_j}{\pi_{j-1}}\lambda\sum_{i=j}^{\infty}\frac{\pi_{i-1}}{\lambda + a_i} + \frac{\pi_\infty}{1-\pi_\infty}\sum_{j=1}^{\infty}\frac{\eta_j}{\pi_{j-1}}\lambda\sum_{i=1}^{\infty}\frac{\pi_{i-1}}{\lambda + a_i}$$
$$= \sum_{j=1}^{\infty}\frac{\eta_j}{\pi_{j-1}}(\pi_{j-1} - \pi_\infty) + \frac{\pi_\infty}{1-\pi_\infty}\sum_{j=1}^{\infty}\frac{\eta_j}{\pi_{j-1}}(1 - \pi_\infty) = \sum_{i=1}^{\infty}\eta_i.$$

This establishes convergence of the resolvents $(\lambda - A_n)^{-1}$, $n \geq 1$.

3.3.6 *Return to* $i = 1$ *after explosion: The limit semigroup*

By the Trotter–Kato Theorem 1.4.2, our analysis shows that the semigroups generated by A_n's converge to a limit semigroup, defined perhaps on a subspace of l^1. We will show that this semigroup is in fact defined on the entire l^1 and that its generator extends $D + O$.

Let G_1 be the operator defined on the domain $\mathcal{D}(G_1)$ equal to $\mathcal{D}(G)$ of (3.25) by the formula $G_1\,(\xi_i)_{i\ge 1} = (\eta_i)_{i\ge 1}$, where

$$\eta_1 = -a_1\xi_1 + h, \qquad \eta_i = -a_i\xi_i + a_{i-1}\xi_{i-1}, \quad i \ge 2. \tag{3.35}$$

Here $h := \lim_{n\to\infty} a_n\xi_n$ exists since absolute convergence of the series from the definition of $\mathcal{D}(G_1)$ implies this series's convergence.

This definition implies that $A_n\,(\xi_i)_{i\ge 1}$ differs from $G_1\,(\xi_i)_{i\ge 1}$ by $a_n\xi_n - h$ in the first coordinate, and is the same at coordinates $i = 2, \ldots, n$. It follows that

$$\|A_n\,(\xi_i)_{i\ge 1} - G_1\,(\xi_i)_{i\ge 1}\,\| = |a_n\xi_n - h| + \sum_{i=n+1}^{\infty} |a_i\xi_i - a_{i-1}\xi_{i-1}|.$$

Since this converges to 0, as $n \to \infty$, the extended limit of A_n's extends G_1. Moreover, $\mathcal{D}(G_1)$ is dense in l^1, implying that so is the domain of the extended limit. Therefore, by the Sova–Kurtz version of the Trotter–Kato Theorem 1.4.3, the limit semigroup is defined on the whole of l^1. Moreover, its generator, being equal to the extended limit of A_n's, extends G_1. On the other hand, by (1.25), each member of the domain of the generator is of the form (3.30)–(3.31) for some $(\eta_i)_{i\ge 1} \in l^1$. Then, (3.31) implies that $(a_i\xi_i - a_{i-1}\xi_{i-1})_{i\ge 1}$ is the difference of two members of l^1: $(\eta_i)_{n\ge 1}$ and $\lambda\,(\xi_i)_{i\ge 1}$. Thus, it is a member of l^1 and $(\xi_i)_{i\ge 1}$ is a member of $\mathcal{D}(G_1)$. This proves that the generator of the limit semigroup is G_1.

Finally, the domain of $D + O$ is

$$\mathcal{D}(D + O) = \mathcal{D}(D) = \{(\xi_i)_{i\ge 1} \in l^1;\, (a_i\xi_i)_{i\ge 1} \in l^1\},$$

and for $(\xi_i)_{i\ge 1}$ in $\mathcal{D}(D + O)$,

$$(D + O)\,(\xi_i)_{i\ge 1} = (-a_i\xi_i + a_{i-1}\xi_{i-1})_{i\ge 1}\,,$$

where for notational convenience we agree that $a_0\xi_0 = 0$. Clearly, $\mathcal{D}(D + O) \subset \mathcal{D}(G_1)$. Moreover, $G_1\,(\xi_i)_{i\ge 1}$ and $(D + O)\,(\xi_i)_{i\ge 1}$ differ only in the first coordinate, the first coordinate of $G_1\,(\xi_i)_{i\ge 1}$ being $-a_1\xi_1 + h$ and the first coordinate of $(D+O)\,(\xi_i)_{i\ge 1}$ being $-a_1\xi_1$. However, for $(\xi_i)_{i\ge 1} \in \mathcal{D}(D+O)$, $h = \lim_{n\to\infty} a_n\xi_n = 0$. It follows that G_1 extends $D + O$. In particular, the semigroup generated by G_1 dominates the minimal semigroup of Kato's Theorem.

To summarize, the operator G_1 defined in (3.25)–(3.35) is the generator of a Markov semigroup related to the process in which a pure birth Markov chain after each explosion returns to the state $i = 1$. In agreement with Kato's Theorem, G_1 was shown to extend $D + O$. We also note the following relation between G, the generator of the minimal semigroup, and G_1: their domains coincide, and for $x = (\xi_i)_{i\ge 1} \in \mathcal{D}(G) = \mathcal{D}(G_1)$,

$$\boxed{G_1 x = Gx + h(x)e_1,} \tag{3.36}$$

where

$$h(x) = \lim_{n\to\infty} a_n \xi_n$$

is well defined on $\mathcal{D}(G)$.

3.3.7 Return to $i = 2$ after explosion: Construction

The semigroup describing the pure birth Markov process returning to $i = 1$ after explosion, constructed in the previous sections, can be used as a building block for the semigroup describing this chain's return to $i = 2$. Here are the steps one needs to take.

(i) First, we construct the semigroup describing return to $i = 1$ for the shifted sequence $a_n, n \ge 1$, that is, for $a_n' := a_{n+1}$. We call this semigroup $\{T(t), t \ge 0\}$ and denote its generator G_1'.

(ii) The subspace $l^\diamond \subset l^1$ composed of $(\xi_i)_{i\ge 1}$ with $\xi_1 = 0$ is isometrically isomorphic to l^1 with isomorphism $I : l^\diamond \to l^1$ being the shift to the left: $I\,(\xi_i)_{i\ge 1} = (\xi_{i+1})_{i\ge 1}$. The formula $T^\diamond(t) = I^{-1}T(t)I$ defines the isomorphic image of $\{T(t), t \ge 0\}$ in $l^\diamond$. The generator of this semigroup is $G_1^\diamond = I^{-1}G_1' I$ with domain $I^{-1}\mathcal{D}(G_1')$; see Section 1.1.10.

(iii) Let G_2 be the operator in l^1 with domain composed of vectors of the form $\xi_1 e_1 + x$ where $\xi_1 \in \mathbb{R}$ and $x \in \mathcal{D}(G_1^\diamond) \subset l^\diamond$, defined as

$$G_2(\xi_1 e_1 + x) = -a_1\xi_1 e_1 + a_1 \xi_1 e_2 + G_1^\diamond x.$$

G_2 is the generator of the semigroup we wanted to construct.

Intuitively, the semigroup $\{T(t), t \ge 0\}$ describes the process returning to $i = 1$, but the intensity of escaping from $i \ge 2$ is that belonging to the state $i + 1$. By shifting in step (ii) we make intensities right and at the same time force the process to return to $i = 2$ after explosion. However, this does not provide the rules of behavior for the process starting at $i = 1$, so we fill this gap in step (iii).

3.3.8 Return to $i = 2$ after explosion: Analysis

First of all, G_2 is a Markov generator. For, in solving its resolvent equation we are searching for $\xi_1 \in \mathbb{R}$ and $x \in \mathcal{D}(G_1^\diamond) \subset l^\diamond$ such that

$$\lambda(\xi_1 e_1 + x) + a_1\xi_1 e_1 - a_1\xi_1 e_2 - G_1^\diamond x = y$$

where $\lambda > 0$ and $y = (\eta_i)_{i\ge 1} \in l^1$ are given. Since $G_1^\diamond x \in l^\diamond$, this task comes down to solving

$$(\lambda + a_1)\xi_1 e_1 = \eta_1 e_1 \qquad \text{and} \qquad \lambda x - G_1^\diamond x = y^\diamond + a_1\xi_1 e_2,$$

where $y^\diamond := y - \eta_1 e_1 \in l^\diamond$. This system has the unique solution:

$$\xi_1 = \frac{\eta_1}{\lambda + a_1} \qquad \text{and} \qquad x = \left(\lambda - G_1^\diamond\right)^{-1}(y^\diamond + a_1\xi_1 e_2), \tag{3.37}$$

and it is clear that this solution is nonnegative provided y is nonnegative (since $\left(\lambda - G_1^\diamond\right)^{-1}$ is a positive operator). Since $\mathcal{D}(G_2)$ is clearly dense in l^1 and $\Sigma G_2(\xi_1 e_1 + x) = 0 + \Sigma G_1^\diamond x = 0$, all conditions of the Hille–Yosida Theorem 2.4.3 are satisfied, proving the claim.

Here is a more detailed description of G_2: we have

$$G_1'\,(\xi_i)_{i\ge 1} = (-a_2\xi_1 + \lim_{n\to\infty} a_{n+1}\xi_n, a_2\xi_1 - a_3\xi_2, a_3\xi_2 - a_4\xi_3, \dots)$$

provided $\sum_{i=1}^\infty |a_i\xi_{i-1} - a_{i+1}\xi_i| < \infty$; ($\xi_0 := 0$). Hence, under the same condition,

$$\begin{aligned} G_1^\diamond(0, \xi_2, \xi_3, \dots) &= I^{-1}G_1'(\xi_2, \xi_3, \dots) \\ &= I^{-1}(-a_2\xi_2 + h, a_2\xi_2 - a_3\xi_3, a_3\xi_3 - a_4\xi_4, \dots) \\ &= (0, -a_2\xi_2 + h, a_2\xi_2 - a_3\xi_3, a_3\xi_3 - a_4\xi_4, \dots) \end{aligned}$$

where, as before, $h = h\,(\xi_i)_{i\ge 1} = \lim_{n\to\infty} a_n\xi_n$. It follows that $\mathcal{D}(G_2) = \mathcal{D}(G)$ and

$$\begin{aligned} G_2(\xi_1, \xi_2, \dots) &= -a_1\xi_1 e_1 + a_1\xi_1 e_2 + G_1^\diamond(0, \xi_2, \xi_3, \dots) \\ &= (-a_1\xi_1, a_1\xi_1 - a_2\xi_2 + h, a_2\xi_2 - a_3\xi_3, \dots). \end{aligned} \tag{3.38}$$

In other words,

$$\boxed{G_2 x = Gx + h(x)e_2,} \qquad x \in \mathcal{D}(G). \tag{3.39}$$

As in 3.3.6, we argue that G_2 extends $D + O$. The definitions of $G_2 x$ and $(D + O)x$ differ in the second coordinate. The domain of $D + O$ is a subset of $\mathcal{D}(G_2) = \mathcal{D}(G)$ and on this subset the term $h = h(x)$, by which these coordinates differ is 0.

Finally, the semigroups generated by G_1 (of Section 3.3.6) and by G_2 do not coincide. This is clear since they have different generators, but may also be seen from (3.30) and (3.37). Indeed, (3.30) shows that the upper left entry in the matrix representing the resolvent of the former semigroup is $\frac{1}{1-\pi_\infty}\frac{1}{\lambda+a_1}$, whereas the same entry in the matrix representing the resolvent of G_2 is, by (3.37), equal to $\frac{1}{\lambda+a_1}$. The former is larger than the latter. This is because in 3.3.6, with nonzero probability the process after leaving $i = 1$ returns there. This possibility is ruled out in the process governed by G_2, and $\frac{1}{\lambda+a_1}$ is the Laplace transform of the probability that the process is 'still at $i = 1$.'

3.3.9 Exercise

Given sequences $(c_n)_{n\geq 1}$, $(d_n)_{n\geq 1}$, and $(\eta_n)_{n\geq 1}$, consider the recurrence

$$\xi_{n+1} = c_n\xi_n + d_n\eta_n, \qquad n \geq 1.$$

Check to see that this recurrence has infinitely many solutions, each of them of the form $\xi_1 = p$,

$$\xi_n = \left(\prod_{i=1}^{n-1} c_i\right) p + \sum_{i=1}^{n-1}\left(\prod_{j=i+1}^{n-1} c_j\right) d_i\eta_i, \qquad n \geq 2,$$

where p is a parameter.

3.3.10 Exercise

Check that equations in the second line of (3.27) imply that

$$\xi_i = (\lambda + a_1)\frac{\pi_{i-1}}{\lambda + a_i}\xi_1 + \frac{\pi_{i-1}}{\lambda + a_i}\sum_{j=2}^{i}\frac{\eta_j}{\pi_{j-1}}, \qquad i = 2, \ldots, n, \tag{3.40}$$

and then deduce the formula for ξ_1 given in (3.28) from the first line in (3.27).

3.3.11 Exercise

(a) Prove (3.33) by direct induction argument. (b) The operator A_n from Section 3.3.4 generates a Markov semigroup. Hence, $\Sigma\lambda\,(\lambda - A_n)^{-1}\,e_1 = 1$. Use this to deduce (3.33) from (3.40) and (3.28) .

3.3.12 Exercise

Check to see that the norm of the operator $(\eta_i)_{i\ge 1} \mapsto (\xi_i)_{i\ge 1}$ given by (3.24) is $(1-\pi_\infty)\lambda^{-1}$.

3.3.13 Exercise

By modifying the construction of Sections 3.3.7 and 3.3.8, show that, for any $i \ge 3$, the operator

$$G_i x = Gx + h(x)e_i, \qquad x \in \mathcal{D}(G)$$

is a Markov generator; the related process after explosion starts all over again from the state i.

3.4 On the Domain of the Minimal Semigroup Generator

Besides the issues already discussed, the examples of generators presented in Section 3.3 clarify another important matter. To explain, given an intensity matrix $Q = \left(q_{i,j}\right)_{i,j\in\mathbb{I}}$, let us define the domain of the related operator, say, $\mathfrak{Q}$, to be the set of $(\xi_i)_{i\ge 1} \in l^1$ such that the series $\sum_{j\in\mathbb{I}} \xi_j q_{j,i}$, $i \in \mathbb{I}$ converge absolutely and

$$\sum_{i\in\mathbb{I}} \left| \sum_{j\in\mathbb{I}} \xi_j q_{i,j} \right| < \infty,$$

and let $\mathfrak{Q}$ be defined by the formula

$$\mathfrak{Q}(\xi_i)_{i\in\mathbb{I}} = \left(\sum_{j\in\mathbb{I}} \xi_j q_{j,i} \right)_{i\in\mathbb{I}}. \tag{3.41}$$

Intuitively, this operator is in a sense maximal, and Theorem 6.20 in [6] confirms this intuition. However, one may be tempted to think that $\mathfrak{Q}$ is maximal because it extends generators of all possible postexplosion processes or, put otherwise, that to describe a particular postexplosion process, that is, to obtain a generator, one needs to restrict this maximal operator to a smaller domain, as was the case in the second Kolmogorov–Kendall–Reuter semigroup. In reality, the situation is more complex.[4]

[4] Interestingly, in the dual perspective of l^∞, postexplosion processes are described precisely by such restrictions of domains. See Sections 5.7 and 5.8.

For, in the case of the pure birth process, $\mathcal{D}(\mathfrak{Q})$ is composed of $(\xi_i)_{i\geq 1}$ such that $\sum_{i=1}^{\infty} |a_i\xi_i - a_{i-1}\xi_{i-1}| < \infty$, and thus coincides with the domain $\mathcal{D}(G)$ of the generator of the minimal semigroup of Section 3.3.2; in fact $\mathfrak{Q}$ is G. Moreover, neither G_1 or G_2 are obtained by restricting the domain of $\mathfrak{Q} = G$, but rather by modifying values of G on $\mathcal{D}(G) \setminus \{x \in \mathcal{D}(G); \Sigma Gx = 0\}$.

As we will see later, the general rule is that discrete exit boundary does not change the domain of $\mathfrak{Q}$ but rather modifies the way this operator acts; its points are expressed as additional terms accompanying the generator of the minimal semigroup (see (3.92)). It is the entrance boundary that indeed does restrict the domain of $\mathfrak{Q}$ (see Section 4.4), and the second Kolmogorov–Kendall–Reuter process involves an entrance boundary. Hence, when both an entrance and exit boundary are faced, both the domain of $\mathfrak{Q}$ and the way it acts may be altered.

Needless to say, in general, the generator G of the minimal semigroup need not coincide with $\mathfrak{Q}$. Our main goal in this section is to establish that in the birth and death process example of Sections 2.4.11 and 3.2.7 all three inclusions in the chain

$$\mathcal{D}(D + O) \subset \mathcal{D}(\overline{D + O}) \subset \mathcal{D}(G) \subset \mathcal{D}(\mathfrak{Q}) \tag{3.42}$$

are simultaneously proper. Before doing that, however, we note that the pure birth process exemplifies the fact that the first two inclusions may be simultaneously proper.

3.4.1 *Inclusions (3.42) in the pure birth process*

Since the pure birth process example is explosive, $\overline{D + O} \neq G$ (see 3.2.1), that is, the second inclusion in (3.42) is proper. We will make this statement even more explicit by recalling that

$$\mathcal{D}(D + O) = \{(\xi_i)_{i\geq 1} \in l^1; \sum_{i=1}^{\infty} a_i|\xi_i| < \infty\} \tag{3.43}$$

and showing that

$$\mathcal{D}(\overline{D + O}) = \{(\xi_i)_{i\geq 1} \in \mathcal{D}(G); \lim_{i\to\infty} a_i\xi_i = 0\}, \tag{3.44}$$

where $\mathcal{D}(G)$ is given by (3.25). The fact that $\mathcal{D}(\overline{D + O})$ is contained in the set, say, $\mathcal{D}$, on the right-hand side of (3.44) is a special case of (3.9) because, by (3.25),

$$\Sigma Gx = -\lim_{n\to\infty} a_n\xi_n, \qquad x \in \mathcal{D}(G). \tag{3.45}$$

To check the reverse inclusion, consider an $x := (\xi_i)_{i\ge 1} \in \mathcal{D}$ and let $x_n := (\xi_1, \dots, \xi_n, 0, 0, \dots), n \ge 1$. Then $x_n \in \mathcal{D}(D + O)$ and $\lim_{n\to\infty} x_n = x$. Moreover, Gx and $(D + O)x_n$ differ by $-a_{n+1}\xi_{n+1}$ on the $(n + 1)$st coordinate, and by $a_{i-1}\xi_{i-1} - a_i\xi_i$ on the ith coordinate, where $i \ge n + 2$. Thus $\|Gx - (D + O)x_n\| = |a_{n+1}\xi_{n+1}| + \sum_{i=n+2}^{\infty} |a_{i-1}\xi_{i-1} - a_i\xi_i|$ converges to zero, as $n \to \infty$. This shows that $x \in \mathcal{D}(\overline{D + O})$, completing the proof of (3.44).

It is now clear that $\mathcal{D}(D + O)$ is a proper subset of $\mathcal{D}(\overline{D + O})$ because $\sum_{i\ge 1} |a_i\xi_i| < \infty$ implies $\sum_{i\ge 1} |a_{i-1}\xi_{i-1} - a_i\xi_i| < \infty$ and $\lim_{i\to\infty} a_i\xi_i = 0$, but not vice versa (take, e.g., $\xi_i = \frac{1}{ia_i}$). Similarly, $\mathcal{D}(\overline{D + O})$ is a proper subset of $\mathcal{D}(G)$ because the functional defined on $\mathcal{D}(G)$ by (3.45) is non-zero: condition $\sum_{i\ge 1} |a_{i-1}\xi_{i-1} - a_i\xi_i| < \infty$ does not imply $\lim_{i\to\infty} a_i\xi_i = 0$ (take, e.g., $\xi_i := a_i^{-1}\sum_{k=1}^{i} \frac{1}{k^2}$).

We turn to the case of the birth and death process of Sections 2.4.11 and 3.2.7. Our key to proving that all inclusions in (3.42) are proper in this case is the fact that the generator A of Section 2.4.11 coincides with the generator of the minimal semigroup. We establish this fact in Section 3.4.4; Sections 3.4.2 and 3.4.3 prepare a way for this proof.[5]

3.4.2 $\mathfrak{Q}$ *extends* G

In the case of the birth and death process under consideration, it is rather clear that $\mathfrak{Q}$ extends G. For, here the domain $\mathcal{D}(D + O) = \mathcal{D}(D)$ is composed of $(\xi_i)_{i\ge 1}$ such that

$$\sum_{i=1}^{\infty} 3^i |\xi_i| < \infty, \tag{3.46}$$

and the ith coordinate of $(D + rO)\,(\xi_i)_{i\ge 1}$, where $(\xi_i)_{i\ge 1} \in \mathcal{D}(D)$, is $9r\xi_2 - 6\xi_1$ or $3^{i-1}(9r\xi_{i+1} - 9\xi_i + 2r\xi_{i-1})$ depending on whether $i = 1$ or $i \ge 2$. Also, given any $x \in \mathcal{D}(G)$ and any sequence $(r_n)_{n\ge 1}$ converging to 1 from the left, we may find $x_n \in \mathcal{D}(D)$ such that $\lim_{n\to\infty} x_n = x$ and $\lim_{n\to\infty}(D + r_nO)x_n = Gx$. Combining this with the fact that convergence in norm implies convergence of coordinates, we obtain that the ith coordinate of Gx is either $9\xi_2 - 6\xi_1$ (for $i = 1$) or $3^{i-1}(9\xi_{i+1} - 9\xi_i + 2\xi_{i-1})$ (for $i \ge 2$). This means that Gx coincides with $\mathfrak{Q}x$ for $x \in \mathcal{D}(G)$.

[5] Sections 3.4.3–3.4.4 are based on the idea I owe to Jacek Banasiak; see also Lemma 7.16 in [6].

3.4.3 Eigenvectors of $\mathfrak{Q}$

Let $\lambda > 0$. For every $p \in \mathbb{R}$, there is precisely one solution $(\rho_i)_{i\geq 1}$ to the system

$$\begin{aligned} \lambda\rho_1 &= 9\rho_2 - 6\rho_1, \\ \lambda\rho_i &= 3^{i+1}\rho_{i+1} - 3^{i+1}\rho_i + 2\cdot 3^{i-1}\rho_{i-1}, \qquad i \geq 2, \end{aligned} \tag{3.47}$$

such that $\rho_1 = p$. (This solution may, but need not be a member of l^1.) For $p = 0$ the solution is trivial. Also, if $p \geq 0$ then

$$\rho_i > \left(\frac{4}{9}\right)^{i-1} \rho_1, \qquad i \geq 1. \tag{3.48}$$

The first two claims are clear: if $\rho_1 = p$ is given, then the first relation in (3.47) determines ρ_2, and the second implies that two consecutive coordinates of $(\rho_i)_{i\geq 1}$ determine the next coordinate. Also, for $p = 0$ all these coordinates are seen to be zeros.

To show (3.48), it suffices to establish that

$$\rho_i > \frac{4}{9}\rho_{i-1}, \qquad i \geq 2, \tag{3.49}$$

and to this end, we use an induction argument. First of all, by the first equation in (3.47) it is obvious that (3.49) is true for $i = 2$. Next, suppose that this inequality is true for all indices up to some i. Since this implies $\rho_i > 0$, the second relation in (3.47) says that $9\rho_{i+1} > 9\rho_i - 2\rho_{i-1}$. Thus, by the induction assumption,

$$9\rho_{i+1} > 9\rho_i - 2\rho_{i-1} > 9\rho_i - \frac{9}{2}\rho_i > 4\rho_i.$$

It follows that (3.49) is true also when i is replaced by $i + 1$. This completes our proof.

3.4.4 $A = G$

To establish that the operator A of Section 2.4.11 coincides with the minimal semigroup generator, it suffices to show that for any $\lambda > 0$ and any nonnegative $y \in l^1$,

$$(\rho_i)_{i\geq 1} := (\lambda - A)^{-1}\, y - (\lambda - G)^{-1}\, y$$

is zero. From 3.2.7 we know that this vector is nonnegative. Moreover, since $(\lambda - A)^{-1}\, y$ and $(\lambda - G)^{-1}\, y$ are solutions to the resolvent equations for A and G, respectively, with the same right-hand side (equaling y), and since both

A and G are restrictions of $\mathfrak{Q}$ (for G this was proven above, for A it is obvious by definition: cf. (3.50) and (2.37)), $(\rho_i)_{i\ge 1}$ is a solution to the system (3.47).

Suppose that $\rho_1 > 0$, and let $(\xi_i)_{i\ge 1} = (\lambda - A)^{-1}$. Then $(\xi_i)_{i\ge 1} = (\lambda - G)^{-1} y + (\rho_i)_{i\ge 1} \ge (\rho_i)_{i\ge 1}$, and, by (3.48),

$$3^i \xi_i > 3 \left(\frac{4}{3}\right)^{i-1} \rho_1 .$$

This, however, contradicts that fact that for $(\xi_i)_{i\ge 1} \in \mathcal{D}(A)$ the limit $\lim_{i\to\infty} 3^i \xi_i$ exists. It follows that the only choice we have is $\rho_1 = 0$ and this implies $(\rho_i)_{i\ge 1} = 0$, as desired.

3.4.5 *All inclusions are proper*

It is finally easy to see that in the birth and death example of Sections 2.4.11 and 3.2.7 all inclusions in (3.42) are proper. Indeed, the domain $\mathcal{D}(\mathfrak{Q})$ of the maximal operator $\mathfrak{Q}$ related to the intensity matrix of the birth and death process of Section 3.2.7 is the set of all $(\xi_i)_{i\ge 1}$ such that

$$\sum_{i=2}^{\infty} 3^{i-1} |3(3\xi_{i+1} - \xi_i) - 2(3\xi_i - \xi_{i-1})| < \infty, \tag{3.50}$$

and we have just proved that the domain $\mathcal{D}(G)$ of the generator of the minimal semigroup is the set of all $(\xi_i)_{i\ge 1}$ such that

$$\sum_{i=2}^{\infty} 3^i |3\xi_i - \xi_{i-1}| < \infty. \tag{3.51}$$

Hence, $\mathcal{D}(G)$ is a proper subset of $\mathcal{D}(\mathfrak{Q})$, because (3.51) implies (3.50) but not vice versa. Indeed, for the sequence $(\xi_i)_{i\ge 1}$ defined by the recurrence

$$\xi_{i+1} = \frac{1}{3}\xi_i + \frac{1}{3^{i+1}}\left(2^i - \frac{1}{2^i}\right), \qquad i \ge 1$$

with arbitrary ξ_1, we have $\sum_{i=1}^{\infty} |3^{i+1}\xi_{i+1} - 3^i \xi_i| = \sum_{i=1}^{\infty} \left(2^i - \frac{1}{2^i}\right) = \infty$, whereas

$$\sum_{i=1}^{\infty} |(3^{i+1}\xi_{i+1} - 3^i \xi_i) - 2(3^i \xi_i - 3^{i-1}\xi_{i-1})| = \sum_{i=1}^{\infty} \left(\frac{1}{2^{i-2}} - \frac{1}{2^i}\right) < \infty.$$

(The reader who, rightly, wonders if the $(\xi_i)_{i\ge 1}$ considered here is a member of l^1, should consult Section 3.4.6.)

To see that the other two inclusions in (3.42) are proper, it suffices to recall that $\mathcal{D}(D+O) = \mathcal{D}(D)$ is the set of $(\xi_i)_{i\ge 1}$ satisfying (3.46), and prove that

$$\mathcal{D}(\overline{D+O}) = \{(\xi_i)_{i\ge 1}\,;\,(3.51)\text{ holds and }\lim_{i\to\infty} 3^i\xi_i = 0\}. \tag{3.52}$$

Examples of elements of $\mathcal{D}(G)\setminus\mathcal{D}(\overline{D+O})$ and $\mathcal{D}(\overline{D+O})\setminus\mathcal{D}(D+O)$ may be constructed as in 3.4.1.

The proof of (3.52) also follows the lines of 3.4.1. Since, as shown in 2.4.11, $\Sigma G\,(\xi_i)_{i\ge 1} = -\lim_{i\to\infty} 3^i\xi_i$, $(\xi_i)_{i\ge 1}\in\mathcal{D}(G)$, the fact that $\mathcal{D}(\overline{D+O})$ is contained in the right-hand side of (3.52) is a special case of inclusion (3.9). Conversely, given any $x = (\xi_i)_{i\ge 1}$ satisfying (3.51) and $\lim_{i\to\infty} 3^i\xi_i = 0$, there are $x_n\in\mathcal{D}(D)$ such that $\lim_{n\to\infty} x_n = x$ and $\lim_{n\to\infty}(D+O)x_n = Gx$. Indeed, take $x_n = (\xi_1,\dots,\xi_n,0,0,\dots)$, $n\ge 1$. Clearly, $x_n\in\mathcal{D}(D+O)$ and $\lim_{n\to\infty} x_n = x$. Moreover, for $n\ge 2$, $(D+O)x_n$ and Gx are the same at the first $n-1$ coordinates, and differ by (a) $3^{n+1}\xi_{n+1}$ on the nth coordinate (b) $3^{n+2}\xi_{n+2} - 3^{n+2}\xi_{n+1}$ on the $(n+1)$-st coordinate, and (c) $3^{k-1}(9\xi_{k+1} - 9\xi_k + 2\xi_{k-1})$ on the kth coordinate as long as $k\ge n+2$. It follows that

$$\begin{aligned}\|(D+O)x_n - Gx\| &\le 3^{n+1}|\xi_{n+1}| + 3^{n+2}|\xi_{n+2}| + 3\cdot 3^{n+1}|\xi_{n+1}|\\ &\quad + \sum_{k=n+2}^{\infty} 3^{k-1}|9\xi_{k+1} - 9\xi_k + 2\xi_{k-1}|.\end{aligned}$$

Since the right-hand side here converges to 0, as $n\to\infty$, we are done.

3.4.6 *A clarification on* $\mathcal{D}(\Omega)$

Is it clear that the set of $(\xi_i)_{i\ge 1}$ satisfying (3.50) coincides with $\mathcal{D}(\Omega)$? By definition, $\mathcal{D}(\Omega)$ is the set of $(\xi_i)_{i\ge 1}\in l^1$ satisfying (3.50). Hence, the question is, does (3.50) imply $(\xi_i)_{i\ge 1}\in l^1$?

Here is an argument showing that the answer is in the affirmative. Let $(\xi_i)_{i\ge 1}\in\mathcal{D}(\Omega)$. Calculating as in our solution to Exercise 2.4.18, we check that the limit $\lim_{i\to\infty} 3^i(3\xi_{i+1} - 2\xi_i)$ exists. It follows that there is a positive constant c such that

$$|3\xi_{i+1} - 2\xi_i| \le \frac{c}{3^i}, \qquad i\ge 1. \tag{3.53}$$

Next, take an $\alpha\in(\frac{2}{3},1)$. Since $3\alpha > 2$, there is an integer i_0 such that $(3\alpha-2)(3\alpha)^i \ge c$ for $i\ge i_0$. Let d be the maximum of the numbers $1, c$ and $\alpha^{-i}|\xi_i|$, $i = 1,\dots,i_0$. I claim that

$$|\xi_i| \le \alpha^i d, \qquad i\ge 1; \tag{3.54}$$

this condition clearly implies $(\xi_i)_{i\ge 1}\in l^1$.

To prove (3.54), we proceed by induction. For $i = 1, \ldots, i_0$, (3.54) is true by the definition of d. Also, assuming that (3.54) holds for some $i \ge i_0$, we see, by (3.53), that

$$|\xi_{i+1}| \le \left|\xi_{i+1} - \frac{2}{3}\xi_i\right| + \frac{2}{3}|\xi_i| \le \frac{c}{3^{i+1}} + \frac{2}{3}\alpha^i d.$$

Moreover, since $d \ge 1$ and $i \ge i_0$, we have $\frac{c}{3^{i+1}} \le (\alpha - \frac{2}{3})\alpha^i d$. It follows that $|\xi_{i+1}|$ does not exceed $d\alpha^{i+1}$, that is, that (3.54) holds with i replaced by $i+1$. This completes the proof.

3.5 Beyond Kato's Theorem

We come back to the results established in Section 3.3. Formulae (3.36) and (3.39) show an interesting way the information on return of the process after explosion is contained in the generator. They suggest that if we want the process to return to $i \ge 1$ then we add $h(x)e_i$ to the minimal semigroup generator G; this claim may be proved by modifying the argument of Sections 3.3.7 and 3.3.8 (see Exercise 3.3.13).

Here is a bolder hypothesis:[6] let $0 \neq u = (\upsilon_i)_{i\ge 1} \in l^1$ be nonnegative with $\Sigma u \le 1$. Then the operator $H = H_u$ with domain $\mathcal{D}(H) = \mathcal{D}(G)$ given by

$$\boxed{Hx = Gx + h(x)u} \tag{3.55}$$

is the generator of a sub-Markov semigroup related to the process which after explosion starts all over again at i with probability υ_i. If $\Sigma u = 1$, that is, if u is a density, H generates a Markov semigroup.

We prove this hypothesis in the next two sections.

3.5.1 *H is a sub-Markov generator*

For $x = (\xi_i)_{i\ge 1} \in \mathcal{D}(G)$,

$$\Sigma Gx = \lim_{n\to\infty} \sum_{i=1}^{n} (a_{i-1}\xi_{i-1} - a_i\xi_i) = -\lim_{n\to\infty} a_n\xi_n = -h(x). \tag{3.56}$$

Thus, $\Sigma Hx = \Sigma Gx + h(x)\Sigma u \le 0$ (with equality if $\Sigma u = 1$). Since $\mathcal{D}(G)$ is dense in l^1, in view of the Hille–Yosida Theorem 2.4.3, we are left with

[6] For a different motivation for (3.55), see Section 4.2 and formula (4.14) in particular.

proving that for any nonnegative y and any $\lambda > 0$ there is a unique solution x of the resolvent equation

$$\lambda x - Gx - h(x)u = y, \tag{3.57}$$

and that this solution is nonnegative.

First, we will show that y and λ determine $h(x)$. For, applying $\Sigma G R_\lambda$ (recall that $R_\lambda = (\lambda - G)^{-1}$) to both sides of the equation and using (3.56), we obtain

$$h(x)[1 - h(R_\lambda u)] = h(R_\lambda y).$$

Moreover, since $GR_\lambda u = \lambda R_\lambda u - u$,

$$1 - h(R_\lambda u) = 1 + \Sigma G R_\lambda u = 1 + \Sigma \lambda R_\lambda u - \Sigma u \ge \Sigma \lambda R_\lambda u > 0,$$

because $\Sigma R_\lambda u = 0$ would imply $\|R_\lambda u\| = 0$, that is, $u = 0$. Thus $h(x) = \frac{h(R_\lambda y)}{1 - h(R_\lambda u)}$. It follows that the solution of (3.57) must be of the form

$$x = R_\lambda y + \frac{h(R_\lambda y)}{1 - h(R_\lambda u)} R_\lambda u. \tag{3.58}$$

To see that this x is a true solution to (3.57), we check first that $h(x) = \frac{h(R_\lambda y)}{1 - h(R_\lambda u)}$ and then that

$$(\lambda - G)x - h(x)u = y + \frac{h(R_\lambda y)}{1 - h(R_\lambda u)} u - h(x)u = y.$$

This completes the proof, because it is obvious that x is nonnegative if y is.

3.5.2

To support the intuition that the process related to H returns, after explosion, to i with probability υ_i, we proceed analogously as in Section 3.3.3. Given u one may construct nonnegative $u_n = \left(\upsilon_{n,i}\right)_{i \ge 1}$ such that $\lim_{n\to\infty} u_n = u$, $\upsilon_{n,i} = 0$ for $i \ge n$, and, if u is a density, then so are u_n. Then one may modify the operators A_n of Section 3.3.3 by replacing the last row in (3.26) by

$$(a_n \upsilon_{n,1}, a_n \upsilon_{n,2}, \ldots, a_n \upsilon_{n,n-1}, -a_n).$$

Let B_n be the so-modified A_n. The process related to B_n goes down the ladder, spends an exponential time with parameter a_n at the lowest rung $i = n$, and then jumps to $j \in \{1, \ldots, n-1\}$ with probability $\upsilon_{n,j}$. If $\Sigma u_n = 1$, $n \ge 2$, the related semigroups are composed of Markov operators.

Now, A_n and B_n are related by the formula

$$B_n x = A_n x + a_n \xi_n (u_n - e_1), \qquad x = (\xi_i)_{i \ge 1} \in l^1.$$

Moreover, we know from Section 3.3.6 that

$$\lim_{n\to\infty} A_n x = G_1 x = Gx + h(x)e_1, \qquad x \in \mathcal{D}(G).$$

It follows that

$$\lim_{n\to\infty} B_n x = Gx + h(x)u, \qquad x \in \mathcal{D}(G).$$

By the Sova–Kurtz Theorem (see Section 1.4.1) this proves that the semigroups generated by B_n converge to that generated by H, thus supporting our claim.

3.5.3 *Mea culpa*

I admit, the results just proved are much more elegant and general than those of Section 3.3, and the proof is much simpler here. So, why did we bother going through all those tiresome and troublesome calculations of resolvents and stuff? Well, if we start from an abstract theorem like that, how do we build our intuition?

3.5.4 *An abstract version of 3.5.1: The key functional*

The main idea of 3.5.1 goes far beyond the context of pure birth process and may be applied to any Kolmogorov matrix. This, however, requires defining an abstract functional on $\mathcal{D}(G)$, playing the role of $h\,(\xi_i)_{i\ge1} = \lim_{n\to\infty} a_n\xi_n$, the leading actor of Section 3.3.

Let G be the generator of Kato's minimal semigroup for a given Kolmogorov matrix. On $\mathcal{D}(G)$ we define:

$$\boxed{h(x) = -\Sigma G x,}$$

and note that $h(x) \ge 0$ provided $x \ge 0$ (formula (3.56) reveals that in the case of the pure birth process, h coincides with our old friend).

In what follows the following property of h will turn out to be crucial:

$$h_{|\mathcal{D}(D)} = 0. \tag{3.59}$$

We have encountered this fact in Section 3.1.7 (as formula (3.8)), where it has been laboriously derived from (3.2). For a slightly more direct proof (yet using the same calculation that lies behind (3.2)), let $x = (\xi_i)_{i\ge1} \in \mathcal{D}(D)$. We have

$$\sum_{i=1}^{\infty}\sum_{j\ne i} |\xi_j q_{j,i}| = \sum_{j=1}^{\infty} |\xi_j| \sum_{i\ne j} q_{j,i} = \sum_{j=1}^{\infty} |\xi_j| q_j = \|Dx\|,$$

and so

$$\sum_{i=1}^{\infty}\sum_{j=1}^{\infty}|\xi_j q_{j,i}| = \sum_{i=1}^{\infty}\sum_{j\neq i}|\xi_j q_{j,i}| + \sum_{i=1}^{\infty}|\xi_i|q_i = 2\|Dx\| < \infty.$$

It follows that changing the order of summation in the following calculation is justified:

$$-h(x) = \Sigma Gx = \sum_{i=1}^{\infty}\sum_{j=1}^{\infty}\xi_j q_{j,i} = \sum_{j=1}^{\infty}\xi_j\sum_{i=1}^{\infty}q_{j,i} = 0.$$

This completes the proof.

We recall, finally, that if Q is explosive, the functional h, despite (3.59), is not zero on the entire $\mathcal{D}(G)$ (see 3.2.1, point (h)). It follows that the generator H considered below does not coincide with G.

3.5.5 *An abstract version of 3.5.1*

Let G be the generator of Kato's minimal semigroup for a Kolmogorov matrix, and let h be the functional defined on $\mathcal{D}(G)$ in the previous section. For any nonnegative $u \neq 0$ in l^1 such that $\Sigma u \leq 1$, the operator

$$\boxed{Hx = Gx + h(x)u}$$

defined on $\mathcal{D}(G)$ is a sub-Markov generator. If u is a density, H is a Markov generator. H coincides with G and $D + O$ on $\mathcal{D}(D)$, but $G \neq H$.

Proof We argue as in 3.5.1; all the preparatory work has been done in the previous section; for instance, $H = G = D + O$ on $\mathcal{D}(D)$ because of (3.59).

(a) $\mathcal{D}(G)$ is dense in l^1, G being the generator.

(b) For $x \geq 0$, $x \in \mathcal{D}(G)$, we have

$$\Sigma Hx = \Sigma Gx + h(x)\Sigma u \leq 0$$

since h is a nonnegative functional.

(c) In the resolvent equation $\lambda x - Gx - h(x)u = y$, $h(x)$ is determined by y and λ: we must have $h(x)[1 - h(R_\lambda u)] = h(R_\lambda y)$, and since

$$1 - h(R_\lambda u) = 1 + \Sigma\lambda R_\lambda u - \Sigma u \geq \Sigma\lambda R_\lambda u > 0,$$

this implies $h(x) = \frac{h(R_\lambda y)}{1-h(R_\lambda u)}$. This in turn yields

$$x = R_\lambda y + \frac{h(R_\lambda y)}{1 - h(R_\lambda u)}R_\lambda u, \tag{3.60}$$

and such x is the only true solution to the resolvent equation. This solution is nonnegative if y is nonnegative, because h is a nonnegative functional.

Finally, $H \neq G$, since h is nonzero. □

3.5.6 *Is this all?*

Is this all one can do? Are the operators H of the previous section the only possible extensions of $D + O$ that generate sub-Markov or Markov semigroups? Absolutely not (at least not in the general case). To explain, after explosion, the process described by an $H = H_u$ starts all over again at a state i with probability being equal to the ith coordinate of u, and this does not depend on how explosion came about. Sometimes one may but introduce different rules for 'different explosions.'

The last sentence may seem peculiar. Are there different types of explosion? In a sense, yes. If we think again of the pure birth process as going down a ladder, then there might be many different, so to say, disjoint, ladders, and to each of them one may attach a different set of rules of starting over again. So, one may construct Markov and sub-Markov semigroups that command the process to start all over again after explosion, but depending on the type of explosion to provide different starting distributions. In the next sections we construct such a family of semigroups for two birth processes going down two different ladders.

3.5.7 *Two infinite ladders: The generator of the minimal semigroup*

Let $(a_n)_{n\geq 1}$ satisfy (3.22). We consider a process which, starting at i waits at this state for an exponential time with parameter a_i and then jumps to $i + 2$; thus we go down a ladder either using only even-numbered rungs or only odd-numbered rungs. Then

$$\mathcal{D}(D) = \{(\xi_i)_{i\geq 1}\,;\,(a_i\xi_i)_{i\geq 1} \in l^1\},$$

$(\lambda - D)^{-1}\, e_i = \frac{1}{\lambda+a_i} e_i$ and $Oe_i = a_i e_{i+2}$, $i \geq 1$ so that

$$B_\lambda^n e_i = \left(\prod_{k=0}^{n-1} \frac{a_{i+2k}}{\lambda + a_{i+2k}}\right) e_{i+2n}, \qquad n \geq 0.$$

Let $l^1_{\rm e}$ and $l^1_{\rm o}$ be the subspaces of l^1 composed of vectors with all odd (even) coordinates vanishing, respectively. The space l^1 is a **direct sum** of $l^1_{\rm e}$ and $l^1_{\rm o}$:

$$l^1 = l^1_{\rm e} \oplus l^1_{\rm o}.$$

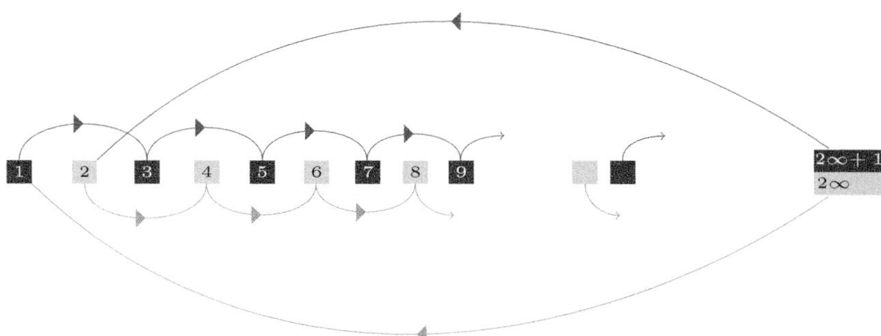

Figure 3.3 An example of 'two infinite ladders': here, the process after reaching infinity via the odd ladder returns to $i = 2$; after reaching infinity via the even ladder, it returns to $i = 1$.

This means that each $x \in l^1$ may be uniquely represented as $x = x_{\mathrm{e}} + x_{\mathrm{o}}$ where $x_{\mathrm{e}} \in l^1_{\mathrm{e}}$ and $x_{\mathrm{o}} \in l^1_{\mathrm{o}}$; the uniqueness of representation is a consequence of the fact that $l^1_{\mathrm{e}} \cap l^1_{\mathrm{o}} = \{0\}$. Moreover, B_λ leaves both these subspaces **invariant**: if $x \in l^1_{\mathrm{e}}$ ($x \in l^1_{\mathrm{o}}$) then $B_\lambda x \in l^1_{\mathrm{e}}$ ($B_\lambda x \in l^1_{\mathrm{o}}$), and so does $(\lambda - D)^{-1}$. Since l^1_{e} and l^1_{o} are closed, for R_λ of (3.5), we obtain

$$R_\lambda x = R_\lambda x_{\mathrm{e}} + R_\lambda x_{\mathrm{o}}$$

with $R_\lambda x_{\mathrm{e}} \in l^1_{\mathrm{e}}$ and $R_\lambda x_{\mathrm{o}} \in l^1_{\mathrm{o}}$.

Now, $I_{\mathrm{e}} : l^1_{\mathrm{e}} \to l^1$ given by

$$I_{\mathrm{e}}(0, \xi_2, 0, \xi_4, 0, \xi_6, \dots) = (\xi_2, \xi_4, \xi_6, \dots)$$

establishes an isometric isomorphism of l^1_{e} and l^1. Therefore, restrictions of R_λ, $(\lambda - D)^{-1}$ and B_λ to l^1_{e} have their isomorphic images in l^1. Denoting by B^{e}_λ the image of B_λ we have, for instance,

$$\begin{aligned}(B^{\mathrm{e}}_\lambda)^n e_i &= I_{\mathrm{e}} B^n_\lambda I_{\mathrm{e}}^{-1} e_i = I_{\mathrm{e}} B^n_\lambda e_{2i} = I_{\mathrm{e}} \left(\prod_{k=0}^{n-1} \frac{a_{2i+2k}}{\lambda + a_{2i+2k}} \right) e_{2i+2n} \\ &= \left(\prod_{k=0}^{n-1} \frac{a'_{i+k}}{\lambda + a'_{i+k}} \right) e_{i+n}\end{aligned}$$

where $a'_i = a_{2i}$. The last expression here is identical to the right-hand side of (3.13) with $(a_n)_{n\ge1}$ replaced by $\left(a'_n\right)_{n\ge1}$. Consulting 3.3.2 we come to the conclusion that R^{e}_λ, the image of the restriction of R_λ to l^1_{e}, is the resolvent of the operator given by (3.25) and the formula following it, with $(a_n)_{n\ge1}$

replaced by $(a'_n)_{n\ge 1}$. Coming back to the original space we see that for $x \in l^1_{\mathrm{e}}$, $R_\lambda x = (\lambda - G_{\mathrm{e}})^{-1} x$ where

$$\mathcal{D}(G_{\mathrm{e}}) = \{(\xi_i)_{i\ge 1} \in l^1_{\mathrm{e}};\ \sum_{i=1}^{\infty} |a_{2i-2}\xi_{2i-2} - a_{2i}\xi_{2i}| < \infty\},$$

$$G_{\mathrm{e}}\,(\xi_i)_{i\ge 1} = I_{\mathrm{e}}^{-1}\,(a_{2i-2}\xi_{2i-2} - a_{2i}\xi_{2i})_{i\ge 1}\,.$$

Similarly one argues that for $x \in l^1_{\mathrm{o}}$, $R_\lambda x = (\lambda - G_{\mathrm{o}})^{-1} x$ where

$$\mathcal{D}(G_{\mathrm{o}}) = \{(\xi_i)_{i\ge 1} \in l^1_{\mathrm{o}};\ \sum_{i=1}^{\infty} |a_{2i-3}\xi_{2i-3} - a_{2i-1}\xi_{2i-1}| < \infty\},$$

$$G_{\mathrm{o}}\,(\xi_i)_{i\ge 1} = I_{\mathrm{o}}^{-1}\,(a_{2i-3}\xi_{2i-3} - a_{2i-1}\xi_{2i-1})_{i\ge 1}\,.$$

Here, by convention, $a_{-1}\xi_{-1} = 0$, and I_{o} is the isomorphism of l^1_{o} and l^1 defined as follows:

$$I_{\mathrm{o}}(\xi_1, 0, \xi_3, 0, \xi_5, \dots) = (\xi_1, \xi_3, \xi_5, \dots).$$

Combining these results we conclude that $R_\lambda = (\lambda - G)^{-1}$, where $\mathcal{D}(G)$ is composed of $x \in l^1$ such that $x_{\mathrm{e}} \in \mathcal{D}(G_{\mathrm{e}})$ and $x_{\mathrm{o}} \in \mathcal{D}(G_{\mathrm{o}})$, and

$$Gx = G_{\mathrm{e}}x_{\mathrm{e}} + G_{\mathrm{o}}x_{\mathrm{o}}.$$

In other words, G thus defined is the generator of Kato's minimal semigroup.

3.5.8 *Two infinite ladders: Many possible (happy) returns*

Armed with the information of the previous section, we may construct a semigroup related to a process which features two different types of explosion and, consequently, two different types of behavior after explosions. To this end, let nonzero u and v be nonnegative elements of l^1 such that $\Sigma u, \Sigma v \le 1$. For $(\xi_i)_{i\ge 1} \in \mathcal{D}(G)$ we define

$$\begin{aligned} h_{\mathrm{e}}\,(\xi_i)_{i\ge 1} &= -\Sigma G_{\mathrm{e}}\left[(\xi_i)_{i\ge 1}\right]_{\mathrm{e}} = \lim_{n\to\infty} a_{2n}\xi_{2n}, \\ h_{\mathrm{o}}\,(\xi_i)_{i\ge 1} &= -\Sigma G_{\mathrm{o}}\left[(\xi_i)_{i\ge 1}\right]_{\mathrm{o}} = \lim_{n\to\infty} a_{2n-1}\xi_{2n-1}, \end{aligned} \tag{3.61}$$

and claim that $H = H_{u,v}$ defined by

$$\boxed{Hx = Gx + h_{\mathrm{e}}(x)u + h_{\mathrm{o}}(x)v,} \qquad x \in \mathcal{D}(G)$$

extends $D + O$ and is a sub-Markov generator. If u and v are densities, H is a Markov generator.

Proof [7] The domain $\mathcal{D}(H) = \mathcal{D}(G)$ is dense in l^1 and

$$\begin{aligned} \Sigma Hx &= \Sigma G_e x_e + \Sigma G_o x_o + h_e(x)\Sigma u + h_o(x)\Sigma v \\ &= h_e(x)(\Sigma u - 1) + h_o(x)(\Sigma v - 1) \le 0, \qquad x \in \mathcal{D}(G) \end{aligned}$$

(with equality if u and v are densities), since h_e and h_o are nonnegative functionals. Hence, our task reduces to studying properties of solution(s) to the resolvent equation for H:

$$\lambda x - Hx = y$$

where $y \ge 0$ and $\lambda > 0$ are given. Since $l^1 = l^1_e \oplus l^1_o$, this equation is equivalent to the system

$$\begin{aligned} \lambda x_e - G_e x_e - h_e(x)u_e - h_o(x)v_e &= y_e, \\ \lambda x_o - G_o x_o - h_e(x)u_o - h_o(x)v_o &= y_o. \end{aligned} \tag{3.62}$$

Applying, as in 3.5.1, R_λ and then h_e to both sides of the first equation, and proceeding similarly with the second, we see that the unknown $h_e(x)$ and $h_o(x)$ satisfy

$$\begin{aligned} [1 - h_e(R_\lambda u_e)]h_e(x) \qquad - h_e(R_\lambda v_e)\, h_o(x) &= h_e(R_\lambda y_e), \\ - h_o(R_\lambda u_o)\, h_e(x) + [1 - h_o(R_\lambda v_o)]h_o(x) &= h_o(R_\lambda y_o). \end{aligned} \tag{3.63}$$

Noting that

$$\begin{aligned} h_e(R_\lambda u_e) + h_o(R_\lambda u_o) = h_e(R_\lambda u) + h_o(R_\lambda u) &= -\Sigma G R_\lambda u \\ = \Sigma u - \lambda \Sigma R_\lambda u &=: c \in [0, 1), \end{aligned} \tag{3.64}$$

and, similarly, $h_e(R_\lambda v_e) + h_o(R_\lambda v_o) =: d \in [0, 1)$, we estimate the value of the main determinant of (3.63) as follows:

$$\begin{aligned} W :&= [1 - h_e(R_\lambda u_e)]\,[1 - h_o(R_\lambda v_o)] - [c - h_e(R_\lambda u_e)]\,[d - h_o(R_\lambda v_o)] \\ &\ge (1 - c)[1 - h_o(R_\lambda v_o)]. \end{aligned}$$

This last expression is positive because $1 - h_o(R_\lambda v_o) = 1 - \Sigma v_o + \Sigma \lambda R_\lambda v_o$ is nonnegative and could be zero only if $1 - \|v_o\|$ and $\|\lambda R_\lambda v_o\|$ were equal to zero simulataneously, but the last case is impossible. Also, both coefficient determinants, say, W_1 and W_2, are nonnegative: for example,

$$\begin{aligned} W_1 :&= \begin{vmatrix} h_e(R_\lambda y_e) & -h_e(R_\lambda v_e) \\ h_o(R_\lambda y_o) & 1 - h_o(R_\lambda v_o) \end{vmatrix} \\ &= h_e(R_\lambda y_e)[1 - h_o(R_\lambda v_o)] + h_o(R_\lambda y_o) h_e(R_\lambda v_e) \ge 0. \end{aligned}$$

[7] In 3.7.9, we prove a more general theorem using a reasoning that requires a slightly smaller number of calculations.

It follows that $h_{\rm e}(x)$ and $h_{\rm o}(x)$ are determined by u, v, y and $\lambda > 0$: $h_{\rm e}(x) = W_1 W^{-1}$ and $h_{\rm o}(x) = W_2 W^{-1}$, and are nonnegative. Moreover, the only possible solution of (3.62) is

$$x_{\rm e} = \frac{W_1}{W} R_\lambda u_{\rm e} + \frac{W_2}{W} R_\lambda v_{\rm e} + R_\lambda y_{\rm e},$$
$$x_{\rm o} = \frac{W_1}{W} R_\lambda u_{\rm o} + \frac{W_2}{W} R_\lambda v_{\rm o} + R_\lambda y_{\rm o}$$

or, simply,

$$x = \frac{W_1}{W} R_\lambda u + \frac{W_2}{W} R_\lambda v + R_\lambda y. \tag{3.65}$$

We claim that this x is a true solution of the resolvent equation. To show this, we note first that for this x:

$$h_{\rm e}(x) = \frac{W_1}{W} h_{\rm e}(R_\lambda u) + \frac{W_2}{W} h_{\rm e}(R_\lambda v) + h_{\rm e}(R_\lambda y) = \frac{W_1}{W}.$$

Here, in the second step we used the fact that replacing the unknowns: $h_{\rm e}(x)$ and $h_{\rm o}(x)$ in (3.63) by $W_1 W^{-1}$ and $W_2 W^{-1}$, respectively, results in two equalities. For the same reason $h_{\rm o}(x) = W_2 W^{-1}$. This in turn shows that

$$\begin{aligned}(\lambda - H)x &= (\lambda - G)x - h_{\rm e}(x)u - h_{\rm o}(x)v \\ &= \frac{W_1}{W} u + \frac{W_2}{W} v + y - h_{\rm e}(x)u - h_{\rm o}(x)v = y,\end{aligned}$$

as claimed. We note that x of (3.65) is nonnegative since y is and, as we have seen, $W_1 W^{-1}, W_2 W^{-1} \geq 0$.

To complete the proof, we note that on $\mathcal{D}(D)$ the functionals $h_{\rm e}$ and $h_{\rm o}$ vanish; thus $H_{|\mathcal{D}(D)} = G_{|\mathcal{D}(D)} = D + O$. □

3.5.9 *Some intuition*

Let us compare 3.5.8 with 3.5.5. In the case of two infinite ladders,

$$h(x) = -\Sigma G x = h_{\rm e}(x) + h_{\rm o}(x), \qquad x \in \mathcal{D}(G).$$

Therefore, in this case, the result of the previous section may be deduced from 3.5.5 iff $u = v$.

To explain the difference between the two results in question, it might be useful to think of the two ladders example of Section 3.5.8 as follows. There are two additional states for the related Markov chain: one is the bottom of the 'even' ladder and the other is the bottom of the 'odd' ladder. The process spends no time there since both states are instantaneous, but each of them 'distributes' paths going through it differently: one uses u, the other uses v. Theorem 3.5.5 does not allow for such a discrimination: whether the process reaches the bottom of the 'even' ladder or the bottom of the 'odd' ladder, its distribution right after that is u.

As it transpires, these additional states are in no way imaginary. We will be able to say more about them in the following sections.

3.5.10 Exercise

Check that c of (3.64) is in fact nonzero. To this end, look at (3.24) and use (3.34).

3.5.11 Exercise

Assume that, in addition to $(a_n)_{n\geq 1}$, we are given another sequence $(b_n)_{n\geq 1}$ of nonnegative numbers. A Markov chain with intensity matrix

$$\begin{pmatrix} -a_1 & a_1 & 0 & 0 & 0 & \dots \\ b_2 & -a_2 - b_2 & a_2 & 0 & 0 & \dots \\ 0 & b_3 & -a_3 - b_3 & a_3 & 0 & \dots \\ \vdots & \vdots & \ddots & \ddots & \ddots & \dots \end{pmatrix}$$

is a **birth and death process**. Assuming that $(b_n)_{n\geq 1}$ is bounded, check to see that in this case $h\,(\xi_i)_{i\geq 1} = \lim_{i\to\infty}(a_i\xi_i - b_{i+1}\xi_{i+1}) = \lim_{i\to\infty} a_i\xi_i$.

3.5.12 Exercise

Persuade yourself (or another good mathematician) that (3.32) (together with (3.30)) is a particular case of (3.60).

3.6 Close to the Edge: Lattice $\mathfrak{B}$

The example of two infinite ladders, discussed in the previous section, hinges on constructing the functionals h_{o} and h_{e}. Each of these functionals describes one ladder, and each ladder involves its bottom, a curious place down below which distributes paths going through it as if it were a part of the state-space.

This is a simple, but illustrative, example of an (exit) boundary point for a Markov chain.

It was the idea of Feller [41, 42] to construct the (exit) boundary for a Markov chain governed by a Kolmogorov matrix $Q = \left(q_{i,j}\right)_{i,j\geq 1}$ by means of functionals $f \in l^\infty = (l^1)^*$ such that

$$\boxed{\Pi^* f = f \qquad \text{and} \qquad 0 \leq f \leq \Sigma,} \tag{3.66}$$

where $\Pi \in \mathcal{L}(l^1)$ is the Markov operator related to the so-called **jump chain** matrix

$$\left((1-\delta_{i,j})\frac{q_{i,j}}{q_i}\right)_{i,j\geq 1} \tag{3.67}$$

(hence, $\Pi^* f$ is the product of the matrix Π and of the column-vector f). We note that $(1-\delta_{i,j})\frac{q_{i,j}}{q_i}$ (where, as always, $\delta_{i,j} = 1$ is 1 if $i = j$ and 0 otherwise) is the probability of jumping to state j after sojourn in the state i.

This section is devoted to the description of the set $\mathfrak{B}$ of functionals satisfying (3.66). Here and in what follows, to avoid unnecessary complications, we always assume that

$$q_i > 0, \qquad i \in \mathbb{N};$$

that is to say that we assume that no state is absorbing.

Before proceeding, it is perhaps worth noting that bounded linear functionals satisfying (3.66) were in fact involved in our analysis of two infinite ladders example. For, the definition of the unbounded functionals h_{o} and h_{e} given in (3.61) may be equivalently written as

$$\begin{aligned} h_{\text{e}}\,(\xi_i)_{i\geq 1} &= -\Sigma_{\text{e}} G\,(\xi_i)_{i\geq 1}, \\ h_{\text{o}}\,(\xi_i)_{i\geq 1} &= -\Sigma_{\text{o}} G\,(\xi_i)_{i\geq 1}, \qquad (\xi_i)_{i\geq 1} \in \mathcal{D}(G), \end{aligned} \tag{3.68}$$

where the functionals Σ_{e} and Σ_{o} are represented by the following two bounded sequences:

$$(0, 1, 0, 1, \dots) \qquad (1, 0, 1, 0, \dots). \tag{3.69}$$

In other words, the unbounded functionals h_{o} and h_{e} are constructed by means of the generator G and two bounded functionals just defined (as we will see later, in (3.89), this is a typical way such functionals are constructed). It is also easy to see that both Σ_{e} and Σ_{o} satisfy (3.66), for the action of Π^* amounts to a shift by two coordinates to the left:

$$(\Pi^* f)(i) = f(i+2), \qquad i \in \mathbb{N}$$

(consult 3.6.1, below, if you find this notation mysterious).

I hasten to add that not all functionals satisfying (3.66) are useful in constructing semigroups dominating Kato's minimal semigroup. For instance, if the example of two infinite ladders is modified so that

$$a_{2i} = i, \qquad i \in \mathbb{N} \tag{3.70}$$

(yet still $\sum_{i=1}^{\infty} a_{2i-1}^{-1} < \infty$), then it will take an infinite time for a minimal process to go down the 'even ladder.' Thus, the bottom of this ladder will never be reached, and it can play no role in constructing a postexplosion process. Nevertheless, the operator Π is not affected by such a change and so Σ_e still satisfies (3.66). As we shall see later, in this modified example, Σ_e is a member of the so-called **passive boundary** (see Section 3.7.4). It is one of the aims of the following analysis to distinguish such passive functionals from those forming the 'real' boundary.

3.6.1 *Notation*

Elements of the space l^1 will continue to be denoted $(\xi_i)_{i\geq 1}$, $(\eta_i)_{i\geq 1}$, and so on, whereas elements of l^∞ will be denoted f, g, and so on, and seen as bounded functions on the set of natural numbers. When needed, $f \in l^\infty$ will be identified with the sequence $(f(i))_{i\geq 1}$ where, of course, $f(i)$ is the value of f at $i \in \mathbb{N}$. As often as not, though, we will rightfully see f as a functional on l^1 and write $f(x)$ to denote $\sum_{i=1}^{\infty} f(i)\xi_i$ for $x = (\xi_i)_{i\geq 1}$. In particular, $f(i) = f(e_i)$. However, since double parentheses do not look right, instead of $f((\xi_i)_{i\geq 1})$ we will prefer to write $f\,(\xi_i)_{i\geq 1}$. By $f \geq g$ for $f, g \in l^\infty$ we mean that $f(x) \geq g(x)$ for all nonnegative $x \in l^1$.

3.6.2 $\mathfrak{B}$ *as a lattice*

Let $\mathfrak{B}$ denote the set of functionals satisfying (3.66).

(a) For all $f, g \in \mathfrak{B}$, there exist the largest member of the set of $h \in \mathfrak{B}$ satisfying $h \leq f$ and $h \leq g$. This unique element (which may be equal 0) is denoted $f \cap g$.
(b) For all $f, g \in \mathfrak{B}$, there exist the smallest member of the set of $h \in \mathfrak{B}$ satisfying $h \geq f$ and $h \geq g$. This unique element (which may be equal to Σ) is denoted $f \cup g$.
(c) Suppose that for some $f, g \in \mathfrak{B}$, $f + g \in \mathfrak{B}$, that is, suppose that $f + g \leq \Sigma$. Then

$$f + g = f \cap g + f \cup g. \tag{3.71}$$

Proof

(a) Let $h_0(i) = \min(f(i), g(i))$, $i \in \mathbb{N}$ (note that in general $h_0 \notin \mathfrak{B}$). Since, for any nonnegative $x \in l^1$,

$$h_0(\Pi x) \le f(\Pi x) = (\Pi^* f)(x) = f(x), \text{ and}$$
$$h_0(\Pi x) \le g(\Pi x) = (\Pi^* g)(x) = g(x),$$

we have $h_0(\Pi x) \le h_0(x)$. Thus, the sequence $(h_0(\Pi^n x))_{n\ge 1}$, being nonincreasing and bounded by 0, converges to a nonnegative limit. Since any $x \in l^1$ may be written as a difference of two nonnegative elements of l^1, the limit $h_1(x) = \lim_{n\to\infty} h_0(\Pi^n x)$ exists for all $x \in l^1$, and defines a nonnegative functional satisfying $h_1 \le h_0 \le \Sigma$. The calculation

$$h_1(\Pi x) = \lim_{n\to\infty} h_0(\Pi^{n+1} x) = \lim_{n\to\infty} h_0(\Pi^n x) = h_1(x),$$

shows that $\Pi^* h_1 = h_1$, that is, $h_1 \in \mathfrak{B}$. It is clear that $h_1 \le f$ and $h_1 \le g$. Moreover, if for a certain $h \in \mathfrak{B}$ we have $h \le f$ and $h \le g$, then for any nonnegative $x \in l^1$ and any $n \ge 1$,

$$h(x) = ((\Pi^*)^n h)(x) = h(\Pi^n x) \le f(\Pi^n x)$$

and, similarly, $h(x) \le g(\Pi^n x)$. Thus, $h(x) \le h_0(\Pi^n x)$ and letting $n \to \infty$ we conclude that $h(x) \le h_1(x)$. Uniqueness is clear and we may define $f \cap g := h_1$.

(b) Let $k \in \mathfrak{B}$ be such that $k \ge f$ and $k \ge g$. Since Π is a stochastic matrix, $\Sigma \in \mathfrak{B}$ and we may thus take, for example, $k = \Sigma$. Then, it makes sense to define

$$f \cup g := k - (k - f) \cap (k - g), \tag{3.72}$$

because all the operations needed to calculate the right-hand side are well defined, and the resulting functional belongs to $\mathfrak{B}$. Since $k - f \le k$ and $k - g \le k$,

$$f \cup g \ge k - k \cap (k - g) = k - (k - g) = g.$$

Similarly, $f \cup g \ge f$. On the other hand, suppose $f \le h \le \Sigma$ and $g \le h \le \Sigma$. Then $h' := h \cap k$ satisfies $f \le h' \le k$ and $g \le h' \le k$ and so $0 \le k - h' \le k - f$ and $0 \le k - h' \le k - g$. It follows that

$$f \cup g \le k - (k - h') \cap (k - h') = k - (k - h') = h' \le h.$$

This shows that the functional defined by (3.72) possesses all the required properties stated in (b). In particular, since $f \cup g$ is defined uniquely, the definition (3.72) does not depend on the choice of k.

(c) Under the assumptions stated, one may take $k = f + g$ in (3.72). The latter equation then becomes $f \cup g = f + g - g \cap f$. Since $f \cap g = g \cap f$, this completes the proof. □

3.6.3 Remark

In what follows it will be desirable to be able to calculate the minima and maxima not only for functionals in $\mathfrak{B}$ but also those which, besides being nonnegative and satisfying $\Pi^* f = f$ are merely bounded. (In other words, we relax condition $0 \le f \le \Sigma$, and assume instead that $f \in l^\infty$ and $f \ge 0$.) One way to do that is by

$$f \cap g = c(f_c \cap g_c) \qquad \text{and} \qquad f \cup g = c(f_c \cup g_c),$$

where $f_c = c^{-1} f, g_c = c^{-1} g$ and $c > 0$ is chosen so that $f_c, g_c \in \mathfrak{B}$. It is easy to see that this definition does not depend on c, is consistent with the previous one, and that with this extended definition basic properties of minima and maxima remain unchanged. In particular, $f \cap g$ is the largest element of the set of h such that $h \le f, h \le g$ and $\Pi^* h = h$. Also, formula (3.71) remains true, and the assumption on $f + g$ belonging to $\mathfrak{B}$ is now redundant.

3.6.4 Remark

As a by-product of the proof of (a) in 3.6.2,

$$(f \cap g)(i) \le \min(f(i), g(i)), \qquad i \in \mathbb{N}.$$

Using (3.71) and the obvious equality $f(i) + g(i) = \min(f(i), g(i)) + \max(f(i), g(i))$, we obtain also

$$\max(f(i), g(i)) \le (f \cup g)(i), \qquad i \in \mathbb{N}. \tag{3.73}$$

3.6.5 Remark

It should perhaps be stressed here that both $\mathfrak{B}$ itself and the lattice structure in $\mathfrak{B}$ depend on Π (i.e., on Q). In particular, if f and g belong to two sets $\mathfrak{B}$ related to two Π's, $f \cap g$ in one of these sets may be different than in the other.

3.6.6 *A dual characterization of* (3.66)

Suppose f satisfies (3.66). Then (cf. (3.8))

$$f(Ox) = f(-Dx) \tag{3.74}$$

for all $x \in \mathcal{D}(D)$ (see Sections 3.1.2 and 3.1.3 for appropriate definitions).

Proof In coordinates, (3.66) reads

$$\sum_{j \neq i} q_{i,j} f(j) = q_i f(i), \qquad i \in \mathbb{N},$$

and for $x = (\xi_i)_{i \geq 1} \in \mathcal{D}(D)$ we have $(q_i |\xi_i|)_{i \geq 1} \in l^1$. Therefore,

$$\begin{aligned}\sum_{i=1}^{\infty} \sum_{j \neq i} |q_{j,i} \xi_j f(i)| &= \sum_{j=1}^{\infty} |\xi_j| \sum_{i \neq j} q_{j,i} f(i) = \sum_{j=1}^{\infty} |\xi_j| q_j f(j) \\ &= f\,(q_i |\xi_i|)_{i \geq 1} < \infty.\end{aligned}$$

It follows that the change of the order of summation in the following calculation is justifiable:

$$\begin{aligned}f(Ox) &= \sum_{i=1}^{\infty} \sum_{j \neq i} q_{j,i} \xi_j f(i) = \sum_{j=1}^{\infty} \xi_j \sum_{i \neq j} q_{j,i} f(i) = \sum_{j=1}^{\infty} \xi_j q_j f(j) \\ &= f\,(q_i \xi_i)_{i \geq 1} = f(-Dx),\end{aligned}$$

completing the proof. □

The interplay between $f \in \mathfrak{B}$ and the functional f_λ introduced in the next section is of crucial importance for the entire analysis.

3.6.7 *f* and f_λ

Let f belong to $\mathfrak{B}$ and let $\lambda > 0$. The limit

$$f_\lambda(y) = \lim_{n \to \infty} f(B_\lambda^n y), \qquad y \in l^1$$

exists and defines a nonnegative bounded linear functional such that

$$f = \lambda R_\lambda^* f + f_\lambda, \tag{3.75}$$

where $R_\lambda, \lambda > 0$ is the resolvent of Kato's minimal semigroup. (See (3.4) for the definition of B_λ.)

Proof Let $y \in l^1$. Then

$$x_n := \sum_{k=0}^{n} (\lambda - D)^{-1} B_\lambda^k y$$

is a member of $\mathcal{D}(D)$ and so, by (3.74), $f(Ox_n) = f(-Dx_n)$. Equation (3.10), on the other hand, shows that $f(y) + f(Ox_n) = f(\lambda x_n - Dx_n) + f(B_\lambda^{n+1} y)$, or

$$f(y) = \lambda f(x_n) + f(B_\lambda^{n+1} y).$$

Since $\lim_{n\to\infty} x_n = R_\lambda y$ (see (3.5)), it follows that $\lim_{n\to\infty} f(B_\lambda^n y)$ exists, and (3.75) is true. □

3.6.8 *Characterization of* f_λ

Let $\mathfrak{B}_{\lambda,f}$ be the set of $g \in l^\infty$ such that $B_\lambda^* g = g$ and $0 \le g \le f$. Then f_λ is the largest element of $\mathfrak{B}_{\lambda,f}$: f_λ belongs to $\mathfrak{B}_{\lambda,f}$, and $g \in \mathfrak{B}_{\lambda,f}$ implies $g \le f_\lambda$.

Proof Formula (3.75) shows that $f_\lambda \le f$. Moreover, for any $x \in l^1$,

$$B_\lambda^* f_\lambda(x) = f_\lambda(B_\lambda x) = \lim_{n\to\infty} f(B_\lambda^{n+1} x) = \lim_{n\to\infty} f(B_\lambda^n x) = f_\lambda(x),$$

proving that f_λ belongs to $\mathfrak{B}_{\lambda,f}$. Finally, for a $g \in \mathfrak{B}_{\lambda,f}$ and a nonnegative $x \in l^1$,

$$g(x) = (B_\lambda^*)^n g(x) = g(B_\lambda^n x) \le f(B_\lambda^n x).$$

Thus, letting $n \to \infty$, we obtain $g \le f_\lambda$. This completes the proof. □

3.6.9 Example

Formula (3.75) for $f = \Sigma$ takes the form

$$\lambda R_\lambda^* \Sigma = \Sigma - \Sigma_\lambda \tag{3.76}$$

and reveals that the functional Σ_λ measures how far λR_λ is away from being a Markov operator, that is, how much probability mass is lost in λR_λ because of explosion. The larger is the probability mass loss, the larger is Σ_λ. In particular $\Sigma_\lambda = 0$ iff λR_λ is a Markov operator.

Let's see what Σ_λ is like in the pure birth Markov chain example. Being a bounded linear functional, Σ_λ is determined by its values on $e_i, i \ge 1$. By (3.13),

$$\Sigma_\lambda(e_i) = \lim_{n\to\infty} \Sigma(B_\lambda^n e_i) = \lim_{n\to\infty} \left(\prod_{k=i}^{i+n-1} \frac{a_k}{\lambda + a_k} \right) = \frac{\pi_\infty}{\pi_{i-1}},$$

where $\pi_i = \prod_{k=1}^{i} \frac{a_k}{\lambda + a_k}$. In other words, $\Sigma_\lambda = \left(\frac{\pi_\infty}{\pi_{i-1}}\right)_{i\ge 1}$.

3.6.10 *A functional relation for* $f_\lambda, \lambda > 0$

The functionals $f_\lambda, \lambda > 0$ are not unrelated: for $\lambda, \mu > 0$, applying R^*_μ to (3.75) and using the Hilbert equation, we obtain

$$\begin{aligned}(\lambda - \mu) R^*_\mu f_\lambda &= (\lambda - \mu) R^*_\mu f - \lambda(\lambda - \mu) R^*_\mu R^*_\lambda f \\ &= (\lambda - \mu) R^*_\mu f - \lambda(R^*_\mu - R^*_\lambda) f \\ &= \lambda R^*_\lambda f - \mu R^*_\mu f = (f - f_\lambda) - (f - f_\mu) \\ &= f_\mu - f_\lambda.\end{aligned}$$

Combining this with an analogous calculation of $(\lambda - \mu) R^*_\lambda f_\mu$ yields the following formula

$$f_\mu - f_\lambda = (\lambda - \mu) R^*_\mu f_\lambda = (\lambda - \mu) R^*_\lambda f_\mu. \tag{3.77}$$

3.6.11 *The canonical mapping*

In Section 3.6.7 we have seen a path leading from an $f \in \mathfrak{B}$ to the $f_\lambda \in \mathfrak{B}_{\lambda, f}$. Is there a path leading back? Yes, to some extent, as we shall soon explain.

Let

$$\mathfrak{B}_\lambda := \mathfrak{B}_{\lambda, \Sigma}.$$

For $f \in \mathfrak{B}_\lambda$ the limit

$$f^\diamond(x) = \lim_{n\to\infty} f(\Pi^n x), \qquad x \in l^1 \tag{3.78}$$

exists and defines a bounded linear functional $f^\diamond$. Moreover, $f^\diamond$ is the smallest of all g satisfying the following two conditions

$$f \le g \le \Sigma \qquad \text{and} \qquad \Pi^* g = g. \tag{3.79}$$

Proof Note that the entries of the matrix $\left((1 - \delta_{i,j}) \frac{q_{i,j}}{\lambda + q_i}\right)_{i,j\ge 1}$ representing the operator B_λ do not exceed those of the matrix representing Π. Hence, for any nonnegative $x \in l^1$ and any natural n,

$$f(x) = (B^*_\lambda f)(x) = f(B_\lambda x) \le f(\Pi x) \le \Sigma(\Pi x) = \Sigma(x).$$

Thus, the sequence $(f(\Pi^n x))_{n\ge 1}$, being nondecreasing and bounded by $\Sigma(x)$, converges to an $f^\diamond(x)$ such that

$$f(x) \le f^\diamond(x) \le \Sigma(x).$$

A standard argument allows us now to prove the existence of the limit (3.78) for all $x \in l^1$. As in 3.6.8 we also see that $\Pi^* f^\diamond = f^\diamond$. Thus, $f^\diamond$ belongs to the set of g satisfying (3.79). Finally, suppose g satisfies (3.79). Then, for all nonnegative $x \in l^1$,

$$g(x) = ((\Pi^*)^n g)(x) = g(\Pi^n x) \geq f(\Pi^n x).$$

Letting $n \to \infty$, we obtain $g \geq f^\diamond$. □

3.6.12 Definition

Following W. Feller,

$$f \mapsto f^\diamond$$

is termed the **canonical mapping**[8] from $\mathfrak{B}_\lambda$ to $\mathfrak{B}$. We note that

$$f^\diamond \geq f. \tag{3.80}$$

3.6.13 Theorem

For $f \in \mathfrak{B}_\lambda$,

$$\boxed{(f^\diamond)_\lambda = f.} \tag{3.81}$$

Proof By 3.6.8, $(f^\diamond)_\lambda$ is the largest element in $\mathfrak{B}_{\lambda, f^\diamond}$. Since $f \in \mathfrak{B}_\lambda$ and $f \leq f^\diamond$, we have

$$f \leq (f^\diamond)_\lambda \leq f^\diamond.$$

It follows that, for nonnegative $x \in l^1$ and $n \geq 1$,

$$f(\Pi^n x) \leq (f^\diamond)_\lambda(\Pi^n x) \leq f^\diamond(\Pi^n x) = f^\diamond(x).$$

Letting $n \to \infty$, we conclude that

$$f^\diamond(x) \leq \left((f^\diamond)_\lambda\right)^\diamond(x) \leq f^\diamond(x).$$

This reveals that $\left((f^\diamond)_\lambda\right)^\diamond = f^\diamond$ or that the value of the canonical mapping on the nonnegative functional $(f^\diamond)_\lambda - f$ is zero. However, by (3.80), this is impossible unless $(f^\diamond)_\lambda = f$. □

8 Strictly speaking, there are many canonical mappings: there is one canonical mapping $\mathfrak{B}_\lambda \to \mathfrak{B}$ associated to each $\lambda > 0$.

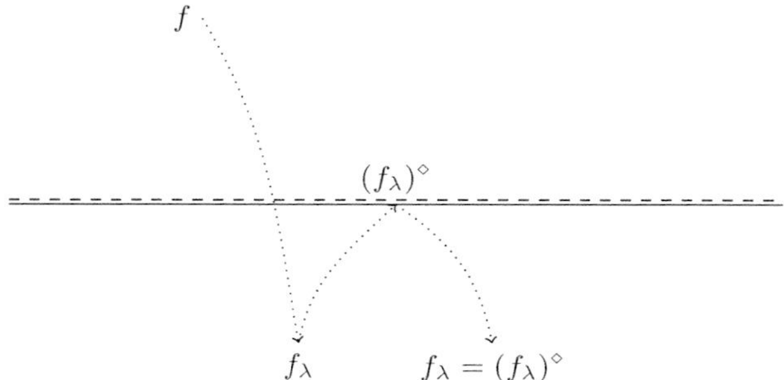

Figure 3.4 Relation between the maps $f \mapsto f_\lambda$ and $f \mapsto f^\diamond$.

3.6.14 Remark

The theorem of the previous section *does not* say that $(f_\lambda)^\diamond = f$ for $f \in \mathfrak{B}$. All we can claim is that

$$f \geq (f_\lambda)^\diamond \tag{3.82}$$

(see Exercise 3.6.19). The nonnegative functional

$$\Sigma_{\text{pass}} := \Sigma - (\Sigma_\lambda)^\diamond = \Sigma - \Sigma_{\max}, \tag{3.83}$$

where $\Sigma_{\max} := (\Sigma_\lambda)^\diamond \leq \Sigma$, will play an important role in what follows. (We will see in particular that its definition does not depend on the choice of λ.)

3.6.15 *Active and passive functionals*

A functional $f \in \mathfrak{B}$ is said to be passive if for some $\lambda > 0$, $f_\lambda = 0$. Relation (3.77) makes it clear that f is passive iff $f_\lambda = 0$ for all $\lambda > 0$. For such f, (3.75) takes the form

$$f = \lambda R_\lambda^* f \tag{3.84}$$

and reveals that passive functionals 'do not see' the loss of probability mass caused by explosion(s) (see also Exercise 3.6.21).

A functional $f \in \mathfrak{B}$ is said to be active if $f = (f_\lambda)^\diamond$ for some $\lambda > 0$. Thus, for active functionals, f_λ is 'so large' that f can be recovered from f_λ by means of Π. The proof of following theorem shows in particular that f is active iff $f = (f_\lambda)^\diamond$ for all $\lambda > 0$.

3.6.16 *Maximal properties of* $\Sigma_{\max}$

Let $f \in \mathfrak{B}$. The following conditions are equivalent:

(a) f is active.
(b) There is no nontrivial passive h such that $h \le f$.
(c) f belongs to the range of the canonical mapping.[9]
(d) $f \le \Sigma_{\max}$.

Moreover, the definition of $\Sigma_{\max}$ does not depend on λ.

Proof Suppose $f \ne (f_\lambda)^\diamond$ for some $\lambda > 0$. Then $h := f - (f_\lambda)^\diamond \le f$ is nontrivial, nonnegative, and passive, since by (3.81)

$$h_\lambda = f_\lambda - ((f_\lambda)^\diamond)_\lambda = f_\lambda - f_\lambda = 0.$$

Conversely, if there is a nontrivial, nonnegative, passive vector $h \le f$, then, by (3.82),

$$f - h \ge ((f - h)_\lambda)^\diamond = (f_\lambda)^\diamond,$$

showing that $f = (f_\lambda)^\diamond$ for no $\lambda > 0$. This proves that f is active iff $(f_\lambda)^\diamond = f$ for all $\lambda > 0$, and iff (b) holds.

Next, we show that (a) is equivalent to (c). An active f belongs to the range of the canonical mapping, being equal to $(f_\lambda)^\diamond$. For the proof of the converse, note that, in view of 3.6.13, each member of $\mathfrak{B}_\lambda$ is of the form g_λ for some $g \in \mathfrak{B}$. Thus, if f is in the range of the canonical mapping, there exists a $g \in \mathfrak{B}$ such that $f = (g_\lambda)^\diamond$. By 3.6.13, it follows that $f_\lambda = g_\lambda$ and so $f = (f_\lambda)^\diamond$.

Before proving that (d) is equivalent to the remaining conditions, we show that the definition of $\Sigma_{\max}$ does not depend on λ. To this end, we note that Σ_λ is the largest element of $\mathfrak{B}_\lambda$. Thus, $\Sigma_{\max} = (\Sigma_\lambda)^\diamond$ is the largest element in the image of $\mathfrak{B}_\lambda$ via the canonical map. However, by the already established equivalence of (b) and (c), this range does not depend on λ. Therefore, neither the largest element in this range depends on λ.

We are left with showing that (a) implies (d), and (d) implies (b). Since $f \le \Sigma$, (a) implies that $f = (f_\lambda)^\diamond \le (\Sigma_\lambda)^\diamond = \Sigma_{\max}$. For the proof of the other implication, we note that, by 3.6.13, $\Sigma_{\max}$ is active. Therefore, if (b) is violated and there is a nontrivial passive $h \le f$, f cannot be $\le \Sigma_{\max}$, for this would imply $h \le \Sigma_{\max}$, which is impossible by the equivalence of (a) and (b). □

3.6.17 *Maximal properties of* Σ_{pass}

(a) The definition of Σ_{pass} does not depend on λ.
(b) A functional $f \in \mathfrak{B}$ is passive iff $f \le \Sigma_{\text{pass}}$.
(c) An intensity matrix is explosive iff $\Sigma_{\text{pass}} \ne \Sigma$.

[9] As we shall see in the proof, the range of all canonical maps $\mathfrak{B}_\lambda \to \mathfrak{B}$ is the same; it is this common range that we have in mind here.

Proof (a) is clear: in the previous section we showed that the definition of $\Sigma_{\max}$ does not depend on λ.

Next we prove (b). First of all, Σ_{pass} itself is passive since, by 3.6.13, $(\Sigma_{\max})_\lambda = \Sigma_\lambda$. Therefore, if $0 \le f \le \Sigma_{\text{pass}}$, then $0 \le f_\lambda \le (\Sigma_{\text{pass}})_\lambda = 0$ implying $f_\lambda = 0$. Conversely, suppose f is passive. Since $\Sigma_{\max}$ is active, there is no nontrivial passive $h \le \Sigma_{\max}$. But any $h \le f$ is passive. Hence, $f \cap \Sigma_{\max} = 0$ and so, by (3.71) extended in Remark 3.6.3,

$$f + \Sigma_{\max} = f \cap \Sigma_{\max} + f \cup \Sigma_{\max} = f \cup \Sigma_{\max} \le \Sigma.$$

This implies $f \le \Sigma - \Sigma_{\max} = \Sigma_{\text{pass}}$.

To prove (c), suppoose $\Sigma = \Sigma_{\text{pass}}$. Then Σ is passive and so, by (3.84), $\Sigma = \lambda R_\lambda^* \Sigma$ for all $\lambda > 0$. It follows that $\Sigma \lambda R_\lambda x = \Sigma x$ for all $x \in l^1$, and this means that Σ is preserved by λR_λ and thus Q is not explosive (by Section 3.2.1). Conversely, since $\Sigma_{\text{pass}} \le \Sigma$, condition $\Sigma_{\text{pass}} \ne \Sigma$ implies (by (b)) that Σ is not passive. Hence, for some (and, in fact, for any) $\lambda > 0$ the functional Σ_λ is nonzero and we have $\Sigma \lambda R_\lambda^* = \Sigma - \Sigma_\lambda$. Thus, $\Sigma \lambda R_\lambda x < \Sigma x$ for some nonnegative $x \ge 0$, implying that λR_λ is not a Markov operator and thus Q is explosive. □

Sections 3.6.16 and 3.6.17 reveal crucial roles played by $\Sigma_{\max}$ and Σ_{pass}. Hence, we complete this section by the following example, where these functionals can be computed explicitly. Before reading it, however, the reader should solve by himself Exercise 3.6.20.

3.6.18 Example

Let us calculate Σ_{pass} and $\Sigma_{\max}$ for the following Markov chain. Starting at $i = 1$, the chain spends an exponential time with parameter 1 there, and then jumps either to $i = 2$ with probability p or to $i = 3$ with probability $p' = 1 - p$. At $i = 2$ it spends another, independent, exponential time with parameter 1 and then jumps either to $i = 1$ with probability q or to $i = 4$ with probability $q' = 1 - q$. To avoid trivialities, we assume $pq \ne 1$. At $i \ge 3$ the chain spends an exponential time with parameter a_i before jumping to $i + 2$, and we assume that

$$\sum_{k=1}^{\infty} a_{2k+1}^{-1} < \infty \qquad \text{whereas} \qquad \sum_{k=1}^{\infty} a_{2k}^{-1} = \infty. \tag{3.85}$$

Since $\Sigma_{\max} = (\Sigma_\lambda)^\diamond$, we begin by calculating Σ_λ. We have $B_\lambda e_i = \frac{a_i}{\lambda + a_i} e_{i+2}$, $i \ge 3$. This implies

$$B_\lambda^n e_{2k+1} = \prod_{i=0}^{n-1} \frac{a_{2(k+i)+1}}{\lambda + a_{2(k+i)+1}} e_{2k+1+2n} = \prod_{j=k}^{k+n-1} \frac{a_{2j+1}}{\lambda + a_{2j+1}} e_{2k+1+2n},$$

which in turn yields, for $k \geq 1$,

$$\Sigma_\lambda(2k+1) = \Sigma_\lambda(e_{2k+1}) = \lim_{n\to\infty} \Sigma(B_\lambda^n e_{2k+1}) = \prod_{j=k}^{\infty} \frac{a_{2k+1}}{\lambda + a_{2j+1}}, \qquad (3.86)$$

with the last product, by (3.85), different from 0. A similar calculation shows that

$$\Sigma_\lambda(2k+2) = 0, \qquad k \geq 1,$$

because the second series in (3.85) diverges.

Since, as we shall see soon, the values of Σ_λ at $i = 1$ and $i = 2$ do not matter in the calculation of $\Sigma_{\max}$, we simply state that

$$\Sigma_\lambda(1) = \frac{p'_\lambda}{1 - p_\lambda q_\lambda} \Sigma_\lambda(3) \qquad \text{and} \qquad \Sigma_\lambda(2) = \frac{p'_\lambda q_\lambda}{1 - p_\lambda q_\lambda} \Sigma_\lambda(3) \qquad (3.87)$$

where

$$p_\lambda = \frac{p}{\lambda+1}, \quad p'_\lambda = \frac{1-p}{\lambda+1}, \quad q_\lambda = \frac{q}{\lambda+1}, \quad q'_\lambda = \frac{1-q}{\lambda+1},$$

leaving verification of these formulae as an exercise for the reader (see Exercise 3.6.22).

Section 3.6.11 reveals that

$$\Sigma_{\max}(i) = \lim_{n\to\infty} (\Pi^*)^n \Sigma_\lambda(i).$$

We note that the first two rows of the matrix representing Π^* are

$$(0, p, p', 0, 0, \dots) \qquad \text{and} \qquad (q, 0, 0, q', 0, 0, \dots),$$

and that for $i \geq 3$, the ith row is e_{i+2}. Therefore,

$$\begin{aligned}
(\Pi^* f)(1) &= pf(2) + p' f(3), \\
(\Pi^* f)(2) &= qf(1) + q' f(4), \\
(\Pi^* f)(i) &= f(i+2), \qquad i \geq 3.
\end{aligned} \qquad (3.88)$$

Combining this with (3.86) we conclude that

$$\Sigma_{\max}(2k+1) = \lim_{n\to\infty} (\Pi^*)^n \Sigma_\lambda(2k+1) = \lim_{n\to\infty} \Sigma_\lambda(2k+1+2n) = 1$$

since the product in (3.86) converges. Analogously, $\Sigma_{\max}(2k+2) = \lim_{n\to\infty}\Sigma_\lambda(2k+2n) = 0$, for $k \geq 1$. Finally, using the first two equations in (3.88) we find that

$$(\Pi^*)^2 f(1) = pqf(1) + pq' f(4) + p' f(5)$$

and, substituting here $(\Pi^*)^n\Sigma$ for f, that

$$(\Pi^*)^{n+2}\Sigma_\lambda(1) = pq(\Pi^*)^n\Sigma_\lambda(1) + pq'(\Pi^*)^n\Sigma_\lambda(4) + p'(\Pi^*)^n\Sigma_\lambda(5).$$

Letting $n \to \infty$ results in

$$\Sigma_{\max}(1) = pq\Sigma_{\max}(1) + pq'\Sigma_{\max}(4) + p'\Sigma_{\max}(5)$$

and, since $\Sigma_{\max}(4) = 0$ and $\Sigma_{\max}(5) = 1$, this shows that $\Sigma_{\max}(1) = \frac{1-p}{1-pq}$. Using the second equation in (3.88) again, we furthermore obtain $\Sigma_{\max}(2) = \frac{(1-p)q}{1-pq}$.

Summarizing,

$$\Sigma_{\max}(i) = \begin{cases} \frac{1-p}{1-pq}, & i = 1, \\ \frac{(1-p)q}{1-pq}, & i = 2, \\ 1, & i = 2k+1, k \geq 1, \\ 0, & i = 2k+2, k \geq 1. \end{cases}$$

By the way, for $p = q = 0$ this solves Exercise 3.6.20, which, however, the reader was supposed to solve by himself. We note that $\Sigma_{\max}(i)$ is thus revealed to be the probability that the minimal process starting at i will explode. Had we known this, we could have calculated these quantities directly – explosion occurs iff the process ever touches one of the rungs of the odd ladder (except for $i = 1$). See Section 4.1 for more on this subject.

3.6.19 Exercise

Show that for $f \in \mathfrak{B}$, $(f_\lambda)^\diamond \leq f$.

3.6.20 Exercise

Show that in the example of two infinite ladders modified according to (3.70),

$$\Sigma_{\text{pass}} = \Sigma_{\text{e}}, \quad \Sigma_{\max} = \Sigma_{\text{o}},$$

where Σ_{e} and Σ_{o} are represented in (3.69).

3.6.21 Exercise

Let f be a passive functional. Show that $f(S(t)x) = f(x)$ for all $x \in l^1, t \geq 0$.

3.6.22 Exercise

Show (3.87).

3.7 The (Discrete, Exit) Boundary

As 3.6.17 (c) reveals, if Q is explosive, there is at least one nonpassive $f \in \mathfrak{B}$ (namely, $f = \Sigma$). In many cases, there are more nonpassive functionals in $\mathfrak{B}$, and the following theorem clarifies the way in which such functionals can be used to construct new sub-Markov generators from the generator of Kato's minimal semigroup. This theorem differs from that of Section 3.5.4 in that, whereas the latter theorem lumps all types of explosion together, here we choose one of them (which, however, may be 'compound,' that is, composed of several other, more basic ways) and command the process to continue after explosion from distribution u only after this particular type of explosion occurs. If there are many infinite ladders for the process to descend, only after reaching the bottom of a particular one of them (or a couple of them) the process is continued (see also Section 3.8.4).

3.7.1 Theorem

Given $f \in \mathfrak{B}$, let h given by

$$\boxed{h(x) = -f(Gx),} \qquad x \in \mathcal{D}(G), \tag{3.89}$$

be an (unbounded) functional with domain equal $\mathcal{D}(G)$.

(a) $h(x) = 0$ for $x \in \mathcal{D}(D)$.
(b) h is nonzero iff f is not passive.
(c) $h \geq 0$.
(d) For any $\lambda > 0$, the functionals h related to f and $(f_\lambda)^\diamond$ coincide.
(e) For any nonzero, nonnegative u such that $\Sigma u \leq 1$, the operator H defined by

$$Hx = Gx + h(x)u, \qquad x \in \mathcal{D}(H) = \mathcal{D}(G) \tag{3.90}$$

is a sub-Markov generator. (By (a), H coincides with G on $\mathcal{D}(D)$, and thus H extends $D + O$. By (b), H and G differ iff f is not passive.)

We do not claim that H is a Markov generator if u is a density, see Section 3.8.4 for an explanation.

Proof

(a) See (3.74).

(b) Fix $\lambda > 0$. A y belongs to $\mathcal{D}(G)$ if there is an $x \in l^1$ such that $y = R_\lambda x$, and then

$$h(y) = h(R_\lambda x) = -f(GR_\lambda x) = f(x) - f(\lambda R_\lambda x) = f_\lambda(x). \tag{3.91}$$

Hence, $h = 0$ iff $f_\lambda = 0$, that is, iff f is passive.

(c) For y of the form $y = R_\lambda x$ with nonnegative x, we have $h(y) \geq 0$ by (3.91). Furthermore, h is clearly continuous in the graph norm of $\mathcal{D}(G)$:

$$\|x\|_{\mathcal{D}(G)} = \|x\| + \|Gx\|.$$

Also, since G is a generator, $x = \lim_{\lambda\to\infty} \lambda R_\lambda x$, for any $x \in l^1$. It follows that for $x \in \mathcal{D}(G)$, $\lim_{\lambda\to\infty} G\lambda R_\lambda x = \lim_{\lambda\to\infty} \lambda R_\lambda Gx = Gx$, and so $\lim_{\lambda\to\infty} \|\lambda R_\lambda x - x\|_{\mathcal{D}(G)} = 0$. Thus we conclude that $h(x) = \lim_{\lambda\to\infty} h(\lambda R_\lambda x) \geq 0$ provided $x \in \mathcal{D}(G)$ is nonnegative.

(d) The functional $f - (f_\lambda)^\diamond$ is passive: by 3.6.13, $(f - (f_\lambda)^\diamond)_\lambda = f_\lambda - ((f_\lambda)^\diamond)_\lambda = 0$.

(e) Assume f is not passive, for otherwise H coincides with G and there is nothing to prove. The argument presented in 3.5.5, showing that, given $\lambda > 0$ and a nonnegative $y \in l^1$, there is precisely one $x \in \mathcal{D}(H)$ satisfying the resolvent equation $\lambda x - Hx = y$, can be repeated with f playing the role of Σ, provided we can show that $h(R_\lambda u) \neq 1$. But (3.91) makes it clear that $0 \leq h(R_\lambda u) \leq 1$, and that $h(R_\lambda u) = 1$ iff $f_\lambda(u) = 1$. Moreover, the latter condition implies $\Sigma_\lambda u = 1$, which in turn yields $\|\lambda R_\lambda u\| = \Sigma\lambda R_\lambda u = \Sigma u - \Sigma_\lambda u \leq 0$, which is impossible, because $R_\lambda, \lambda > 0$ is the resolvent of a generator. Therefore, (3.60) is the unique solution to the resolvent equation, and – since we have seen that $0 \leq h(R_\lambda u) < 1$ and (3.91) shows that $h(R_\lambda y) \geq 0$ – this solution is nonnegative.

It remains to show that $-\Sigma Hx \geq 0$ provided $x \geq 0$. By (c),

$$\begin{aligned} -\Sigma Hx &= -\Sigma Gx - h(x)\Sigma u \geq -\Sigma Gx - h(x) = -\Sigma Gx + f(Gx) \\ &= -(\Sigma - f)(Gx). \end{aligned}$$

The last expression is nonnegative since it is the value of the functional h corresponding to $\Sigma - f$. □

It is an important consequence of Theorem 3.7.1 that if we want to construct new sub-Markov semigroups from Kato's minimal semigroup by means of formulae (3.89) and (3.90), we must work with nonpassive functionals (which, to repeat, exist provided Q is explosive). Moreover, given a nonpassive f we may throw away its 'passive part' $f_{\rm pass} := f - (f_\lambda)^\diamond$ and work with $(f_\lambda)^\diamond$ instead. This reduces our possibilities to the members of the range of the canonical mapping, that is, to active f's.

On the other hand, as suggested by the example of two infinite ladders, our aim should be to prove a generation theorem for an operator H of the form

$$\boxed{Hx = Gx + \sum_{i=1}^{k} h_i(x)u_i,} \tag{3.92}$$

where k is a certain integer, h_i's come from different active functionals f^i, and u_i are also possibly different.[10] However, since we want our theorem to be as general as possible, whereas an active f may lump together a number of types of explosion, we need to work with functionals that are in a sense minimal and, at the same time, to have the sum as large as possible (but not too large): the best would be to have

$$\sum_{i=1}^{k} h_i(x) = -\Sigma Gx = -\Sigma_{\rm max} Gx, \qquad x \in \mathcal{D}(G). \tag{3.93}$$

A natural procedure thus would be to look first for an active $f^1 \le \Sigma_{\rm max}$ that is in a sense minimal, then a minimal $f^2 \le \Sigma_{\rm max} - f^1$ and so on, until equality in (3.93) is reached. Unfortunately, in general existence of such minimal functionals is neither granted nor obvious. Moreover, even if existence of minimal functionals is granted, we need to be sure that the procedure described above produces *distinct* minimal functionals. These considerations lead us to the following definitions.

3.7.2 Definition

An $f \in \mathfrak{B}$ is said to be **extremal** iff $f \cap (\Sigma - f) = 0$. For example, both $\Sigma_{\rm pass}$ and $\Sigma_{\rm max}$ are extremal. For, any $h \le \Sigma_{\rm pass}$ is necessarily passive, but, since $\Sigma_{\rm max}$ is active, there is no nontrivial passive $h \le \Sigma_{\rm max}$.

A nonzero extremal f is said to be **minimal** if there is no other extremal $g \le f$ other than $g = f$ and $g = 0$. An extremal f is said to be **continuous** if there is no minimal extremal $g \le f$.

These definitions are due to W. Feller. However, Feller used the adjective *continuous* in reference to a part of a boundary and not as a description

[10] Here, and in what follows, for extremals f I use superscripts rather than subscripts lest the ith f, denoted f_i, be mistaken with f_i defined in Section 3.6.7.

of a property of a functional. Since the phrase *continuous functional* has an established meaning in mathematics and in functional analysis in particular, in what follows instead of *continuous extremal/functional* I will say a *singular extremal/functional*. Thus, to repeat, an extremal functional is said to be **singular** if there is no minimal extremal smaller than this functional.

3.7.3 Assumption

W. Feller proved that there are at most denumerably many minimal functionals (see Lemma 12.1 in [41]). To make things more manageable (and following the example of Feller [41] and Chung [22]), in what follows we assume that

(1) there is only a finite number of minimal functionals $f \leq \Sigma_{\max}$, and
(2) there are no singular functionals $f \leq \Sigma_{\max}$.

We do not need to assume there are finitely many minimal $f \leq \Sigma_{\text{pass}}$.

3.7.4 Definition

The set $\mathfrak{B}$ of minimal functionals $f \leq \Sigma_{\max}$ (which is finite in our case) is called the **discrete (exit) boundary** for an intensity matrix Q or for a related Markov chain. Extremal functionals $f \leq \Sigma_{\text{pass}}$ form the **passive boundary**.

This definition slightly differs from Feller's. In the Feller's definition, all minimal functionals are members of $\mathfrak{B}$. As explained above, passive functionals introduce no additional possibilities to continue a minimal process after explosion, and I decided to exclude them from the boundary.

3.7.5 *Possibility of decomposition (3.93)*

Under assumptions stated above, there are finitely many minimal functionals summing to $\Sigma_{\max}$: for some $\mathfrak{k}$,

$$\Sigma_{\max} = \sum_{i=1}^{\mathfrak{k}} f^i.$$

(This implies that for the corresponding h_i's equality (3.93) holds.)

For the proof of this result we need the following lemmas.

3.7.6 Lemma

Suppose $g_1, g_2, f \in \mathfrak{B}$. Then

$$f \cap (g_1 + g_2) \leq f \cap g_1 + f \cap g_2.$$

Proof Since

$$f \cap (g_1 + g_2) - g_1 \leq g_1 + g_2 - g_1 = g_2, \text{ and}$$
$$f \cap (g_1 + g_2) - g_1 \leq f - g_1 \leq f,$$

we have $f \cap (g_1 + g_2) - g_1 \leq f \cap g_2$. Combining this with $f \cap (g_1 + g_2) - f \leq 0 \leq f \cap g_2$, we obtain

$$f \cap (g_1 + g_2) - f \cap g_2 \leq f \cap g_1,$$

completing the proof. □

3.7.7 Lemma

(a) If $f \leq \Sigma_{\max}$ is extremal, then so is $g = \Sigma_{\max} - f$.
(b) If f and g are extremal and such that $f + g \leq \Sigma$, then their sum is also extremal.

Proof
(a) By 3.7.6,

$$\begin{aligned} g \cap (\Sigma - g) &= (\Sigma_{\max} - f) \cap (\Sigma_{\text{pass}} + f) \\ &\leq (\Sigma_{\max} - f) \cap \Sigma_{\text{pass}} + (\Sigma_{\max} - f) \cap f \\ &\leq \Sigma_{\max} \cap \Sigma_{\text{pass}} + (\Sigma - f) \cap f = 0. \end{aligned}$$

Since $g \cap (\Sigma - g) \geq 0$, this completes the proof.
(b) By 3.7.6,

$$\begin{aligned} (\Sigma - (f + g)) \cap (f + g) &\leq (\Sigma - (f + g)) \cap f + (\Sigma - (f + g)) \cap g \\ &\leq (\Sigma - f) \cap f + (\Sigma - g) \cap g = 0, \end{aligned}$$

and thus the proof is completed as in (a). □

Proof of *3.7.5* Since $\Sigma_{\max}$ is extremal and there are no singular functionals, there is a minimal $f^1 \leq \Sigma_{\max}$. If $f^1 = \Sigma_{\max}$, there is nothing to prove, hence we assume $f^1 \neq \Sigma_{\max}$.[11] Then $\Sigma_{\max} - f^1$ is nonzero and extremal

[11] The lazy part of me, which is the dominant part, of course, whispers quietly, 'This sentence is a nonsense. This should read, If $f^1 = \Sigma_{\max}$, there is nothing to prove, hence we *do* assume $f^1 = \Sigma_{\max}$.'

by Lemma 3.7.7 (a). Arguing as above we conclude that there is a minimal $f^2 \le \Sigma_{\max} - f^1$. This f^2 is different from f^1: the relation

$$f^1 \cap f^2 \le f^1 \cap (\Sigma_{\max} - f^1) \le f^1 \cap (\Sigma - f^1) = 0 \tag{3.94}$$

rules out the possibility of $f^1 = f^2$. Again, if $f^2 = \Sigma_{\max} - f^1$, we are done. Otherwise, $f^1 + f^2$ is extremal by Lemma 3.7.7 (b), and thus $\Sigma_{\max} - f^1 - f^2$ is nontrivial and extremal, by Lemma 3.7.7 (a). Estimating as in (3.94) we check that a minimal $f^3 \le \Sigma_{\max} - f^1 - f^2$ is distinct from f^1 and f^2. Since, by assumption, there are finitely many minimal functionals, this procedure cannot continue indefinitely: there is a k such that for the kth minimal functional we have $f^k = \Sigma_{\max} - (f^1 + \cdots + f^{k-1})$. □

3.7.8 Remark

There are no other nonzero minimal functionals $f \le \Sigma_{\max}$ but those featuring in the sum $\sum_{i=1}^{k} f^i$. For, if f and g are minimal functionals then necessarily $f \cap g = 0$ or $f \cap g = f = g$ (see Exercise 3.8.5 in the next section). On the other hand, if a nonzero $f \le \Sigma_{\max}$ is minimal, then by 3.7.6 and induction argument

$$f = f \cap \Sigma_{\max} = f \cap \sum_{i=1}^{k} f^i \le \sum_{i=1}^{k} (f \cap f^i)$$

and the last sum cannot be zero, since this would imply $f = 0$. Therefore, precisely one term in the sum is nonzero, and this means that f is one of the f^i's.

We are now ready to state our main theorem in the second part of this chapter. In this theorem there are precisely k ways explosion may come about and to each of them a possibly different distribution after explosion is assigned.

3.7.9 Theorem

Let $f^1, \ldots, f^k$ be the minimal functionals constructed in the proof of 3.7.5 so that f^i's are distinct and $\sum_{i=1}^{k} f^i = \Sigma_{\max}$, and let h_i be the related functionals on $\mathcal{D}(G)$:

$$\boxed{h_i(x) = -f^i(Gx),} \qquad x \in \mathcal{D}(G). \tag{3.95}$$

Also, assume u_i, $i = 1, \ldots, k$ are nonnegative elements of l^1 satisfying $\Sigma u_i \le 1$. Then, formula (3.92) defines a sub-Markov generator such that

(a) $Hx = Gx$ for $x \in \mathcal{D}(D)$,

(b) $H \neq G$ provided at least one u_i is nonzero,
(c) H is a Markov operator if all u_i's are densities.

We still need a lemma.

3.7.10 Lemma

Suppose $\gamma_{i,j}$, $i, j = 1, \ldots, k$ are nonnegative numbers such that

$$\sum_{j=1}^{k} \gamma_{i,j} < 1 \qquad \text{for all } i = 1, \ldots, k.$$

Then, for any $\eta_1, \ldots, \eta_k \in \mathbb{R}$, the system

$$\xi_j - \sum_{i=1}^{k} \gamma_{i,j}\xi_i = \eta_j \qquad \text{for all } j = 1, \ldots, k \tag{3.96}$$

has precisely one solution $x = (\xi_1, \ldots, \xi_k) \in \mathbb{R}^k$. Moreover, $x \geq 0$ provided $y = (\eta_1, \ldots, \eta_k) \geq 0$.

Proof Let $\mathbb{R}^k$ be equipped with the norm $\|x\| = \sum_{i=1}^{k} |\xi_i|$ and let $M \in \mathcal{L}(\mathbb{R}^k)$ be the operator given by

$$M(\xi_i)_{i=1,\ldots,k} = \left(\sum_{i=1}^{k} \gamma_{i,j}\xi_i \right)_{j=1,\ldots,k}.$$

(This is multiplying the matrix $M = (\gamma_{i,j})_{i,j=1,\ldots k}$ by a row-vector from the left.) Then

$$\begin{aligned} \|M\| &\leq \sup_{\|x\|=1} \sum_{j=1}^{k} \sum_{i=1}^{k} \gamma_{i,j} |\xi_i| = \sup_{\|x\|=1} \sum_{i=1}^{k} |\xi_i| \sum_{j=1}^{k} \gamma_{i,j} \\ &\leq \gamma := \max_{i=1,\ldots,k} . \end{aligned}$$

Since, by assumption, $\gamma < 1$, the series $\sum_{n=0}^{\infty} M^n$ converges in the operator norm. Moreover, equation (3.96) may be written as $x - Mx = y$. Therefore, its solution is unique and given by $x = \sum_{n=0}^{\infty} M^n y$. The rest is clear since $M \geq 0$. □

Proof of *3.7.9* Interestingly, only minor modifications of the proof of 3.7.1 are needed. We are apparently well prepared by now, and we have Lemma 3.7.10 at our disposal.

Without loss of generality, in what follows we assume that all $u_i \neq 0$. If this is not the case, several terms in (3.92) vanish, and the argument must be modified by solving (3.97) with smaller $\mathfrak{k}$.

Let us check that H is a sub-Markov generator. In solving the resolvent equation for H,

$$\lambda x - Gx - \sum_{i=1}^{\mathfrak{k}} h_i(x)u_i = y, \tag{3.97}$$

we argue first that $h_i(x)$'s are determined by λ, y and u_i's, via a system of linear equations (and thus may be calculated without prior knowledge of x). To this end, we apply R_λ and then h_j to both sides of (3.97), obtaining

$$h_j(x) - \sum_{i=1}^{\mathfrak{k}} h_j(R_\lambda u_i)h_i(x) = h_j(R_\lambda y), \qquad \text{for all } j = 1, \dots, \mathfrak{k}. \tag{3.98}$$

In this system, $h_j(R_\lambda u_i)$'s and $h_j(R_\lambda y)$'s are treated as known and $h_j(x)$'s are unknowns. Since (see (3.91))

$$\sum_{j=1}^{\mathfrak{k}} h_j(R_\lambda u_i) = \sum_{j=1}^{\mathfrak{k}} f_\lambda^j(u_i) = (\Sigma_{\max})_\lambda(u_i) = \Sigma_\lambda(u_i) \in [0, 1)$$

(we cannot have $\Sigma_\lambda(u_i) = 1$ for this would imply $\Sigma \lambda R_\lambda u_i = 0$, i.e., $u_i = 0$), conditions of Lemma 3.7.10 are satisfied and there is precisely one vector

$$(\xi_1, \dots, \xi_{\mathfrak{k}}) = (h_1(x), \dots, h_{\mathfrak{k}}(x)) \in \mathbb{R}^{\mathfrak{k}}$$

solving (3.98). To repeat, this vector depends merely on λ, u_j's and y. Hence, a solution of (3.97) must be of the form

$$x = R_\lambda y + \sum_{j=1}^{\mathfrak{k}} \xi_j R_\lambda u_j. \tag{3.99}$$

To check that this x is a true solution to (3.97), we apply h_j to both sides of the definition of x. Since replacing $h_j(x)$'s by ξ_j's in (3.98) gives $\mathfrak{k}$ identities, this results in the expected equality

$$h_j(x) = h_j(R_\lambda y) + \sum_{i=1}^{\mathfrak{k}} \xi_i h_j(R_\lambda u_i) = \xi_j,$$

and thus $\lambda x - Gx = y + \sum_{i=1}^{\mathfrak{k}} h_i(x)u_i$, that is, (3.97) holds. If $y \geq 0$, then $h_j(R_\lambda y) = f_\lambda^j(y) \geq 0$, and Lemma 3.7.10 guarantees that then $\xi_j = h_j(x)$, $j = 1, \dots, \mathfrak{k}$ are nonnegative also, and so is x given in (3.99).

Finally, let $x \in \mathcal{D}(G)$ be nonnegative. Since in such a case all $h_i(x)$'s are nonnegative,

$$\Sigma Hx \le \Sigma Gx + \sum_{i=1}^{\mathfrak{k}} h_i(x) = \Sigma Gx - \sum_{i=1}^{\mathfrak{k}} f^i(Gx)$$
$$= \Sigma Gx - \Sigma_{\max} Gx = \Sigma_{\text{pass}} Gx = 0.$$

The first inequality here may be replaced by equality iff $\Sigma u_i = 1$ for all $i = 1, \dots, \mathfrak{k}$. Hence, in the latter case, H is a Markov generator; this proves (c). Since properties (a) and (b) are clear, we are done. □

3.8 More on Extremal and Minimal Functionals

Since our main goal in the previous section was Theorem 3.7.9, immediately after the definition of extremal and minimal elements we proceeded to Assumption 3.7.3 leading to the possibility of decomposition (3.93). Now, that the main theorem is proved, let us pause for a moment and characterize the extremal and minimal elements. This characterization will reveal in particular that Theorem 3.7.1 is in fact contained in Theorem 3.7.9.

3.8.1 *A characterization of extremal functionals*

The following are equivalent.

(1) f is extremal.
(2) If $f \ge \alpha g$ for a $g \in \mathfrak{B}$ and some $\alpha \in (0, 1)$, then $f \ge g$.
(3) If

$$f = \alpha g_1 + (1 - \alpha) g_2 \tag{3.100}$$

for $g_1, g_2 \in \mathfrak{B}$ and $\alpha \in (0, 1)$, then $f = g_1 = g_2$.

Proof

(1) $\Longrightarrow$ (2) If f is extremal and $f \ge \alpha g$, then (by Lemma 3.7.6)

$$g = g \cap (f + \Sigma - f) \le g \cap f + g \cap (\Sigma - f)$$
$$\le g \cap f + (\alpha^{-1} f) \cap (\Sigma - f).$$

Since $(\alpha^{-1} f) \cap (\Sigma - f) \le (\alpha^{-1} f) \cap (\alpha^{-1}(\Sigma - f)) = \alpha^{-1}(f \cap (\Sigma - f)) = 0$, we see that $g \le g \cap f$, that is, $g \le f$.

(2) $\Longrightarrow$ (3) If (3.100) holds, $\alpha g_1 \le f$. By assumption it follows that $g_1 \le f$. Thus, using (3.100) again, we see that $(1 - \alpha) g_2 = f - \alpha g_1 \ge (1 - \alpha) g_1$,

that is, $g_2 \geq g_1$. Since the roles of g_1 and g_2 in (3.100) are symmetric, $g_1 \geq g_2$, implying $g_1 = g_2$. Hence, $f = g_1 = g_2$.

(3) $\Longrightarrow$ (1) Suppose $f \cap (\Sigma - f) = g$ is nonzero. Then $0 \leq f + g \leq f + \Sigma - f = \Sigma$ and $0 \leq f - g \leq f$. Hence, $g_1 := f + g$ and $g_2 := f - g$ are elements of $\mathfrak{B}$ and $f = \frac{1}{2}g_1 + \frac{1}{2}g_2$, even though $g_1 \neq g_2$. □

3.8.2 *A characterization of minimal elements*

(a) Suppose f is minimal and $g \in \mathfrak{B} \setminus \{0, f\}$ is $\leq f$. Then, $g = \alpha f$ for some $\alpha \in (0, 1)$.

(b) Conversely, if all $g \leq f$ other than 0 and f are of the form $g = \alpha f$ for some $\alpha \in (0, 1)$, then a nonzero extremal f is minimal.

Proof The argument for (a) involves techniques that are developed in Section 4.1. Therefore, I postpone the proof till the end of Section 4.1.

(b) For a nonzero f and $\alpha \in (0, 1)$, αf is not extremal:

$$(\alpha f) \cap (\Sigma - \alpha f) \geq (\alpha f) \cap ((1 - \alpha) f) \geq \min(\alpha, 1 - \alpha) f \neq 0.$$

Thus, the assumption implies that all $g \leq f$ other than $g = 0$ and $g = f$ are nonextremal. □

3.8.3 *An $f \in \mathfrak{B}$ is active iff it is of the form $f = \sum_{i=1}^{\Bbbk} \alpha_i f^i$ where $\alpha_i \in [0, 1]$*

For, each f of this form is clearly $\leq \Sigma_{\max}$, and thus is active. Conversely, suppose $f \leq \Sigma_{\max}$, and take an $i = 1, \ldots, \Bbbk$. Since $f \cap f^i \leq f^i$ and f^i is minimal, there is an $\alpha_i \in [0, 1]$ such that $f \cap f^i = \alpha_i f^i$. Hence, by a repeated application of Lemma 3.7.6,

$$f = f \cap \Sigma_{\max} = f \cap \sum_{i=1}^{\Bbbk} f^i \leq \sum_{i=1}^{\Bbbk} (f \cap f^i) = \sum_{i=1}^{\Bbbk} \alpha_i f^i.$$

We will argue that, since $(f \cap f^i) \cap (f \cap f^j) \leq f^i \cap f^j = 0$ (for $i \neq j$), the inequality here can be replaced by equality. To this end, by induction argument, it suffices to show that $g_1 \cap g_2 = 0$ implies that the inequality in Lemma 3.7.6 may be replaced by equality: $f \cap (g_1 + g_2) = f \cap g_1 + f \cap g_2$. Hence, we need to show that

$$f \cap (g_1 + g_2) \geq f \cap g_1 + f \cap g_2, \tag{3.101}$$

provided $g_1 \cap g_2 = 0$. By (3.71), the right-hand side equals $(f \cap g_1) \cup (f \cap g_2) + (f \cap g_1) \cap (f \cap g_2)$ and the second term here is clearly 0. Also, since

$$f \cap (g_1 \cup g_2) \geq f \cap g_1 \qquad \text{and} \qquad f \cap (g_1 \cup g_2) \geq f \cap g_2,$$

and $(f \cap g_1) \cup (f \cap g_2)$ is the smallest of elements of $\mathfrak{B}$ that are larger than both $f \cap g_1$ and $f \cap g_2$, the right-hand side of (3.101) is $\leq f \cap (g_1 \cup g_2)$. Using (3.71) again, we see that $g_1 \cup g_2 = g_1 + g_2$, completing the proof.

3.8.4 *Theorem 3.7.1 is a special case of Theorem 3.7.9*

By 3.8.3, the functional f of Theorem 3.7.1 may be represented as $f = \sum_{i=1}^{k} \alpha_i f^i$. (Recall that, by Theorem 3.7.1 (d), we may restrict ourselves to the case where f is active.) Given u of Theorem 3.7.1, let u_i's in (3.92) be defined by $u_i = \alpha_i u$. Then

$$\sum_{i=1}^{k} h_i(x) u_i = \sum_{i=1}^{k} \alpha_i h_i(x) u = -\sum_{i=1}^{k} \alpha_i f^i(Gx) u = h(x) u,$$

proving that (3.92) reduces to (3.90).

In particular, even if u is a density the process governed by H of Theorem 3.7.9 may be sub-Markov: if the ith type of explosion occurs the process is continued (according to distribution u) with probability $\alpha_i \leq 1$.

3.8.5 Exercise

Use the characterization of 3.8.2 to show that if f and g are minimal functionals then necessarily $f \cap g = 0$ or $f \cap g = f = g$.

3.9 Notes

With the exception of the proof of 3.1.4, Section 3.1 follows Kato's argument presented in [51,54] closely. Feller [42] gives an apparently simpler reasoning, by avoiding the use of the Hille–Yosida Theorem and using instead the classical Widder's Theorem on inverting the Laplace transform. Kato's argument, however, has the advantage of being applicable to more general situations: Voigt [88] uses it to extend Kato's Theorem to L^1 spaces, whereas Banasiak and Lachowicz [7] use it to extend the theorem to Kantorovič–Banach spaces (see also [6, 16]); yet further generalizations may be found in [84] and [93]. Notably, Arlotti & Banasiak [6] (see also [5]) also have a nice chapter on birth and death processes as seen from the perspective of Kato's Theorem.

A thorough probabilistic description of what may happen with a chain after explosion can be found in Chung's little book [22]. The equally intriguing original papers by Feller [41] and [42] possess an additional functional-analytic flavor, and we borrowed a handful of arguments from those papers. However, whereas Feller focuses his attention on the resolvents of the nonminimal processes, we try to work with generators.

The example of a birth and death chain for which all inclusions in (3.42) are proper is due to J. Banasiak (see [6] Proposition 7.22, p. 195). Our argument, however, differs from the one given in [6]: Whereas in [6] the fact that these inclusions are proper is deduced from certain general criteria developed earlier in that book, we focus on this particular example alone, and are able to characterize the domains of $D + O$, $\overline{D + O}$, G and $\mathfrak{Q}$ quite explicitly.

Functionals h of the type introduced in Section 3.5.4 are used to measure mass loss not only in probabilistic context. For instance, in the framework of fragmentation processes with growth and decay, they were originally studied in [8] and [4]. Intuitions developed in these papers led later to a quite satisfying abstract theory in [6] and [68] (see also [9]). In fact, in the general setting of the latter works, the role of h is played by the difference of two functionals, say, $h^\sharp$ and $h^\flat$, where

$$\begin{aligned} h^\flat(x) &= -\Sigma G x, \\ h^\sharp_\lambda(x) &= \sum_{n=0}^{\infty} h^\flat((\lambda - D)^{-1} B^n_\lambda(\lambda - G)x) \end{aligned} \tag{3.102}$$

and the definition of $h^\sharp_\lambda$ is proved to be independent of $\lambda > 0$. The case of Kolmogorov matrices, considered in this book, is simpler, because (3.59) implies that all the terms defining $h^\sharp$ in (3.102) are zero, resulting in $h^\sharp = 0$, and so $h = h^\flat$. The theorem of Section 3.5.5 is due to Chin Pin Wong [92,93] (see the proof of Theorem 2.3.4 in [92] or Theorem 2.3 in [93], cf. Theorem 4.10.28 in [9]), and this result is true in the framework much more general than that provided by Kolmogorov matrices. On the other hand, in the latter framework, results presented by W. Feller in Section 8 of [42] are more general than Chin Pin Wong's (see in particular equation (8.1) and the equation on the top of page 547 in [42], and compare them with our formula (3.60)), and even more so are those of his Section 13. Our Theorem 3.7.9 is a version of these latter results.

The title of Section 3.6 alludes, of course, to the fifth studio album by the English progressive rock band Yes, with 𝔅ill 𝔅ruford on drums (later in King Crimson). The second part of the Close to the Edge suite contained in that album is Total Mass Retain.

4

Boundary Theory Continued

With formula (3.92) and Theorem 3.7.9 our exposition of boundary theory for Markov chains has reached its climax: we know that extremal functionals provide a way for a Markov chain to start afresh after explosion in various ways, and the formula just mentioned shows how such information of postexplosion process is hidden in its generator. There is, however, still much we can learn. In the first section of this chapter, Section 4.1, extremal functionals, somewhat abstract beings, are revealed to be sojourn sets for the minimal Markov chain involved. Hence, in agreement with our intuition, each functional is simply a way to reach infinity in finite time. Next, in Section 4.2, these extremal functionals, that is, sojourn sets, are finally seen as proper elements of the state-space. We will argue that from the perspective of such an extended state-space, phenomena discovered before are even more transparent.

And then, when everything seems to be finally clear and properly explained, in Section 4.3, we come to the example of a Markov chain that shatters our confidence into pieces: that is, P. Lévy's flash. In this example, the information on postexplosion process is encrypted in the domain of the generator; this is in contrast to formula (3.92), where this information is expressed as extra terms accompanying the generator of the minimal chain. But, as is the case with many apparent paradoxes, we will finally learn from this chain something new and precious:[1] besides exit boundary for a Markov chain, there is also its entrance boundary. We will also see, by examining examples, how the information about postexplosion processes involving entrance boundary is reflected in the domain of the generator. We will learn that, if both entrance and exit boundary is present, the generator of a chain which after explosion starts afresh through a so-called entrance law or from a randomly distributed position

[1] As quoted in [1], p. 289, Niels Bohr supposedly said: 'How wonderful that we have met with a paradox. Now we have some hope of making progress.'

in the state-space may differ from the generator of the minimal process both by additional terms and in the domain.

4.1 Sojourn Sets and Sojourn Solutions

Our main goal in this section is, guided by Example 3.6.18, to identify extremal functionals as **sojourn solutions**, that is, vectors of probabilities of sojourning in certain sets.

In analyzing sojourn solutions, it is convenient to refer to the so-called **jump chain**. To explain, let $\{X(t), t \geq 0\}$ be a Markov chain with intensity matrix Q. Think of the value, say, Y_1, of the process at the moment of its first jump (this moment is random and different for different elementary events); then think of the value, say, Y_n, of the process at the time of its nth jump. A look at (3.67) shows that Π gathers all transition probabilities of the discrete Markov chain $Y_1, Y_2, \ldots$:

$$\mathbb{P}[Y_{n+1} = j | Y_n = i] = \pi_{i,j} := (1 - \delta_{i,j})\frac{q_{i,j}}{q_i}; \qquad n \geq 1, i, j \in \mathbb{N}.$$

It is this chain that is known as the jump chain.

4.1.1 *Notational convention*

Let $A \subset \mathbb{N}$ be a set. The spaces $l^1(A)$ and $l^\infty(A)$ may be naturally identified with subspaces of $l^1(\mathbb{N})$ and $l^\infty(\mathbb{N})$: for $(\xi)_{i\in A} \in l^1(A)$ and $f \in l^\infty(A)$ it suffices to agree that $\xi_i = 0$ and $f(i) = 0$ for $i \notin A$.

Furthermore, if $\widetilde{\Pi} = \left(\widetilde{\pi}_{i,j}\right)_{i,j\in\mathbb{N}}$ is a sub-Markov operator in $l^1(\mathbb{N})$ and $B \supset A$ is another subset of $\mathbb{N}$, then the matrix composed of

$$\widetilde{\pi}_{i,j}(A) = \begin{cases} \widetilde{\pi}_{i,j}, & i, j \in A, \\ 0, & \text{otherwise} \end{cases} \tag{4.1}$$

induces a sub-Markov operator, say, $\widetilde{\Pi}_A$, in the space $l^1(B)$, by

$$\widetilde{\Pi}_A\,(\xi_i)_{i\in B} = \left(\sum_{j\in B} \widetilde{\pi}_{j,i}(A)\xi_j\right)_{i\in B} \in l^1(B).$$

Figuratively speaking, $\widetilde{\Pi}_A$ represents a chain in which only jumps between elements of A are possible: the remaining elements of B are 'frozen' and jumps from A to $B \setminus A$ are also disallowed.

With such a notation, it might be unclear whether for $x \in l^1(B)$, $\widetilde{\Pi}_A x$ is a member of $l^1(B)$ or of $l^1(\mathbb{N})$, but the elements in question may be identified (as above).

4.1.2 *Sojourn sets*

Let A be an arbitrary subset of $\mathbb{N}$, Π be the jump chain matrix (3.67), and $\Sigma_A \in l^\infty(A)$ be the functional (playing the role of Σ in $l^1(A)$) given by

$$\Sigma_A\ (\xi_i)_{i\in A} = \sum_{i\in A} \xi_i.$$

For the operator Π_A acting in $l^1(A)$, defined in (4.1), we have $\Pi_A^*\Sigma_A \le \Sigma_A$. Therefore, we may argue as in, for example, Section 3.6.2 to see that the limit

$$\sigma_A(x) := \lim_{n\to\infty} \Sigma_A(\Pi_A^n x)$$

exists for all $x \in l^1(A)$, and defines a member of $\mathfrak{B}_A$, the latter set being composed of functionals $f \in l^\infty(A)$ such that $0 \le f \le \Sigma_A$ and $\Pi_A^* f = f$.

Since $\Pi_A^n e_i$ is the ith row of the matrix Π_A^n, it is clear that $\Sigma_A(\Pi_A^n e_i)$ is the probability of the event that the jump chain starting at $i \in A$ will be in A after n jumps without ever leaving A in the meantime. Since such events form a decreasing sequence, $\sigma_A(i) = \lim_{n\to\infty} \Sigma_A(\Pi_A^n e_i)$ is the probability that the chain starting at $i \in A$ will never leave this set: A is termed a **sojourn set** if $\sigma_A \ne 0$.

4.1.3 *Construction of sojourn solutions*

Let $A \subset \mathbb{N}$ be a set. Since $\Pi x \ge \Pi_A x$ for nonnegative $x \in l^1(\mathbb{N})$, extending σ_A as in 4.1.1, we have

$$\Pi^*\sigma_A(x) = \sigma_A(\Pi x) \ge \sigma_A(\Pi_A x) = \sigma_A(x).$$

Arguing as in Section 3.6.2 again, we conclude that

$$s_A(x) = \lim_{n\to\infty} \sigma_A(\Pi^n x)$$

exists for all $x \in l^1(\mathbb{N})$ and defines a member of $\mathfrak{B}$.

To interpret s_A, we note first that if $x \in l^1(\mathbb{N})$ is a distribution, then $\sigma_A(x)$ is the probability that the jump chain starting at a randomly selected point in $\mathbb{N}$ according to the distribution x starts in fact in A and never leaves this set afterward. Therefore, $\sigma_A(\Pi^n e_i)$ is the probability of the event that the jump chain starting at i lands in A after n steps and from that time on stays in A for ever. Since such events form an increasing sequence,

$$s_A(i) = \lim_{n\to\infty} \sigma_A(\Pi^n e_i)$$

is the probability that the jump chain starting at i will eventually hit A to never leave A afterward.

The functional s_A is termed **sojourn solution**. Clearly,

$$s_A \geq \sigma_A. \tag{4.2}$$

To recall, our main goal in this section is to show that all extremal functionals are sojourn solutions and vice versa. We start with three lemmas. Sections 4.1.8 and 4.1.9, following them, are stepping stones for our main results contained in 4.1.10 and 4.1.11. Section 4.1.11 is also of its own importance to be revealed in Chapter 5.

4.1.4 Lemma

Suppose $A, B \subset \mathbb{N}$ are disjoint sets. Then

$$s_A + s_B = s_A \cup s_B \leq s_{A\cup B}.$$

Proof Both s_A and s_B are smaller than $s_{A\cup B}$. Thus, by definition of $\cup$, $s_A \cup s_B \leq s_{A\cup B}$. Similarly, $s_A \cup s_B \leq s_A + s_B$, whether A and B are disjoint or not: it is the reverse inequality that we really need to prove.

Functionals σ_A and σ_B are defined as elements of $l^\infty(A)$ and $l^\infty(B)$, respectively, and then extended to elements of $l^\infty(\mathbb{N})$, as in Section 4.1.1, by supplying zeros on appropriate coordinates. Since $A \cap B = \emptyset$,

$$(\sigma_A + \sigma_B)(i) = \begin{cases} \sigma_A(i), & i \in A, \\ \sigma_B(i), & i \in B, \\ 0, & \text{otherwise}, \end{cases} \leq \max[s_A(i), s_B(i)] \leq (s_A \cup s_B)(i),$$

with the last inequality being a special case of (3.73). Therefore, for any nonnegative $x \in l^1(\mathbb{N})$,

$$(\sigma_A + \sigma_B)(\Pi^n x) \leq (s_A \cup s_B)(\Pi^n x),$$

and letting $n \to \infty$, yields $s_A + s_B \leq s_A \cup s_B$, as desired. □

4.1.5 *An auxiliary inequality*

Suppose A is a nonempty subset of $\mathbb{N}$, $\widetilde{\Pi} = \left(\widetilde{\pi}_{i,j}\right)_{i,j\in A}$ is a sub-Markov operator in $l^1(A)$, a nonnegative, non-zero $f \in l^\infty(A)$ satisfies $f \geq \widetilde{\Pi}^* f$, and B is a nonempty subset of A. Then, for any $i \in A$,

$$f(i) \geq \sum_{j\in B} \widetilde{\pi}_{i,j} f(j) + \sum_{k\in A\setminus B} \widetilde{\pi}_{i,k} f(k). \tag{4.3}$$

If i belongs to B, and $\widetilde{\pi}_{i,j}(B)$ is defined as in (4.1), this may be written as

$$f(i) \geq \sum_{j\in B} \widetilde{\pi}_{i,j}(B) f(j) + \sum_{j\in B} \widetilde{\pi}^0_{i,j}(B) \sum_{k\in A\setminus B} \widetilde{\pi}_{j,k} f(k), \tag{4.4}$$

because $\widetilde{\pi}^0_{i,j}(B) = 1$ only if $i = j$, and is zero otherwise. I claim that this formula may be generalized as follows: for any $n \geq 1$ and $i \in B$,

$$f(i) \geq \sum_{j\in B} \widetilde{\pi}^n_{i,j}(B) f(j) + \sum_{\ell=0}^{n-1} \sum_{j\in B} \widetilde{\pi}^\ell_{i,j}(B) \sum_{k\in A\setminus B} \widetilde{\pi}_{j,k} f(k), \tag{4.5}$$

where $\left(\widetilde{\pi}^\ell_{i,j}(B)\right)_{i,j\in B}$ is the matrix representing $(\widetilde{\Pi}_B)^\ell$. Since for $n = 1$, (4.5) reduces to (4.4), for an induction argument it suffices to show the induction step. If (4.5) is true for a certain n, and S_n denotes the second sum on the right-hand side of this inequality, then replacing $f(j)$ in (4.5) according to (4.3),

$$f(i) \geq \sum_{j\in B} \widetilde{\pi}^n_{i,j}(B) \sum_{k\in B} \widetilde{\pi}_{j,k} f(k) + \sum_{j\in B} \widetilde{\pi}^n_{i,j}(B) \sum_{k\in A\setminus B} \widetilde{\pi}_{j,k} f(k) + S_n.$$

Here, the first sum on the right equals $\sum_{j\in B} \widetilde{\pi}^{n+1}_{i,j}(B) f(j)$ and the remaining two add up to S_{n+1}. This shows that (4.5) with n replaced by $n+1$ is true, thus completing the induction step, and the entire argument. It is also clear that if $\widetilde{\Pi}^* f = f$, inequality in (4.5) may be replaced by equality.

Let now, for a set $C \subset B (\subset A)$,

$$\widetilde{\pi}^n_{i,C}(B) = \sum_{j\in C} \widetilde{\pi}^n_{i,j}(B)$$

be the probability that the jump chain starting at i is in the set C after n steps and in the meantime has never left B. Instead of $\widetilde{\pi}^n_{i,C}(A)$ we simply write $\widetilde{\pi}^n_{i,C}$.

With this notation, applying (4.5) to $f = \Sigma_A$ (we have $\Pi^*\Sigma_A \leq \Sigma_A$, so that using Σ_A in place of f is legitimate) yields

$$1 \geq \widetilde{\pi}^n_{i,B}(B) + \sum_{\ell=0}^{n-1} \sum_{j\in B} \widetilde{\pi}^\ell_{i,j}(B) \widetilde{\pi}_{j,A\setminus B}, \qquad i \in A. \tag{4.6}$$

We note that the second term here is the sum of probabilities of disjoint events $E_0, \ldots, E_{n-1}$, where E_ℓ holds if the chain related to $\widetilde{\Pi}$ starting at i stays in B for (precisely) ℓ steps and then, in the $(\ell+1)$st step, leaves B; the summation index, j, is the last position of the chain in B, before jumping out of B.

4.1.6 *The crucial lemma*

Let A and $\widetilde{\Pi}$ be as in the previous section. Suppose a nonzero $f \geq 0$ satisfies $\widetilde{\Pi}^* f = f$. For $\delta \in (0, \|f\|)$, let

$$F(\delta) = \{i \in A; f(i) > \delta\},$$

and let $\delta' \in (\delta, \|f\|)$. Then

$$\widetilde{\pi}^n_{i,F(\delta)}(F(\delta)) > \frac{\delta' - \delta}{\|f\| - \delta} \tag{4.7}$$

for all $n \geq 1$ and i such that $f(i) > \delta'$.

Proof Look at equality (!) (4.5) with $B = F(\delta)$. The first term on the right-hand side does not exceed

$$\|f\| \sum_{j \in F(\delta)} \widetilde{\pi}^n_{i,j}(F(\delta)) = \|f\| \widetilde{\pi}^n_{i,F(\delta)}(F(\delta)).$$

Since the k's in the second term are not members of $F(\delta)$, the second term does not exceed

$$\delta \sum_{\ell=0}^{n-1} \sum_{j \in F(\delta)} \widetilde{\pi}^\ell_{i,j}(F(\delta)) \widetilde{\pi}_{j, A \setminus F(\delta)},$$

which, by (4.6), does not exceed $\delta(1 - \widetilde{\pi}^n_{i,F(\delta)}(F(\delta)))$. Thus, if $f(i) > \delta'$, (4.5) renders

$$\delta' < \|f\| \widetilde{\pi}^n_{i,F(\delta)}(F(\delta)) + \delta(1 - \widetilde{\pi}^n_{i,F(\delta)}(F(\delta))).$$

This inequality is equivalent to (4.7). □

4.1.7 *$F(\delta)$ is a sojourn set; nonzero sojourn solutions have norm* 1

Let Π be the jump chain matrix. Suppose that for a set A and $\widetilde{\Pi} = \Pi_A$ there is a nonzero $f \in \mathfrak{B}_A$. Then $\pi^n_{i,F(\delta)}(F(\delta))$ in (4.7) is just another notation for the probability $\Sigma_{F(\delta)}(\Pi^n_{F(\delta)} e_i)$ (see Section 4.1.2). It follows that for i such that $f(i) > \delta'$ (and such i does exist in $F(\delta)$)

$$\sigma_{F(\delta)}(i) \geq \frac{\delta' - \delta}{\|f\| - \delta} > 0. \tag{4.8}$$

Hence, $F(\delta)$ is a sojourn set. Moreover, since an i with $f(i) > \delta'$ may be found in $F(\delta)$ for any $\delta' \in (\delta, \|f\|)$, (4.8) shows that

$$\|\sigma_{F(\delta)}\|_{l^\infty(F(\delta))} = 1.$$

In particular, if A is a sojourn set, Lemma 4.1.6 applies to $\widetilde{\Pi} = \Pi_A$ and $f = \sigma_A$. Similarly, since $\Sigma_A \geq \Sigma_{F(\delta)}$, we see that $\sigma_A \geq \sigma_{F(\delta)}$ and conclude that

$$\|\sigma_A\|_{l^\infty(A)} = 1 \tag{4.9}$$

(which automatically implies $\|s_A\|_{l^\infty(\mathbb{N})} = 1$).

4.1.8 *A and $A(\delta)$ are equivalent*

Let A be a sojourn set and, for a given $\delta \in (0, 1)$, let

$$A(\delta) = \{i \in A; \sigma_A(i) > \delta\}.$$

Then

$$s_A = s_{A(\delta)},$$

which is to say that the sets A and $A(\delta)$ are **equivalent**.

Proof

(i) We are to show that

$$\lim_{n\to\infty} \sigma_A(\Pi^n x) = \lim_{n\to\infty} \sigma_{A(\delta)}(\Pi^n x), \qquad x \in l^1(\mathbb{N}). \tag{4.10}$$

The following considerations reduce this task to proving an equality of certain functionals on $l^1(A)$.

Since, for any nonnegative $x \in l^1(A)$,

$$\Sigma_A(x) \geq \sigma_A(x) = \sigma_A(\Pi_A x) \geq \sigma_{A(\delta)}(\Pi_A x) \geq \sigma_{A(\delta)}(\Pi_{A(\delta)} x) = \sigma_{A(\delta)}(x),$$

the sequence $\left(\sigma_{A(\delta)}(\Pi_A^n x)\right)_{n\geq 1}$ increases and is bounded from above by $\Sigma_A(x)$. Therefore, the limit

$$\sigma_{A(\delta)}^A(x) := \lim_{n\to\infty} \sigma_{A(\delta)}(\Pi_A^n x)$$

exists for all $x \in l^1(A)$ and defines a member of $\mathfrak{B}_A$ such that

$$\sigma_{A(\delta)}^A \geq \sigma_{A(\delta)}.$$

On the other hand, for nonnegative $x \in l^1(\mathbb{N})$ and $n \geq 1$,

$$\sigma_{A(\delta)}(\Pi_A^n x) \leq \sigma_{A(\delta)}(\Pi^n x) \leq s_{A(\delta)}(x),$$

showing that

$$\sigma^A_{A(\delta)} \le s_{A(\delta)}.$$

Combining these two inequalities, we conclude that, for any $n \ge 1$ and nonnegative $x \in l^1(\mathbb{N})$,

$$\sigma_{A(\delta)}(\Pi^n x) \le \sigma^A_{A(\delta)}(\Pi^n x) \le s_{A(\delta)}(\Pi^n x) = s_{A(\delta)}(x).$$

It follows that $s_{A(\delta)}(x) = \lim_{n\to\infty} \sigma^A_{A(\delta)}(\Pi^n x)$, $x \in l^1(\mathbb{N})$. Thus, instead of (4.10), it suffices to show that

$$\sigma_A = \sigma^A_{A(\delta)}.$$

(ii) It is clear that $\sigma_A \ge \sigma^A_{A(\delta)}$, and that $f := \sigma_A - \sigma^A_{A(\delta)}$ belongs to $\mathfrak{B}_A$ (since σ_A and $\sigma^A_{A(\delta)}$ do). We need to show that $\|f\| = 0$.

Suppose that this is not so and consider a $\gamma \in (0, \|f\|)$. Then, by 4.1.7,

$$F(\gamma) = \{i \in A;\ f(i) > \gamma\}$$

is a sojourn set contained in A. Also, let $\delta' \in (\delta, 1)$ be so close to 1 that $\gamma + \frac{\delta'-\delta}{1-\delta} > 1$.

For $i \in A(\delta') \subset A(\delta)$, by (4.8),

$$\sigma^A_{A(\delta)}(i) \ge \sigma_{A(\delta)}(i) > \frac{\delta' - \delta}{1 - \delta}.$$

Therefore, such an i cannot belong to $F(\gamma)$ for this would imply $\sigma_A(i) = f(i) + \sigma^A_{A(\delta)}(i) > 1$. Hence, $F(\gamma) \subset A \setminus A(\delta')$, that is,

$$\sigma_A(i) \le \delta', \qquad i \in F(\delta).$$

Since $\sigma_{F(\gamma)} \le \sigma_A$, this contradicts the fact that $\|\sigma_{F(\gamma)}\|_{l^\infty(F(\gamma))} = 1$. The proof is complete. □

4.1.9 Lemma

Let $\epsilon > 0$ be given and let A be a sojourn set such that

$$s_A(i) > \epsilon, \qquad i \in A. \tag{4.11}$$

Then

$$s_A + s_{A^{\complement}} = \Sigma,$$

where $A^{\complement} = \mathbb{N} \setminus A$. (See also Section 4.1.13.)

Proof For simplicity of notation, let $B = A^{\complement}$. We need to show that

$$f := \Sigma - s_A - s_B$$

is zero. By Section 4.1.4, $0 \le f \le \Sigma$.

Suppose $\|f\| > 0$. There are numbers $\delta \in (0, 1)$ and $\gamma \in (0, \|f\|)$ such that $2\delta > 1$ and $\gamma + \delta > 1$. Then the sets

$$\begin{aligned} A(\delta) &= \{i \in A; s_A(i) > \delta\}, \\ B(\delta) &= \{i \in B; s_B(i) > \delta\}, \\ F(\gamma) &= \{i \in \mathbb{N}; f(i) > \gamma\} \end{aligned}$$

are pairwise disjoint, because for an i in the intersection of any two of these sets $\Sigma(i)$ would exceed 1. Hence, by 4.1.4,

$$s_{A(\delta)} + s_{B(\delta)} + s_{F(\gamma)} \le \Sigma.$$

Next, we claim that $s_{A(\delta)} = s_A$ and $s_B = s_{B(\delta)}$. For the proof of the first equality, let $A'(\delta) = \{i \in A; \sigma_A(i) > \delta\}$. We have $A'(\delta) \subset A(\delta) \subset A$. Therefore, $s_{A'(\delta)} \le s_{A(\delta)} \le s_A$. However, $s_{A'(\delta)} = s_A$ by Section 4.1.8. This proves the first claim. Also, the other claim is true if B is a sojourn set. In the other case, $s_B = 0$ and so $B(\delta)$ is empty, implying $s_{B(\delta)} = 0 = s_B$.

Therefore,

$$s_A + s_B + s_{F(\gamma)} \le \Sigma. \tag{4.12}$$

It follows by the definition of f that $f \ge s_{F(\gamma)}$ and, since $\|s_{F(\gamma)}\| = 1$, that $\|f\| = 1$.

Thus γ can be chosen as close to 1 as we wish and in particular it can be so large that $\gamma > 1 - \epsilon$ for the ϵ featuring in our assumption. For such γ, $F(\gamma)$ and A must be disjoint, because for i in the intersection of these two sets we would have $\Sigma(i) = f(i) + s_A(i) + s_B(i) > \gamma + \epsilon > 1$. Thus, $F(\gamma) \subset B$. But then (4.12) shows that $\Sigma \ge s_B + s_{F(\gamma)} \ge 2s_{F(\gamma)}$. This is impossible, since $\|\Sigma\| = \|s_{F(\gamma)}\| = 1$. □

4.1.10 *Any sojourn solution is an extremal functional*

Proof Let A be a sojourn set. Take any $\epsilon \in (0, 1)$. For $A' := A(\epsilon)$ where $A(\epsilon)$ is defined in 4.1.8, $s_A = s_{A'}$ and, on the other hand, conditions of Lemma 4.1.9 are satisfied. Therefore,

$$s_A \cap (\Sigma - s_A) = s_{A'} \cap (\Sigma - s_{A'}) = s_{A'} \cap s_{\mathbb{N}\setminus A'}.$$

Moreover, by Lemma 4.1.4,

$$s_{A'} + s_{\mathbb{N}\setminus A'} = s_{A'} \cup s_{\mathbb{N}\setminus A'}.$$

Hence, by (3.71), $s_{A'} \cap s_{\mathbb{N}\setminus A'} = 0$, that is,

$$s_A \cap (\Sigma - s_A) = 0,$$

proving that s_A is an extremal functional. □

4.1.11 *Any extremal functional is a sojourn solution*

Proof Let nonzero f be extremal. Take $\delta \in (0, \|f\|)$. Then

$$F(\delta) = \{i \in \mathbb{N};\ f(i) > \delta\}$$

is a sojourn set. We aim at showing that $f = s_{F(\delta)}$.

By definition of $F(\delta)$, $f > \delta\Sigma_{F(\delta)} \geq \delta\sigma_{F(\delta)}$. Therefore, $f \geq \delta s_{F(\delta)}$ and Section 3.8.1 (2) yields $f \geq s_{F(\delta)}$. Our task thus reduces to showing that

$$g := f - s_{F(\delta)} \geq 0 \tag{4.13}$$

is in fact zero.

If $g \neq 0$,

$$A := \left\{i \in \mathbb{N};\ g(i) > \frac{\|g\|}{2}\right\}$$

is a sojourn set and so is

$$A(\delta) := \{i \in A;\ s_A(i) > \delta\}.$$

Moreover, see the proof of 4.1.9, $f \geq g \geq \frac{\|g\|}{2} s_A$ which in turn implies $f \geq s_A$. Thus, for $i \in A(\delta)$, we have $f(i) \geq s_A(i) > \delta$, showing that $A(\delta) \subset F(\delta)$. But then, by (4.13),

$$f \geq \frac{\|g\|}{2} s_A + s_{F(\delta)} \geq \left(\frac{\|g\|}{2} + 1\right) s_{A(\delta)}.$$

This is impossible, because $\|s_{A(\delta)}\| = 1$ and $f \in \mathfrak{B}$. This contradiction shows that $g = 0$, completing the proof. □

We complete this section by providing the missing part of the proof of characterization of minimal elements.

4.1.12 *Proof of 3.8.2(a)*

Suppose $g \leq f$ for a nonzero $g \in \mathfrak{B}$. Then,

$$A(\delta) = \{i \in \mathbb{N};\ g(i) > \delta\}$$

is a sojourn set for any $\delta \in (0, \|g\|)$. By the definition of $A(\delta)$, $f \ge g \ge \delta\Sigma_{A(\delta)} \ge \delta\sigma_{A(\delta)}$. Hence, $f \ge g \ge \delta s_{A(\delta)}$ and so, by 3.8.1 (2), $f \ge s_{A(\delta)}$. Since $s_{A(\delta)}$ is extremal and f is minimal, $f = s_{A(\delta)}$. Thus, $g \ge \delta f$ for any $\delta \in (0, \|g\|)$. It follows that

$$g \ge \|g\| f.$$

The same reasoning applied to $f - g$ instead of g renders

$$f - g \ge \|f - g\| f \ge (1 - \|g\|) f,$$

because any extremal functional, being a sojourn solution, has norm 1. These two inequalities result in $g = \|g\| f$. Since $0 \le g \le f$, we have $\|g\| \in [0, \|f\|]$, but extremal values are disqualified by assumption.

4.1.13 *Two warnings*

(i) The seemingly superfluous assumption (4.11) of Lemma 4.1.9 is in fact quite crucial. Feller refers to sojourn sets satisfying (4.11) as **representative**, and notes that for nonrepresentative sojourn sets Lemma 4.1.9 need not be true. Here is an example.

Suppose natural numbers are arranged in four infinite rows as in Figure 4.1. For instance, the first row counted from above, denoted R_1, may be composed of natural numbers divisible by 4, the second, R_2, of those that are divisible by four with remainder 1, and so on. Suppose also that when in the first or fourth row, the jump chain always moves one unit to the right, so that it never leaves this row. Suppose further that when in the kth column and second row it moves either to the first or third row in the $(k+1)$st column with probabilities $\frac{1}{k^2+1}$ and $\frac{k^2}{k^2+1}$, respectively. Similarly when in the kth column and third row

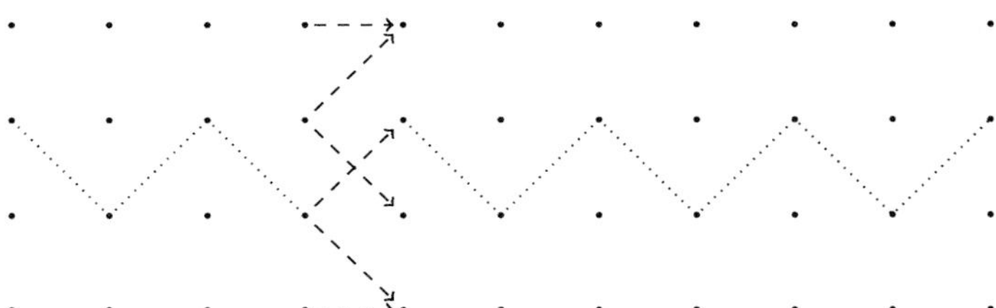

Figure 4.1 Nonrepresentative sojourn sets. The chain jumps from the inner rows to the outer (representative) rows with smaller and smaller probabilities as it moves further to the right. The probability of oscillating between the second and third rows is then nonzero.

it moves either to the fourth or the second row in the $(k+1)$st column with probabilities $\frac{1}{k^2+1}$ and $\frac{k^2}{k^2+1}$, respectively.

Then, R_1 and R_4 are sojourn sets (they are in fact representative), and so are $A = R_1 \cup R_2$ and $B = R_3 \cup R_4$. Moreover, for any point p, $s_A(p) = s_{R_1}(p)$ is the probability that the chain starting at p will after finitely many steps reach the first row; hence, if p is in the kth column and second row, $s_A(p)$ does not exceed $\sum_{j=0}^{\infty} \frac{1}{(k+2j)^2+1} < \sum_{j=k}^{\infty} \frac{1}{j^2+1}$, and the latter quantity is as small as we wish provided k is large enough. It follows that A is not representative, and the same applies to B. We will show that the conclusion of Lemma 4.1.9 is not valid.

It suffices to show that $s_A(p) + s_B(p) < 1$ for the point p in the second row and first column. We note that this sum is the probability that a chain starting at this point will after finitely many steps reach either the first or the fourth row (and then stay there for ever). However, with probability $\prod_{k=1}^{\infty} \frac{k^2}{k^2+1}$, the process starting there will for ever oscillate between the second and third rows (this infinite product is nonzero by Criterion 2.4.6), showing that the former probability is smaller than 1.

(ii) Sojourn solutions (even those that are representative) may correspond to passive functionals and thus be useless in constructing post-explosion processes. For instance, if in the example presented above all sojourn times are exponential with parameter 1, all sojourn sets are passive. If, on the other hand, sojourn times in the kth column are exponential with parameter k^{-2}, all sojourn solutions are active. We are simply repeating the obvious fact that the jump chain matrix can be derived from an intensity matrix (if no states are instantaneous) but not vice versa.

4.2 A Broader Perspective

Now that we have constructed the discrete (exit) boundary for a Markov chain and have been able to identify elements of this boundary as sojourn sets for the chain, we are ready to see a panoramic view from a larger space. From this perspective the semigroups considered before transpire to be shadows, or should I say, spectra, of certain other quite natural semigroups.

4.2.1 *Stopped process*

Let's start by looking again at the minimal semigroup $\{S(t), t \geq 0\}$ for a given Kolmogorov matrix Q. The related process is defined only up to a random time of explosion, and as a result, $\{S(t), t \geq 0\}$ is composed of sub-Markov operators. A simple way to make a Markov semigroup out of $\{S(t), t \ge 0\}$ is to think of an additional state, say, a, for the minimal process and agree that after explosion the minimal process goes to this state and stays there for ever.[2] We stress that this new process is clearly not identical to the minimal process: the former has a larger state-space and stays in the extra point of this space whereas the latter is undefined.

The new semigroup, let us denote it $\{S_a(t), t \ge 0\}$, is defined in the space l^1_a of absolutely summable sequences indexed by elements of $\mathbb{N}_a := \mathbb{N} \cup \{a\}$. The original space $l^1 = l^1(\mathbb{N})$ may be identified with the subspace of $x = (\xi_a, \xi_1, \xi_2, \dots) \in l^1_a$ such that $\xi_a = 0$, and it will be convenient to grant the latter elements dual citizenship, and treat them as members of both l^1 and l^1_a. With this agreement, $S_a(t)$ can be written explicitly:

$$S_a(t)x = (\xi_a + \Sigma Lx - \Sigma S(t)Lx)e_a + S(t)Lx, \qquad x \in l^1_a,$$

where

- $x = (\xi_a, \xi_1, \xi_2, \dots)$,
- $Lx = (0, \xi_1, \xi_2, \dots)$,
- $e_a = (1, 0, 0, \dots) \in l^1_a$,
- and Lx is treated as an element of l^1 so that $S(t)Lx$ makes sense, whereas $S(t)Lx$ is seen as a member of l^1_a.

This formula says what it is supposed to say: if merely the states $1, 2, 3, \dots$ are observed, the process is identical to the minimal one:

$$LS_a(t)x = S(t)x, \qquad x \in l^1.$$

However, the probability mass lost in the minimal process (if x is a distribution, $\Sigma Lx - \Sigma S(t)Lx$ is this probability mass), is now being accumulated at the state a.

The generator, say, G_a, of the semigroup $\{S_a(t), t \ge 0\}$ can also be calculated explicitly: the difference quotients (note that $\xi_a e_a = x - Lx$)

$$\frac{S_a(t)x - x}{t} = \frac{\Sigma Lx - \Sigma S(t)Lx}{t} e_a + \frac{S(t)Lx - Lx}{t}$$

converge as $t \to 0+$ iff Lx belongs to $\mathcal{D}(G)$, where G is the generator of $\{S(t), t \ge 0\}$; this is because the terms in this formula are linearly independent. For such x,

$$\begin{aligned} G_a x &:= \lim_{t \to 0+} \frac{S_a(t)x - x}{t} = -(\Sigma GLx)e_a + GLx \\ &= h(Lx)e_a + GLx. \end{aligned} \tag{4.14}$$

[2] Many a witty author refers to this state as the cemetery.

This form of the generator, in fact, does not surprise us a bit, because we remember Section 3.5.5 (we also borrowed notation for h from the latter section). Didn't we here, as in 3.5.5, command the process to start all over again after explosion? The only difference is that now we have commanded it to start at a newly created state a, which is absorbing and its connections with the other states in $\mathbb{N}_a$ can hardly be called bilateral, and so talking about a new beginning is a bit of an exaggeration.

And vice versa, looking back at the labour pains of Chapter 3, we realize that the little calculation we have made here could have helped us in guessing the form of the generator of a postexplosion process we discovered in (3.36), (3.39), and (3.55).

4.2.2 *A better bookkeeping*

The approach presented in the previous section is an example of a rather poor accounting of lost probability mass. If we know that there are $\mathfrak{k}$ extremal sojourn sets (boundary points), we could accumulate the probability mass lost in the minimal process in $\mathfrak{k}$ containers, using one container (boundary point) for one sojourn set.

Here are the details. First, we identify the $\mathfrak{k}$ elements $f^1, \dots, f^{\mathfrak{k}}$ of the discrete boundary $\mathfrak{B}$ with integers $-1, -2, \dots, -\mathfrak{k}$. Next, we consider the space $l^1(\mathbb{I})$ of absolutely summable sequences indexed by

$$i \in \mathbb{I} := \mathbb{N} \cup \mathfrak{B} = \mathbb{N} \cup \{-1, \dots, -\mathfrak{k}\};$$

members of $l^1(\mathbb{I})$ are thus of the form

$$x = (\xi_{-\mathfrak{k}}, \dots, \xi_{-1}, \xi_1, \dots). \tag{4.15}$$

Again, elements of the original space $l^1 = l^1(\mathbb{N})$ can be identified with $(\xi_i)_{i \in \mathbb{I}} \in l^1(\mathbb{I})$ such that $\xi_{-1} = \xi_{-2} = \cdots = \xi_{-\mathfrak{k}} = 0$. We also need a new version of L:

$$L(\xi_{-\mathfrak{k}}, \dots, \xi_{-1}, \xi_1, \xi_2, \dots) = (0, \dots, 0, \xi_1, \xi_2, \dots).$$

Then, the generator of the searched-for semigroup has the following form (cf. (3.92)):

$$G_{\text{stop}} x = \sum_{i=1}^{k} h_i(Lx) e_{-i} + GLx, \qquad x \in \mathcal{D}(G_{\text{stop}}), \tag{4.16}$$

where h_i is given by (3.95), $e_{-i} \in l^1(\mathbb{I})$ is given by (2.9), and $\mathcal{D}(G_{\text{stop}})$ is composed of $x \in l^1(\mathbb{I})$ such that $Lx \in \mathcal{D}(G)$. Although, formally speaking, the fact that G_{stop} is a Markov generator in $l^1(\mathbb{I})$ is not a direct consequence of Theorem 3.7.9, the proof of this fact differs from the proof of Theorem 3.7.9 only in insignificant details. (See also our solution to Exercise 4.2.6.)

In the Markov chain governed by G_{stop} a particle 'reaching infinity' through the sojourn set related to the ith extremal functional, is commanded to land at the integer $-i$ and from that moment onward to stay there for ever. By contrast, in the chain governed by G_a of the previous section, after explosion, all particles, regardless of the way they 'reached infinity,' land at the 'aggregate' state a.

4.2.3 *Processes in* $\mathbb{N} \cup \mathfrak{B}$ *and their shadows*

Formula (4.16) significantly changes the role of boundary points $f^1, \ldots, f^{\Bbbk}$. In (3.92) they are merely transit hubs, distributing paths after explosion to various states with various probabilities. In (4.16) they gain the status of true state-space points; newcomers, undoubtedly, but occupying their rightful places. These new points differ from other, original, states only by the fact that they cannot be reached by a simple jump from an $i \in \mathbb{N}$; to reach them an infinity of (very particular) jumps is needed.

In the enlarged state-space $\mathbb{N} \cup \mathfrak{B}$ new, more general rules of behavior for a chain may be introduced, by expanding the original matrix Q. We may, for example, allow direct jumps from an $i \in \mathbb{N}$ to a $j \in \mathfrak{B}$ and vice versa. We may also allow direct communication between points of $\mathfrak{B}$. In devising such rules, however, we need to be careful, if we want to have processes that are meaningful from the perspective of the original space. For, if we observe only the states contained in $\mathbb{N}$, we will see particles disappearing after explosion and then after some random time spent at the boundary reappearing again; we need to make sure that such returns do not violate Markovian nature of the process (cf. [72], p. 6).

For example, given densities $u_1, \ldots, u_{\Bbbk} \in l^1(\mathbb{N})$ and positive numbers $r_1, \ldots, r_{\Bbbk}$ (playing the role of escape rates), we may think of the operator

$$G_{\text{er}}x = G_{\text{stop}}x + \sum_{i=1}^{\Bbbk} \xi_{-i} r_i (u_i - e_{-i}), \qquad x \in \mathcal{D}(G_{\text{er}}), \tag{4.17}$$

where $\mathcal{D}(G_{\text{er}})$ is composed of x's such that $Lx \in \mathcal{D}(G)$. Here 'er' stands for 'elementary return.' In the related process a particle reaching, after explosion, a boundary point $i \in \mathfrak{B}$ neither stays there for ever (as in the process described by (4.16)) nor leaves it immediately (as in the process described by (3.92)): it stays there for an exponential time with parameter r_i, and then returns to one of

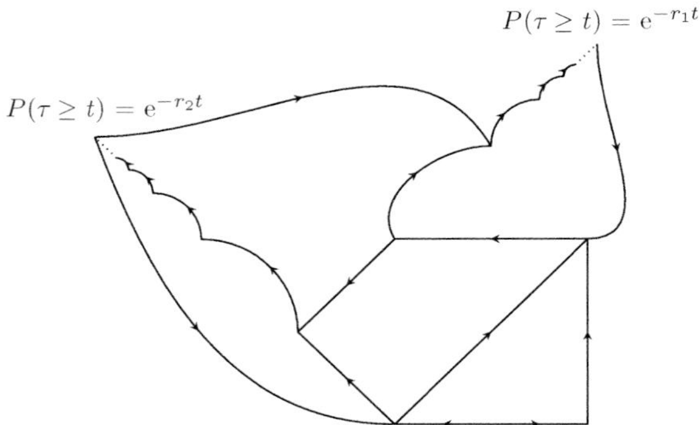

Figure 4.2 An example of elementary return from two (exit) boundary points; τ is the time spent at the boundary. Points at the boundary cannot be reached directly but only via an infinite number of jumps through the regular points.

the states in $\mathbb{N}$, the probability of choosing $j \in \mathbb{N}$ being equal to the jth coordinate in u_i. From the perspective of $\mathbb{N}$, after explosion, a particle disappears for an exponential time (with parameter depending on how explosion came about) and then reappears at a random point according to the rules described above.

That fact that G_{er} is a Markov generator will be proved in the next section.

4.2.4 G_{er} is a Markov generator

Let

$$T(t)x = Lx + \sum_{i=1}^{\mathcal{k}} \xi_{-i}(\mathrm{e}^{-r_i t} e_{-i} + (1 - \mathrm{e}^{-r_i t})u_i), \quad x \in l^1(\mathbb{I}), t \geq 0. \tag{4.18}$$

It is clear that this formula defines a strongly continuous semigroup $\{T(t), t \geq 0\}$ of Markov operators in $l^1(\mathbb{I})$. In the related process, a particle starting at $-i \in \mathfrak{B}$ stays there for an exponential time with parameter r_i and then jumps to a $j \in \mathbb{N}$ with probability equal to the jth coordinate of u_i; all the states $j \in \mathbb{N}$ are absorbing.

It is easy to see that the limit $\lim_{t\to 0+} t^{-1}(T(t)x - x)$ exists for all $x \in l^1(\mathbb{I})$ and equals $\sum_{i=1}^{\mathcal{k}} \xi_{-i} r_i (u_i - e_{-i})$. It follows that the bounded linear operator

$$Rx = \sum_{i=1}^{\mathcal{k}} \xi_{-i} r_i (u_i - e_{-i}) \tag{4.19}$$

is the generator of $\{T(t), t \geq 0\}$.

On the other hand, $G_{\text{er}} = G_{\text{stop}} + R$. Therefore, since G_{stop} is a generator and R is bounded, G_{er} is a generator by the Phillips Perturbation Theorem. Since G_{stop} and R are Markov generators, Trotter's Product Formula implies that G_{er} is a Markov generator as well.

4.2.5 *Instantaneous return from the boundary*

In our discussion of the chain related to (3.92) we frequently described boundary points as, somewhat imaginary, instantaneous states working as transit hubs by distributing particles passing through them to 'true' points of the state-space. The following approximation procedure provides arguments for such a description.

Let $(r_n)_{n\ge 1}$ be a sequence of positive rates with $\lim_{n\to\infty} r_n = \infty$, and let for any $n \ge 1$ an operator A_n be given by

$$A_n x = G_{\text{stop}} x + r_n \sum_{i=1}^{k} \xi_{-i}(u_i - e_{-i}), \tag{4.20}$$

on the common domain composed of x's such that $Lx \in \mathcal{D}(G)$; as before u_i's are fixed densities in $l^1(\mathbb{N})$. Hence, $A_n, n \ge 1$ are Markov generators in $l^1(\mathbb{I})$, and the related Markov chains feature elementary return from the boundary. For a fixed n, return rates of the chain are the same for all boundary points, but generally increase with n. At the same time, distributions after return are different for different boundary points but do not change with n. We will find the limit of $\{e^{tA_n}, t \ge 0\}$.

Formula (4.20) is of the form (1.26) with $A_0 = G_{\text{stop}}$ and B given by (cf. (4.19))

$$Bx = \sum_{i=1}^{k} \xi_{-i}(u_i - e_{-i}). \tag{4.21}$$

Recalling that the operator in (4.19) is the generator of the semigroup (4.18), we also see that

$$\lim_{t\to\infty} e^{tR} x = Px, \qquad \text{where} \qquad Px = Lx + \sum_{i=1}^{k} \xi_{-i} u_i.$$

The definition of P given above implies in particular $Px = x$ for $x \in l^1(\mathbb{N})$ and $Pe_{-i} = u_i, i = 1, \ldots, k$. It follows that

$$PG_{\text{stop}} x = Hx, \qquad x \in \mathcal{D}(H) = \mathcal{D}(G) = \mathcal{D}(G_{\text{stop}}) \cap l^1(\mathbb{N}),$$

where H is given by (3.92).

Thus, since we know that H is a Markov generator in $l^1(\mathbb{N})$, Kurtz's Singular Perturbation Theorem 1.4.4 is in force, and we conclude that

$$\lim_{n\to\infty} \mathrm{e}^{tA_n}x = \mathrm{e}^{tH}Px, \qquad t > 0. \tag{4.22}$$

This formula says that the semigroups $\{\mathrm{e}^{tA_n}, t \ge 0\}$ converge, in a sense, to $\{\mathrm{e}^{tH}, t \ge 0\}$ (I need to add 'in a sense' because $\{\mathrm{e}^{tA_n}, t \ge 0\}$ are defined in a different space than $\{\mathrm{e}^{tH}, t \ge 0\}$) allowing us to infer properties of the process related to H from those of the processes related to A_n's.

To this end, think of a boundary point $-i \in \mathfrak{B}$. In the process related to A_n, the time spent at $-i$ is exponential with parameter r_n, and the distribution of the process after that time is u_i. As $n \to \infty$, the time spent at $-i$ is shorter and shorter but the distribution u_i remains the same. Hence, we expect that in the limit the process 'spends no time at $-i$' but immediately jumps from there to a random point in $\mathbb{N}$, according to the distribution u_i. (In this respect, $-i$ differs from the instantaneous state of the first Kolmogorov–Kendall–Reuter example.) In fact, (4.22) reveals that the time spent at the boundary is so short that the limit process does not see the boundary points as 'true' elements of the state-space (this was also our point of view initially); its state-space is $\mathbb{N}$ not $\mathbb{I}$. We see from our formula that a common initial distribution $x \in l^1(\mathbb{I})$ of the approximating chains must in the limit be replaced by $Px \in l^1(\mathbb{N})$. In particular, e_{-i} must be replaced by u_i; all the probability mass concentrated at the boundary point $-i$ must be immediately distributed across $\mathbb{N}$. This supports our intuition of $-i$ as a transit hub in a process related to H.

4.2.6 Exercise

Show that G_{stop} is a Markov generator.

4.2.7 Exercise

Find an explicit formula for $\{\mathrm{e}^{tG_{\text{stop}}}, t \ge 0\}$.

4.3 P. Lévy's Flash

As we are coming to the end of this chapter, the reader might be led to believe that the general picture has already been drawn, and it is perhaps only details that are missing. If this is the case, the next, simple example may come as a shock: the generator of the semigroup considered there is not of the form

(3.92), even though the underlying process seems not to differ much from those considered before. In order to understand what is going on we need to change our point of view on the state-space $\mathbb{I}$: whereas in the preceding sections seeing $\mathbb{I}$ as no different from $\mathbb{N}$ worked well, here the basic example for $\mathbb{I}$ is $\mathbb{Z}$, the set of integers, which has two natural 'infinities'; each of them will turn out to play a significantly different role.

4.3.1 *P. Lévy's flash* $\rightsquigarrow$

Let $l^1 = l^1(\mathbb{Z})$ be the space of absolutely summable sequences $(\xi_i)_{i\in\mathbb{Z}}$ and let $a_i, i \in \mathbb{Z}$ be positive numbers such that

$$\sum_{i\in\mathbb{Z}} a_i^{-1} < \infty. \tag{4.23}$$

Also, let $\mathfrak{Q}$ be the operator in l^1 given by

$$\mathfrak{Q}\,(\xi_i)_{i\in\mathbb{Z}} = (a_{i-1}\xi_{i-1} - a_i\xi_i)_{i\in\mathbb{Z}}\,,$$

on the domain composed of $(\xi_i)_{i\in\mathbb{Z}}$ such that $(a_{i-1}\xi_{i-1} - a_i\xi_i)_{i\in\mathbb{Z}}$ belongs to l^1. (This is a particular case of the operator $\mathfrak{Q}$ of Section 3.4.) We note that convergence of the series $\sum_{i\in\mathbb{Z}} |a_{i-1}\xi_{i-1} - a_i\xi_i|$ implies existence of

$$\mathfrak{l}_+\,(\xi_i)_{i\in\mathbb{Z}} := \lim_{i\to\infty} a_i\xi_i \qquad \text{and} \qquad \mathfrak{l}_-\,(\xi_i)_{i\in\mathbb{Z}} := \lim_{i\to\infty} a_{-i}\xi_{-i}\,.$$

In particular, it implies, by assumption (4.23), that $(\xi_i)_{i\in\mathbb{Z}} \in l^1$, that is, that the domain of $\mathfrak{Q}$ is contained in l^1. We also have

$$\Sigma\mathfrak{Q}x = \mathfrak{l}_-x - \mathfrak{l}_+x. \tag{4.24}$$

With these preparations out of the way, I claim that the operator A defined as the restriction of $\mathfrak{Q}$ to the domain $\mathcal{D}(A)$ of $x \in \mathcal{D}(\mathfrak{Q})$ such that

$$\mathfrak{l}_+x = \mathfrak{l}_-x, \tag{4.25}$$

is a Markov generator: It is easy to see, using (4.24), that two conditions of the Hille–Yosida Theorem are satisfied, and in Section 4.3.4 we will prove the third of them. Here I would like to explain how the Markov chain related to A behaves.

To this end, let us consider the following sequence of approximating chains (see Figure 4.3[3]). The nth chain is 'active' only in the range of $\{-n, \dots, n\}$; the remaining states are absorbing. At $i \in \{-n, \dots, n-1\}$, the chain stays for

3 Sources I would rather not reveal claim that this figure depicts a cloud from which the flash of Lévy comes.

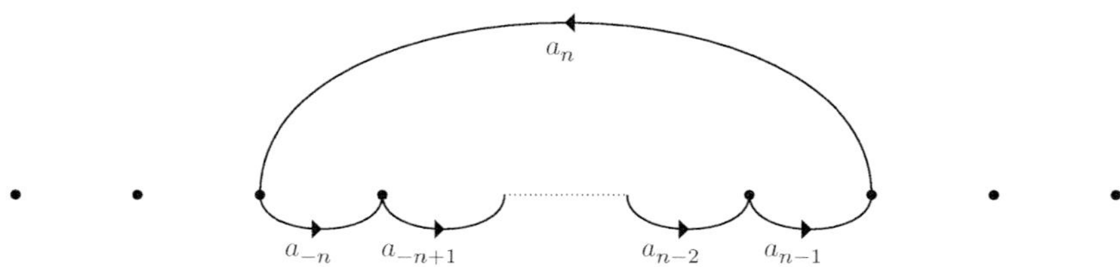

Figure 4.3 A chain approximating P. Lévy's flash.

an exponential time with parameter a_i before jumping to $i + 1$, and after an exponential time with parameter a_n spent at n, it jumps back to $-n$.

To see that the semigroups describing these chains converge to the semigroup generated by A, think of the generator A_n of the nth semigroup. For any $x = (\xi_i)_{i\in\mathbb{Z}} \in \mathcal{D}(A)$, the ith coordinate of $A_n x$ is the same as that of Ax provided $i \in \{-n+1,\ldots,n\}$. Moreover, the $(-n)$th coordinate of $A_n x$ is $a_n\xi_n - a_{-n}\xi_{-n}$ whereas that of Ax is $a_{-n-1}\xi_{-n-1} - a_{-n}\xi_{-n}$. Since all the remaining coordinates of $A_n x$ are zero,

$$\|A_n x - Ax\| = |a_n\xi_n - a_{-n-1}\xi_{-n-1}| + \sum_{i\notin\{-n,\ldots,n\}} |a_{i-1}\xi_{i-1} - a_i\xi_i|.$$

Thus, because of the boundary condition (4.25), $\lim_{n\to\infty} A_n x = Ax$, $x \in \mathcal{D}(A)$, proving convergence of the semigroups by the Sova–Kurtz Theorem.

Hence, the chain related to A runs through all integers, spending exponential time with parameter a_i at the state i and, because of assumption (4.23) the voyage through $\mathbb{Z}$ is completed in a flash; the boundary condition (4.25) then commands the chain to start all over again from $-\infty$. This chain is called **P. Lévy's flash**.

4.3.2 *Entrance points versus exit points*

The difference between the chain of the previous section and those described by (3.92) is that, whereas in (3.92) the chain is commanded to start afresh from a randomly selected but regular point of the state-space, here the chain starts afresh from $-\infty$, which is an example of a point of an **entrance boundary**. (Of course, $+\infty$ is an exit boundary.) It is intriguing that, as exemplified by the boundary condition (4.25), for a description of an entrance boundary a perturbation of the domain of the generator is needed. This is in contrast to exit boundary: formula (3.92) says that an exit boundary introduces additional terms in the generator, but does not change its domain. An entrance boundary does not change the way the generator acts but rather its domain.

Before we study this difference between entrance and exit points in more detail, let us have a closer look at the minimal chain corresponding to P. Lévy's flash.

4.3.3 *The minimal chain and its generator*

The minimal chain described by the intensity matrix hidden in the generator of P. Lévy's flash is a pure birth chain which runs through all integers and after reaching infinity is left undefined. Let us find its generator, G.

As in 3.3.2, we will use (3.5) which is a formula for the resolvent of the generator. We note that (3.13) is still in force but, because there are now infinitely many indices smaller than a given i, formula (3.24) giving the ith coordinate of $R_\lambda\,(\eta_i)_{i\in\mathbb{Z}}$, must be replaced by

$$\xi_i = \frac{\pi_{i-1}}{\lambda + a_i} \sum_{j=-\infty}^{i} \frac{\eta_j}{\pi_{j-1}}, \tag{4.26}$$

where this time $\pi_j = \prod_{k=-\infty}^{j} \frac{a_k}{\lambda+a_k} \neq 0$; the infinite product converges to a nonzero limit because of assumption (4.23), and the series converges because the numbers π_j, $j \in \mathbb{Z}$ are bounded away from 0.

It is a direct consequence of (4.26) that

$$\lim_{i\to-\infty} a_i\xi_i = \lim_{i\to-\infty} \pi_i \sum_{j=-\infty}^{i} \frac{\eta_j}{\pi_{j-1}} = 1 \times 0 = 0.$$

Since R_λ maps l^1 onto $\mathcal{D}(G)$, we obtain that

$$\mathfrak{l}_- x = 0, \qquad x \in \mathcal{D}(G).$$

We will show that the latter is also a sufficient condition for an $x \in \mathcal{D}(\mathfrak{Q})$ to belong to $\mathcal{D}(G)$. In other words (see Section 4.3.1),

$$G \text{ is } \mathfrak{Q} \text{ restricted to the kernel of } \mathfrak{l}_-.$$

To this end, it suffices to show that $\mathfrak{Q}$ restricted to the kernel of $\mathfrak{l}_-$ has the resolvent given by (4.26). Now, in coordinates, the resolvent equation for $\mathfrak{Q}$ reads

$$(\lambda + a_i)\xi_i - a_{i-1}\xi_{i-1} = \eta_i, \qquad i \in \mathbb{Z}. \tag{4.27}$$

Rewriting this as $\xi_i = \frac{1}{\lambda+a_i}(\eta_i + a_{i-1}\xi_{i-1})$ and using induction argument, we obtain that for all $k \geq 1$,

$$\xi_i = \frac{\pi_{i-1}}{\lambda + a_i}\left(\sum_{j=i+1-k}^{i} \frac{\eta_j}{\pi_{j-1}} + \frac{a_{i-k}\xi_{i-k}}{\pi_{i-k}}\right). \tag{4.28}$$

Therefore, letting $k \to \infty$ and using $\mathfrak{l}_- x = 0$, we see that ξ_i is given by the right-hand side of (4.26). This completes the proof.

4.3.4 *Operator* A *is a Markov generator*

Let us complete the proof that A of Section 4.3.1 is a Markov generator, by showing condition (ii) in the Hille–Yosida Theorem. In coordinates the resolvent equation for A is precisely the same as the resolvent equation for $\mathfrak{Q}$, and as we have seen this forces coordinates of its solution to satisfy (4.28). Letting $k \to \infty$, yields thus

$$\xi_i = \frac{\pi_{i-1}}{\lambda + a_i}\left(\sum_{j=-\infty}^{i} \frac{\eta_j}{\pi_{j-1}} + \mathfrak{l}_- x\right), \tag{4.29}$$

and this in turn forces

$$\mathfrak{l}_+ x = \pi_\infty \left(\sum_{j=-\infty}^{\infty} \frac{\eta_j}{\pi_{j-1}} + \mathfrak{l}_- x\right),$$

where, of course, $\pi_\infty = \lim_{i\to\infty} \pi_i = \prod_{i=-\infty}^{\infty} \frac{a_i}{\lambda + a_i}$. It follows that $(\xi_i)_{i\in\mathbb{Z}}$ given by (4.29) belongs to $\mathcal{D}(A)$ iff

$$\mathfrak{l}_- x = \frac{\pi_\infty}{1-\pi_\infty}\sum_{j=-\infty}^{\infty} \frac{\eta_j}{\pi_{j-1}},$$

that is, iff

$$\xi_i = \frac{\pi_{i-1}}{\lambda + a_i}\left(\sum_{j=-\infty}^{i} \frac{\eta_j}{\pi_{j-1}} + \frac{\pi_\infty}{1-\pi_\infty}\sum_{j=-\infty}^{\infty} \frac{\eta_j}{\pi_{j-1}}\right). \tag{4.30}$$

It is perhaps still unclear, though, that the so-defined $(\xi_i)_{i\in\mathbb{Z}}$ belongs to $\mathcal{D}(\mathfrak{Q})$, that is, that $(a_{i-1}\xi_{i-1} - a_i\xi_i)_{i\in\mathbb{Z}}$ belongs to l^1. Fortunately, it is easy to see that $(\xi_i)_{i\in\mathbb{Z}}$ satisfies (4.27). Therefore, it suffices to show that $(\xi_i)_{i\in\mathbb{Z}} \in l^1$ (because $(\eta_i)_{i\in\mathbb{Z}}$ belongs to l^1 by assumption) and a look at (4.30) reduces our problem to showing that

$$\sum_{i\in\mathbb{Z}} \frac{\pi_{i-1}}{\lambda + a_i}\sum_{j=-\infty}^{i} \frac{|\eta_j|}{\pi_{j-1}} < \infty.$$

Indeed, the second summand in the parentheses does not depend on i, and $\sum_{i\in\mathbb{Z}} \frac{\pi_{i-1}}{\lambda+a_i} < \infty$. Finally, because of

$$\frac{\pi_{i-1}}{\lambda + a_i} = \frac{\pi_{i-1} - \pi_i}{\lambda}, \tag{4.31}$$

the sum in question equals

$$\sum_{j\in\mathbb{Z}} \frac{|\eta_j|}{\pi_{j-1}} \sum_{i=j}^{\infty} \frac{\pi_{i-1}}{\lambda + a_i} = \sum_{j\in\mathbb{Z}} \frac{|\eta_j|}{\pi_{j-1}} \frac{\pi_{j-1} - \pi_\infty}{\lambda} < \frac{1-\pi_\infty}{\lambda} \sum_{j\in\mathbb{Z}} \frac{|\eta_j|}{\pi_{j-1}} < \infty,$$

completing the proof.

4.3.5 *An entrance law*

Let us rewrite (4.30) as follows:

$$(\lambda - A)^{-1}\, y = R_\lambda y + \left(\frac{\pi_\infty}{1-\pi_\infty} \sum_{j=-\infty}^{\infty} \frac{\eta_j}{\pi_{j-1}} \right) z_\lambda,$$

where

$$z_\lambda := \left(\frac{\pi_{i-1}}{\lambda + a_i} \right)_{i\in\mathbb{Z}} = \lambda^{-1} \left(\pi_{i-1} - \pi_i \right)_{i\in\mathbb{Z}}$$

is independent of y (recall that π_i, on the other hand, does depend on $\lambda > 0$). Noting that $\pi_\infty \sum_{j=-\infty}^{\infty} \frac{\eta_j}{\pi_{j-1}}$ is, by (4.26), $\mathfrak{l}_+(R_\lambda y)$, we obtain (cf. (3.60))

$$(\lambda - A)^{-1}\, y = R_\lambda y + \frac{\mathfrak{l}_+(R_\lambda y)}{1-\pi_\infty} z_\lambda. \tag{4.32}$$

The map $\lambda \mapsto z_\lambda$ is a crucial player here, and we need to devote some time to its description. First, as it is easy to see, z_λ belongs to $\mathcal{D}(\mathfrak{Q})$ and

$$\mathfrak{Q} z_\lambda = \lambda z_\lambda. \tag{4.33}$$

It follows that

$$(\lambda - \mathfrak{Q})(z_\lambda - z_\mu) = -(\lambda - \mathfrak{Q}) z_\mu = (\mu - \lambda) z_\mu, \qquad \lambda, \mu > 0.$$

Moreover, since $\mathfrak{l}_- z_\lambda = \lim_{i\to-\infty} \pi_i = 1$, we have $z_\lambda - z_\mu \in \mathcal{D}(G)$. Thus, $\mathfrak{Q}$ above may be replaced by G, and we obtain

$$z_\lambda - z_\mu = (\mu - \lambda) R_\lambda z_\mu, \qquad \lambda, \mu > 0. \tag{4.34}$$

A look back at (3.77) now reveals that z_λ's play a dual role to f_λ's of Section 3.6.7 (compare also (4.33) to (5.63)). Moreover, by definition, this equation tells us that $\lambda \mapsto z_\lambda$ is the Laplace transform of an **entrance law**. Instead of going back to the definition (see, e.g., [70], [42], Section XV or Exercise 4.4.8), however, let us look at the determining function of $\lambda \mapsto z_\lambda$.

To this end, let T_i, $i \in \mathbb{Z}$ be independent, exponentially distributed random variables such that $\mathbb{E}\, T_i = a_i^{-1}$. Then, as in 2.7.5, we check that

$$\int_0^\infty \mathrm{e}^{-\lambda t} \mathbb{P}\left[\sum_{j=-\infty}^{i} T_j \le t \right] \mathrm{d}t = \lambda^{-1} \pi_i,$$

and thus

$$z_\lambda = \int_0^\infty \mathrm{e}^{-\lambda t} w_t \, \mathrm{d}t,$$

where

$$w_t = \left(\mathbb{P}\left[\sum_{j=-\infty}^{i-1} T_j \le t \right] - \mathbb{P}\left[\sum_{j=-\infty}^{i} T_j \le t \right] \right)_{i \in \mathbb{Z}} \in l^1. \tag{4.35}$$

(Here, $\int_0^\infty \mathrm{e}^{-\lambda t} w_t \, \mathrm{d}t$ is not a Riemann integral as in Sections 1.1.11 and 1.2.1; rather it is calculated coordinate by coordinate.) It is clear that the ith coordinate of w_t is the probability that a chain 'starting at $-\infty$' has reached the state i before time t, but has not left it. Hence, the map $t \mapsto w_t$ fully deserves the name of 'entrance law': it describes the way the process enters the regular part of the state-space 'through $-\infty$' (see also (4.37)). Remarkably, all coordinates of w_t are nonzero, showing that before any, however small, time elapses, the particle may reach any point, though perhaps with small probability.[4]

4.3.6 *Probabilistic interpretation of (4.32)*

The first summand in formula (4.32) comes from the minimal chain, and its meaning is rather clear. Interestingly, the second term can also be nicely interpreted in probabilistic terms. To this end, we note that its ith coordinate equals

$$\lambda^{-1} \left(\sum_{j=-\infty}^{\infty} \frac{\pi_\infty \eta_j}{\pi_{j-1}} \right) \sum_{n=0}^{\infty} \pi_\infty^n (\pi_{i-1} - \pi_i).$$

Here, $\lambda^{-1} \sum_{j=-\infty}^{\infty} \frac{\pi_\infty \eta_j}{\pi_{j-1}}$ is the Laplace transform of

$$t \mapsto \sum_{j=-\infty}^{\infty} \eta_j \mathbb{P}[T_j + T_{j+1} + \cdots \le t],$$

[4] The reader might remember that a simliar effect was observed in the second Kolmogorov–Kendall–Reuter example.

which, if y is a density, is the probability that the chain starting at i with probability η_i has exploded before time t. This term, when multiplied by $\pi_\infty^n(\pi_{i-1} - \pi_i)$, speaks of the probability that before time t, P. Lévy's flash, which after the first explosion restarted at $-\infty$, managed to race through all the integers n times, and then to reach the state i but did not manage to jump to $i + 1$.

4.3.7 *Reconciliation of generators*

From Section 3.5.5 we know that the generator of the process which after explosion returns to one of randomly chosen states distributed according to a $u \in l^1$ is of the form

$$Hy = Gy + h(y)u$$

on the domain equal to $\mathcal{D}(G)$. Now, the generator of P. Lévy's flash is

$$Ay = \mathfrak{Q}y$$

and is defined on the domain composed of $y \in \mathcal{D}(\mathfrak{Q})$ such that $\mathfrak{l}_+ y = \mathfrak{l}_- y$. Can these two, apparently different types of generators describing apparently similar phenomena, be reconciled? The answer is positive: we will show that the resolvent of the operator A is the limit of the resolvents of operators of type H, that is, that A is the extended limit of the operators of type H. (Of course, by the Trotter–Kato Theorem this also implies convergence of the related semigroups, but this is not the main point here.)

More specifically, for $n \geq 1$ let A_n be the generator of the process which after explosion starts afresh at the state $-n$, so that

$$A_n y = Gy + h(y)e_{-n}, \qquad y \in \mathcal{D}(G).$$

We aim to show that

$$\lim_{n\to\infty} (\lambda - A_n)^{-1} = (\lambda - A)^{-1}.$$

To this end, we recall that, by (3.60),

$$(\lambda - A_n)^{-1} y = R_\lambda y + \frac{h(R_\lambda y)}{1 - h(R_\lambda e_{-n})} R_\lambda e_{-n}.$$

It is thus clear that to establish convergence of $(\lambda - A_n)^{-1}$ one needs to study $\lim_{n\to\infty} \frac{1}{1-h(R_\lambda e_{-n})} R_\lambda e_{-n}$. I claim that

$$\lim_{n\to\infty} h(R_\lambda e_{-n}) = \pi_\infty \tag{4.36}$$

and

$$\lim_{n\to\infty} R_\lambda e_{-n} = z_\lambda. \tag{4.37}$$

(Incidentally, the latter formula confirms the fact that z_λ is the Laplace transform of the law ruling the way probability mass enters through $-\infty$.)

Let $R_\lambda e_{-n} = \left(\xi_{i,n}\right)_{i\in\mathbb{Z}}$, $n \geq 1$. Then, by (4.26),

$$\xi_{i,n} = \begin{cases} \frac{\pi_{i-1}}{\lambda+a_i}\frac{1}{\pi_{-n-1}}, & i \geq -n, \\ 0, & i < -n. \end{cases} \tag{4.38}$$

Since, by (4.24), $h(y) = -\Sigma G y = \mathfrak{l}_+ y$ for $y \in \mathcal{D}(G)$, it follows that

$$h(R_\lambda e_{-n}) = \mathfrak{l}_+(R_\lambda e_{-n}) = \lim_{i\to\infty} \frac{\pi_i}{\pi_{-n-1}} = \frac{\pi_\infty}{\pi_{-n-1}}.$$

This establishes the first claim, because $\lim_{n\to\infty} \pi_{-n-1} = 1$.

As for the second claim, combining (4.38) with (4.31), we obtain $\|R_\lambda e_{-n}\| = \frac{\pi_{-n-1}-\pi_\infty}{\lambda\pi_{-n-1}}$, resulting in $\lim_{n\to\infty}\|R_\lambda e_{-n}\| = \lambda^{-1}(1-\pi_\infty)$. On the other hand, by the definition of z_λ, $\|z_\lambda\| = \lambda^{-1}(1-\pi_\infty)$. Therefore, because of Scheffé's Theorem, it suffices to check that $R_\lambda e_{-n}$ converges to z_λ in coordinates. This, however, is clear from (4.38).

We have established that

$$\lim_{n\to\infty} (\lambda - A_n)^{-1}\, y = R_\lambda y + \frac{h(R_\lambda y)}{1-\pi_\infty} z_\lambda, \qquad y \in l^1;$$

the right-hand side here is $(\lambda - A)^{-1}\, y$ (given in (4.32)), because, as we have seen, $h(y) = \mathfrak{l}_+ y$, $y \in \mathcal{D}(G)$.

4.4 The Joy of Entrance Laws

The study of entrance laws involves families of vectors, say, as in the previous section, z_λ, such that (see (4.33))

$$\mathfrak{Q} z_\lambda = \lambda z_\lambda, \qquad \lambda > 0.$$

The theory is to some extent analogous, or dual (but not quite analogous) to that presented in Section 3.6 (see Section 10 in [42]) – look again at equations (3.77) and (4.34). Instead of going into details of this theory, in this section we intend to play a bit with entrance laws to see how they influence the generators.

4.4.1 *A combination of returns*

Starting a process afresh after explosion through $-\infty$ according to an entrance law may of course be combined with starting at regular state-space points: this leads to generators that are characterized by both (a) a domain differing from that of the generator of the minimal chain, and (b) an extra term in the generator itself. For example, we claim that for any $p \in [0, 1]$ and density $u \in l^1$, the operator

$$Ax = \mathfrak{Q}x + (1 - p)\mathfrak{l}_+(x)u, \qquad x \in \mathcal{D}(A) \tag{4.39}$$

with $\mathcal{D}(A)$ composed of $x \in \mathcal{D}(\mathfrak{Q})$ such that

$$\mathfrak{l}_- x = p\mathfrak{l}_+ x,$$

is a Markov generator.

The chain related to A after explosion chooses one of two options: with probability p it starts afresh from $-\infty$ and with probability $1 - p$ it starts afresh from one of the regular points of the state-space; conditional on choosing the other option, the probability of starting at i is the ith coordinate of u. In particular, for $p = 1$, A is the generator of P. Lévy's flash, and for $p = 0$, the domain of A coincides with $\mathcal{D}(G)$ and A is of the form considered in 3.5.5.

For the proof that A is a generator, we note first that its domain is dense in l^1 since all e_i's belong to $\mathcal{D}(A)$. Moreover, because of the boundary condition and (4.24), $\Sigma\mathfrak{Q}x = \mathfrak{l}_- x - \mathfrak{l}_+ x = (p - 1)\mathfrak{l}_+ x$, for $x \in \mathcal{D}(A)$. It follows that

$$\Sigma Ax = \Sigma\mathfrak{Q}x + (1 - p)\mathfrak{l}_+(x)\Sigma u = (p - 1)\mathfrak{l}_+(x) + (1 - p)\mathfrak{l}_+(x) = 0$$

for all $x \in \mathcal{D}(A)$. Therefore, we are left with showing condition (ii) in the Hille–Yosida Theorem of Section 2.4.3.

By (4.33), the resolvent equation for A (i.e., the equation $\lambda x - Ax = y$) may equivalently be written as

$$(\lambda - \mathfrak{Q})(x - \mathfrak{l}_-(x)z_\lambda) - (1 - p)\mathfrak{l}_+(x)u = y.$$

Since, as we have seen in 4.3.5, $\mathfrak{l}_- z_\lambda = 1$, the vector $x - \mathfrak{l}_-(x)z_\lambda$ is a member of $\mathcal{D}(G)$ and thus $\mathfrak{Q}$ in this equation may be replaced by G. Therefore, applying R_λ to both sides yields

$$x - \mathfrak{l}_-(x)z_\lambda - (1 - p)\mathfrak{l}_+(x)R_\lambda u = R_\lambda y.$$

Recalling that $\mathfrak{l}_+ z_\lambda = \pi_\infty (= \pi_\infty(\lambda))$ and using the boundary condition we obtain

$$\mathfrak{l}_+(x)(1 - p\pi_\infty - (1 - p)\mathfrak{l}_+(R_\lambda u)) = \mathfrak{l}_+(R_\lambda y).$$

Writing the expression in parentheses as $p(1-\pi_\infty)+(1-p)(1-\mathfrak{l}_+(R_\lambda u))$ and recalling that (a) $\mathfrak{l}_+(x)=h(x)$ for $x\in\mathcal{D}(G)$, and $R_\lambda u$ belongs to $\mathcal{D}(G)$, and (b) from Section 3.5.5 we know that $1-h(R_\lambda u)>0$, we conclude that this expression is positive regardless of the choice of p. Therefore,

$$\mathfrak{l}_+(x)=\frac{\mathfrak{l}_+(R_\lambda y)}{1-p\pi_\infty-(1-p)\mathfrak{l}_+(R_\lambda u)},$$

showing that $\mathfrak{l}_+(x)$ in the resolvent equation is determined by u, y, p and λ. It follows that the only possible solution to this equation is

$$x=R_\lambda y+\frac{\mathfrak{l}_+(R_\lambda y)}{1-p\pi_\infty-(1-p)\mathfrak{l}_+(R_\lambda u)}(pz_\lambda+(1-p)R_\lambda u), \tag{4.40}$$

and it is clear that $x\geq 0$ provided $y\geq 0$. Since checking that this x is a true solution to the resolvent equation is straightforward, we are done.

4.4.2 Remark

Formula (4.40) is worth comparing to (3.99). In the terminology introduced in Exercise 4.4.8 (see the end of this chapter) the latter involves Laplace transforms of trivial entrance laws, that is, functions $\lambda\mapsto R_\lambda u_j$, whereas in the former we have a combination of two laws: one trivial ($\lambda\mapsto R_\lambda u$) and one nontrivial ($\lambda\mapsto z_\lambda$). In (4.42) we will have a combination of two trivial and two nontrivial laws. The pattern for further generalizations is clear.

4.4.3 *A combination of returns (continued)*

The probabilistic description of the process related to A, given in the previous section, is confirmed by the following approximation. Given p and u as above, let A_n, $n\geq 1$ be defined by

$$A_n x=Gx+h(x)[pe_{-n}+(1-p)u],\qquad x\in\mathcal{D}(G).$$

These are Markov generators of the type considered in 3.5.5. Since $pe_{-n}+(1-p)u$ is the distribution of the process right after explosion, it is intuitively clear that, if the semigroups $\{\mathrm{e}^{tA_n},t\geq 0\}$ converge, the limit process should behave as explained in 4.4.1. We will show that

$$\lim_{n\to\infty}\mathrm{e}^{tA_n}=\mathrm{e}^{tA}.$$

To this end, by the Trotter–Kato Theorem, it suffices to show that $\lim_{n\to\infty}(\lambda-A_n)^{-1}=(\lambda-A)^{-1}$, and an explicit formula for $(\lambda-A)^{-1}$ is given in (4.40). On the other hand, recalling (3.60) and the fact that $\mathfrak{l}_+(R_\lambda y) = h(R_\lambda y)$ for $y \in l^1$, we see that $(\lambda - A_n)^{-1}\, y$ equals

$$R_\lambda y + \frac{\mathfrak{l}_+(R_\lambda y)}{1 - \mathfrak{l}_+(pR_\lambda e_{-n} + (1-p)R_\lambda u)} R_\lambda(pe_{-n} + (1-p)u).$$

Since in Section 4.3.7 we have established that $\lim_{n\to\infty} \mathfrak{l}_+(R_\lambda e_{-n}) = \pi_\infty$, the denominator here converges to the denominator in (4.40), and (4.37) completes the proof.

4.4.4 *Two entrance laws: Chain's description*

It is probably amply clear that there could be many entrance laws. Each (nontrivial) entrance law, very much like elements of the exit boundary, may be seen as an additional point of the state-space (cf. Section 4.2) and such an entrance law forces a boundary condition on the operator $\mathfrak{Q}$ of Section 3.4. We will look at an example with two entrance laws, leading to two boundary conditions: see, for example, [56] (flash of flashes) for an example with infinitely many entrance laws.

The chain we have in mind is composed of two flashes on two copies of $\mathbb{Z}$, communicating through two minus infinities. For its construction, along with the sequence $a_i, i \in \mathbb{Z}$ of (4.23) we need an analogous one,

$$\sum_{i\in\mathbb{Z}} (a_i')^{-1} < \infty,$$

to rule the chain's behavior on the second copy of $\mathbb{Z}$: When on the first copy of $\mathbb{Z}$, our process is a pure birth chain with intensities $a_i, i \in \mathbb{Z}$; when on the second copy, denoted $\mathbb{Z}'$, it is a pure birth chain with intensities $a_i', i \in \mathbb{Z}'$.

Our chain differs from the minimal chain in that after exploding through $\mathbb{Z}$ it has three possibilities: (a) with probability p it may enter $\mathbb{Z}$ through $-\infty$ as in the preceding sections, (b) with probability q it may start afresh at one of the points of $\mathbb{Z}$ (conditionally on choosing this option, the probability of starting at an $i \in \mathbb{Z}$ is the ith coordinate of a given density u), or (c) with probability $1 - p - q \geq 0$ it may enter $\mathbb{Z}'$ with entrance law defined as in (4.35), but with $a_i, i \in \mathbb{Z}$ replaced by $a_i', i \in \mathbb{Z}'$. (It would not harm to allow the density u to have nonzero entries on $\mathbb{Z}'$, but it is more fun to restrict the roads of communication between $\mathbb{Z}$ and $\mathbb{Z}'$ to those that are 'indirect,' that is, lead through boundaries.) Analogously, after explosion through $\mathbb{Z}'$ the process may either (with probability p') return to $\mathbb{Z}'$ via $-\infty$ with entrance law just described, return to $\mathbb{Z}'$ with the help of density u' (with probability q'), or (with probability $1 - p' - q'$) enter $\mathbb{Z}$ via $-\infty$ according to w_t of (4.35).

4.4.5 *Two entrance laws: The minimal chain*

The description given in the previous section, fully characterizes the matrix of intensities, and the operator $\mathfrak{Q}$ of Section 3.4: the ith coordinate of $\mathfrak{Q}\,(\xi_i)_{i\in\mathbb{Z}\cup\mathbb{Z}'}$ is either $a_{i-1}\xi_{i-1} - a_i\xi_i$ (if $i \in \mathbb{Z}$) or $a'_{i-1}\xi_{i-1} - a'_i\xi_i$ (if $i \in \mathbb{Z}'$). It follows that for $x = (\xi_i)_{i\in\mathbb{Z}\cup\mathbb{Z}'} \in \mathcal{D}(\mathfrak{Q})$ the following four limits exist:

$$\mathfrak{l}_+(x) := \lim_{i\in\mathbb{Z}, i\to\infty} a_i\xi_i, \qquad \mathfrak{l}_-(x) := \lim_{i\in\mathbb{Z}, i\to\infty} a_{-i}\xi_{-i},$$

and

$$\mathfrak{l}'_+(x) := \lim_{i\in\mathbb{Z}', i\to\infty} a'_i\xi_i, \qquad \mathfrak{l}'_-(x) := \lim_{i\in\mathbb{Z}', i\to\infty} a'_{-i}\xi_{-i}.$$

There are obviously two sojourn sets here: $\mathbb{Z}$ and $\mathbb{Z}'$, with corresponding functionals h, h' (see Theorem 3.7.1) given by

$$h(x) = -\sum_{i\in\mathbb{Z}}(a_{i-1}\xi_{i-1} - a_i\xi_i) = \mathfrak{l}_+(x) - \mathfrak{l}_-(x)$$

and

$$h'(x) = -\sum_{i\in\mathbb{Z}'}(a'_{i-1}\xi_{i-1} - a'_i\xi_i) = \mathfrak{l}'_+(x) - \mathfrak{l}'_-(x).$$

Moreover, the domain of the generator G of Kato's minimal semigroup is characterized by

$$\mathfrak{l}_-x = \mathfrak{l}'_-x = 0, \qquad x \in \mathcal{D}(G). \tag{4.41}$$

In particular,

$$h(x) = \mathfrak{l}_+(x) \qquad \text{and} \qquad h'(x) = \mathfrak{l}'_+(x), \qquad x \in \mathcal{D}(G).$$

4.4.6 *Two entrance laws: Approximation*

Let us look at the following approximation. For each $n \geq 1$, let A_n be the Markov generator of type (3.92):

$$A_n x = Gx + h(x)u_n + h'(x)u'_n, \qquad x \in \mathcal{D}(G)$$

where

$$u_n := pe_{-n} + qu + (1-p-q)e_{-n'},$$
$$u'_n := p'e_{-n'} + q'u' + (1-p'-q')e_{-n},$$

and $-n'$ is $-n$, seen as a member of $\mathbb{Z}'$. This chain differs from the one described in Section 4.4.4 in that, after explosion, instead of going directly to a distributing hub at one of two minus infinities, it starts afresh at $-n$ or $-n'$. Of course, we hope that as $n \to \infty$ the chains related to A_n's converge to our chain.

From the proof of Theorem 3.7.9 we know that the resolvent of A_n is given by

$$(\lambda - A_n)^{-1} y = R_\lambda y + \alpha_n R_\lambda u_n + \alpha'_n R_\lambda u'_n$$

where

$$(\alpha_n, \alpha'_n) = (h(R_\lambda y), h'(R_\lambda y)) \cdot (I - M_n)^{-1}$$

and M_n is the 2×2 matrix

$$M_n := \begin{pmatrix} h(R_\lambda u_n) & h'(R_\lambda u_n) \\ h(R_\lambda u'_n) & h'(R_\lambda u'_n) \end{pmatrix}.$$

Using (4.37) and arguing similarly for $\lim_{n\to\infty} R_\lambda e_{-n'}$ we see that

$$\lim_{n\to\infty} R_\lambda u_n = p z_\lambda + q R_\lambda u + (1 - p - q) z'_\lambda,$$

$$\lim_{n\to\infty} R_\lambda u'_n = p' z'_\lambda + q' R_\lambda u' + (1 - p' - q') z_\lambda$$

where, not surprisingly,

$$z'_\lambda := \left(\frac{\pi'_{i-1}}{\lambda + a'_i} \right)_{i \in \mathbb{Z}'} \qquad \text{and} \qquad \pi'_i = \prod_{j=-\infty}^{i} \frac{a'_j}{\lambda + a'_j}.$$

Hence, it is clear that convergence of $(\lambda - A_n)^{-1}$ hinges on convergence of α_n's.

On the other hand, using (4.36), arguing similarly to see that

$$\lim_{n\to\infty} h'(R_\lambda e_{-n'}) = \pi'_\infty (:= \lim_{n\to\infty} \pi'_n),$$

and noting that $h'(R_\lambda e_{-n}) = h(R_\lambda e_{-n'}) = 0$, we conclude that the matrices M_n converge to

$$M = \begin{pmatrix} p\pi_\infty + q h(R_\lambda u) & q h'(R_\lambda u) + (1 - p - q)\pi'_\infty \\ q' h(R_\lambda u') + (1 - p' - q')\pi_\infty & p'\pi'_\infty + q' h'(R_\lambda u') \end{pmatrix}.$$

From the proof of Theorem 3.7.9 we know also that $h(R_\lambda u) + h'(R_\lambda u) < 1$. Since π_∞ and π'_∞ are smaller than 1, the sum of elements in the first row here is $< p + q + (1 - p - q) = 1$, and analogously we show that also the sum of elements of the second row is smaller than 1. Hence, as in the proof of Lemma 3.7.10, M can be thought of as a sub-Markov operator in $\mathbb{R}^2$ with norm smaller than 1. In particular, $(I - M)^{-1}$ exists, and continuity of taking inverse implies that $\lim_{n\to\infty} \alpha_n$ and $\lim_{n\to\infty} \alpha'_n$ converge to α and α', respectively, where

$$(\alpha, \alpha') = (h(R_\lambda y), h'(R_\lambda y)) \cdot (I - M)^{-1}.$$

We have thus established that $\lim_{n\to\infty}(\lambda - A_n)^{-1}\, y$ exists, and equals

$$R_\lambda y + \alpha(pz_\lambda + qR_\lambda u + (1-p-q)z'_\lambda) + \alpha'(p'z'_\lambda + q'R_\lambda u' + (1-p'-q')z_\lambda).$$

Rearranging yields

$$\begin{aligned}\lim_{n\to\infty}(\lambda - A_n)^{-1}\, y = {}& R_\lambda y + q\alpha R_\lambda u + q'\alpha' R_\lambda u' \\ &+ (p\alpha + (1-p'-q')\alpha')z_\lambda \\ &+ (p'\alpha' + (1-p-q)\alpha)z'_\lambda. \end{aligned} \tag{4.42}$$

4.4.7 *Two entrance laws: The generator*

The limit formula (4.42) probably does not look very informative, not at a first glance anyway, especially if we realize that α and α' are given implicitly: It is clear that, since z_λ and z'_λ do not belong do $\mathcal{D}(G)$, in the range of the limit pseudo-resolvent (we have not yet proved that in fact it is a resolvent!) relations (4.41) hold no longer, and that perhaps some other relations between $\mathfrak{l}_+$, $\mathfrak{l}_-$, $\mathfrak{l}'_+$ and $\mathfrak{l}'_-$ can be discovered. Nevertheless, it is difficult to guess these relations from (4.42).

Fortunately, probabilistic intuition tells us that they should read

$$\begin{aligned}\mathfrak{l}_-(x) &= p\mathfrak{l}_+(x) + (1-p'-q')\mathfrak{l}'_+(x),\\ \mathfrak{l}'_-(x) &= (1-p-q)\mathfrak{l}_+(x) + p'\mathfrak{l}'_+(x)\end{aligned} \tag{4.43}$$

(think of $\mathfrak{l}_-$ and $\mathfrak{l}'_-$ as of the 'probabilities' that the process starts afresh at $-\infty$ and $-\infty'$, and $\mathfrak{l}_+$ and $\mathfrak{l}'_+$ as of the probabilities that the process explodes through ∞ and ∞', respectively). Therefore, we venture to claim that the resolvents $(\lambda - A_n)^{-1}$, $n \ge 1$ converge to $(\lambda - A)^{-1}$ where

$$Ax = \mathfrak{Q}x + q\mathfrak{l}_+(x)u + q'\mathfrak{l}'_+(x)u',$$

and the domain of A is composed of $x \in \mathcal{D}(\mathfrak{Q})$ satisfying these two conditions. Even if $q, q' < 1$, this operator is a Markov generator in contrast to the operator given by the same formula, but defined on $\mathcal{D}(G)$. A chain related to the latter, as we know from Section 3.7.9, may lose some probability mass while going through explosion, its probability of starting afresh being q or q' depending on which way explosion came about.

For the proof we note first that A is densely defined, because all e_i, $i \in \mathbb{Z}\cup\mathbb{Z}'$ belong to $\mathcal{D}(A)$. Moreover, for $x \in \mathcal{D}(A)$,

$$\begin{aligned}\Sigma Ax &= \Sigma\mathfrak{Q}x + q\mathfrak{l}_+(x) + q'\mathfrak{l}'_+(x)\\ &= \mathfrak{l}_-(x) - \mathfrak{l}_+(x) + \mathfrak{l}'_-(x) - \mathfrak{l}'_+(x) + q\mathfrak{l}_+(x) + q'\mathfrak{l}'_+(x),\end{aligned}$$

which, because of the boundary conditions, reduces to zero. Thus, we are left with studying the resolvent equation

$$\lambda x - \mathfrak{A} x - q\mathfrak{l}_+(x)u - q'\mathfrak{l}'_+(x)u' = y.$$

As in the previous proofs of this type, we aim to show that $\mathfrak{l}_+(x)$ and $\mathfrak{l}_-(x)$ are determined by y, u, u', λ, q and q' only. To this end, we recall that $\mathfrak{l}_-(z_\lambda) = \mathfrak{l}'_-(z'_\lambda) = 1$, $\mathfrak{l}'_-(z_\lambda) = \mathfrak{l}_-(z'_\lambda) = 0$, and $\mathfrak{A} z_\lambda = \lambda z_\lambda$ and $\mathfrak{A} z'_\lambda = \lambda z'_\lambda$. Therefore, the resolvent equation can be rewritten as

$$(\lambda - \mathfrak{A})(x - \mathfrak{l}_-(x)z_\lambda - \mathfrak{l}'_-(x)z'_\lambda) - q\mathfrak{l}_+(x)u - q'\mathfrak{l}'_+(x)u' = y,$$

or, since $x - \mathfrak{l}_-(x)z_\lambda - \mathfrak{l}'_-(x)z'_\lambda$ belongs to $\mathcal{D}(G)$, as

$$x - \mathfrak{l}_-(x)z_\lambda - \mathfrak{l}'_-(x)z'_\lambda - q\mathfrak{l}_+(x)R_\lambda u - q'\mathfrak{l}'_+(x)R_\lambda u' = R_\lambda y. \tag{4.44}$$

Applying $\mathfrak{l}_+$ to both sides of this equality, using $\mathfrak{l}_+(z_\lambda) = \pi_\infty$, $\mathfrak{l}_+(z'_\lambda) = \pi'_\infty$, the first of the relations following (4.41) and the boundary conditions, we see that

$$(1 - p\pi_\infty - qh(R_\lambda u))\mathfrak{l}_+(x) - (q'h(R_\lambda u') + (1 - p' - q')\pi_\infty)\mathfrak{l}'_+(x) = h(R_\lambda y).$$

Analogously, applying $\mathfrak{l}'_+(x)$, we obtain

$$(1 - p'\pi'_\infty - q'h'(R_\lambda u'))\mathfrak{l}'_+(x) - (qh'(R_\lambda u) + (1 - p - q)\pi'_\infty)\mathfrak{l}_+(x) = h'(R_\lambda y).$$

These two relations show that $\mathfrak{l}_+(x)$ and $\mathfrak{l}'_+(x)$ satisfy the system

$$(\mathfrak{l}_+(x), \mathfrak{l}'_+(x)) \cdot (I - M) = (h(R_\lambda y), h'(R_\lambda y))$$

for the matrix M of the previous section. Since it was established there that $I - M$ is invertible, these two quantities may indeed be determined without prior knowledge of x, and in fact are identical to α and α' considered there. Hence, by (4.44), the only possible solution to the resolvent equation is

$$x = R_\lambda y + q\alpha R_\lambda u + q'\alpha' R_\lambda u' + \mathfrak{l}_-(x)z_\lambda + \mathfrak{l}'_-(x)z'_\lambda,$$

where unknown $\mathfrak{l}_-(x)$ and $\mathfrak{l}'_-(x)$ must be replaced by linear combinations of the known $\mathfrak{l}_+(x) = \alpha$ and $\mathfrak{l}'_+(x) = \alpha'$, according to the prescription of the boundary conditions:

$$\mathfrak{l}_-(x) = p\alpha + (1 - p' - q')\alpha',$$
$$\mathfrak{l}'_-(x) = (1 - p - q)\alpha + p'\alpha'.$$

This shows that x is identical to the vector of equation (4.42). Checking that x belongs to $\mathcal{D}(A)$ and is a true solution to the resolvent equation is straightforward if perhaps tedious.

4.4.8 Exercise

Let $\{S(t), t \geq 0\}$ be a minimal semigroup for a given Kolmogorov matrix. A bounded function $(0, \infty) \ni t \mapsto w_t \in (l^1)^+$ taking nonnegative values is called an entrance law if $S(s)w_t = w_{s+t}, s, t \in (0, \infty)$.

(a) Check that if $u \in l^1$ is nonnegative, then $w_t = S(t)u$ is an entrance law. (Such laws are sometimes referred to as **trivial**.)

(b) It may be proved that any entrance law is continuous in $t \in (0, \infty)$ and the limit $\lim_{t \to 0+} w_t$ in the sense of convergence of coordinates exists (see [71]). This allows thinking of the Laplace transform of w_t (with the integral in the sense of coordinate convergence). Check that the Laplace transform z_λ of an entrance law satisfies (4.34).

4.4.9 Exercise

Let $\mathfrak{Q}$ be the operator of Section 3.4 related to the intensity matrix (2.53). Check that

$$z_\lambda := (0, \pi_2, \pi_3, \dots), \qquad \lambda > 0,$$

where $\pi_i, i \geq 2$ are defined below (2.33), satisfies $\mathfrak{Q}z_\lambda = \lambda z_\lambda$. This means that $z_\lambda, \lambda > 0$ is the Laplace transform of an entrance law.

4.4.10 Exercise

Prove convergence of the semigroups considered in 4.4.3 using the Sova–Kurtz Theorem (and not the Trotter–Kato Theorem). More precisely, without alluding to explicit forms of $(\lambda - A_n)^{-1}$, given $x \in \mathcal{D}(A)$ find $x_n \in \mathcal{D}(A_n)$ such that $\lim_{n\to\infty} x_n = x$ and $\lim_{n\to\infty} A_n x_n = Ax$. Can you do the same for the semigroups of Sections 4.4.6 and 4.4.7?

4.5 Notes

Section 4.1 is a variation of the material presented in [41]. The flash of Section 4.3.1 comes from P. Lévy's paper [63] (p. 366, example 3°); in the context of semigroups of operators it has been analyzed in [56].

5

The Dual Perspective

Up until now, we focused on the analysis of the family of operators

$$x \mapsto x \cdot P(t), \qquad t \geq 0, \tag{5.1}$$

where $x \in l^1(\mathbb{I})$ is thought of as a row-vector, and $P(t)$ is the matrix of transition probabilities of a given chain. Since nonnegative x's in $l^1(\mathbb{I})$ can be identified with measures on $\mathbb{I}$, we were thus studying dynamics of distributions of the Markov chain in time. Under a mild assumption on continuity of $P(\cdot)$, that is, under assumption (2.8), the family in question is a strongly continuous semigroup, and thus we were able to describe it by means of a single closed operator in l^1, that is, its generator. We have also learned quite a bit on how the information on the process is encrypted in this generator. In this chapter, we would like to look at this dynamics from a dual perspective.

More specifically, we will study the operators

$$f \mapsto P(t) \cdot f, \qquad t \geq 0 \tag{5.2}$$

where f is a bounded column-vector, that is, a member of $l^\infty(\mathbb{I})$. Analytically, the map (5.2) is dual to (5.1); probabilistically, it gives dynamics of expected values (see Section 5.1.2). In distinction to (5.1), however, the operators (5.2) rarely form a strongly continuous semigroup in $l^\infty(\mathbb{I})$: D. Williams's Theorem says that this happens under very restrictive assumptions on the related matrix of intensities.

Fortunately, the sun-dual space $\mathbb{X}^\odot \subset l^\infty(\mathbb{I})$ for the underlying semigroup is often sufficiently large, and thus we are able to recover the map (5.1) from (5.2) restricted to $\mathbb{X}^\odot$. Therefore, we are again able to describe the process by means of a single operator; this time this operator is the generator of the sun-dual semigroup (see our Sections 5.6–5.8, for instance). Interestingly, the information on the process which in l^1 was expressed in additional terms of the generator (like in (3.92)) in $l^\infty(\mathbb{I})$ quite often impacts the domain of the dual of the generator rather than the way the latter operator acts, and through its domain – the 'shape' of the sun-dual space (cf. Theorem 1.5.6). In other words, the 'shape' of the sun-dual space may contain information on the process.

A successful analysis with semigroup-analytic tools is also possible if the sun-dual space is not fully known. Namely, it suffices to find a sufficiently large subspace of $\mathbb{X}^{\odot}$ where the formula (5.2) leads to a strongly continuous semigroup – then the dynamics is again encrypted in a single operator, that is, the generator of the semigroup defined in this subspace. (This subspace must be sufficiently large so that (5.1) can be recovered from (5.2).) It was the idea of W. Feller that the role of this subspace may be successfully played by c_0, the space of sequences converging to zero. This fundamental idea was crucial for studying Markov processes with state-space more complicated than a countable set $\mathbb{I}$, that is, with a state-space that is locally compact. In fact, in the latter spaces studying *directly* dynamics of distributions (i.e., studying an analogue of (5.2)) by means of semigroups of operators is quite impossible: even the simplest processes do not lead to strongly continuous semigroups in the space of measures (see, e.g., Section 1.5.2) – the norm in this space is usually too strong to be compatible with delicate stochastic phenomena. (See, however, e.g., [23, 29, 49, 62, 94] and the references given there for nontrivial examples of stochastic processes that can be treated in this norm.)

After a short discussion of Williams's Theorem (in Section 5.1), we commence this chapter with a study of semigroups of Markov chains with Feller property (Sections 5.2 – 5.4). As it transpires, the requirement that c_0 be an invariant subspace for $\{P^*(t), t \geq 0\}$ and that $\{P^*(t), t \geq 0\}$, as restricted to c_0, be strongly continuous is not too limiting: a rather large class of Markov chains possesses the Feller property. Such chains are more regular than other chains; in particular, no state in a chain with Feller property is instantaneous, and the entire chain may be described by a natural operator built by means of the corresponding intensity matrix. These chains are in fact so regular that restarting after explosion is too strange a phenomenon for them: all a Feller semigroup may describe is the minimal chain. Manipulating the generators of Feller semigroups is also comparatively simple: we discuss a version of the Hille–Yosida Theorem for such semigroups, and – in Section 5.5 – explicit formulae for two types of perturbed Feller semigroups, devised by R. Feynman and M. Kac and by V. A. Volkonskii.

Next, in Section 5.6, we come back to the examples of Kolmogorov, Kendall and Reuter. The related chains do not have the Feller property, but we are able to explicitly characterize their sun-dual semigroups. We discuss how the basic information on the chains is hidden in the shape of the sun-dual space and in the domain of generator of the sun-dual semigroup.

The same theme is developed in Sections 5.7–5.9, devoted to sun-duals of the semigroups dominating Kato's minimal semigroup. In particular, we discuss in detail the curious fact that a discrete exit boundary point, expressed as an additional term in the generator of a Markov semigroup (see (3.92)), in the dual space does not affect the way the generator acts but rather the domain of the master operator Q, introduced in Section 5.7.1. Therefore, an additive perturbation of the operator in l^1, in the dual space l^∞ becomes a perturbation of the domain, thus influencing the shape of the sun-dual space. Perhaps even more surprisingly, entrance boundary points, which in l^1 affect the domain, here modify the domain as well: in a fully symmetrical world one could expect that a perturbation of the boundary would become an additive perturbation of the generator (as, e.g., in [17]; see also [16], Chapter 50). This seems indeed to be the case, as exemplified by the second Kolmogorov–Kendall–Reuter semigroup (see Section 5.6.5), if an entrance boundary is not coupled with an exit boundary. If both types of boundary are present, however, all sun-dual semigroups of the semigroups dominating the minimal semigroup are obtained as suitable restrictions of Q. This is in contrast to the situation observed in l^1: there, to describe nonminimal semigroups one needs to either modify the way the 'master operator' $\mathfrak{Q}$ of Section 3.4 acts (as in (3.92)), or modify its domain (as in P. Lévy's flash), or both at the same time (as in Section 4.4.7).

5.1 Preliminaries

5.1.1 *Reminder of notational conventions*

As in the previous chapter, we focus on the case where the Markov chain's state-space is $\mathbb{N}$. By the Steinhaus Theorem (see, e.g., [14], Sections 5.2.3 and 5.2.16), the dual space to $l^1 := l^1(\mathbb{N})$ is $l^\infty = l^\infty(\mathbb{N})$. As before, elements of the former space will be denoted $(\xi_i)_{i\geq 1}$, $(\eta_i)_{i\geq 1}$, and so on, whereas elements of the latter space will be denoted f, g, and so on, and seen as bounded functions on the set of natural numbers. When needed, $f \in l^\infty$ will be identified with the sequence $(f(i))_{i\geq 1}$ where, of course, $f(i)$ is the value of f at $i \in \mathbb{N}$. As often as not, though, we will rightfully see f as a functional on l^1 and write $f(x)$ to denote $\sum_{i=1}^{\infty} f(i)\xi_i$ for $x = (\xi_i)_{i\geq 1}$. However, since double parenthesis does not look right, instead of $f((\xi_i)_{i\geq 1})$ we will prefer to write $f\ (\xi_i)_{i\geq 1}$.

The spaces c and c_0 of convergent sequences, and of sequences converging to 0, respectively, are seen as natural subspaces of l^∞. For $f \in c$, we will write $f(\infty)$ to denote $\lim_{i\to\infty} f(i)$:

$$f(\infty) = \lim_{i\to\infty} f(i).$$

The functional Σ, playing so important a role in the previous chapters, as seen as an element of c is simply a constant function equal to 1 for all i. In this new context it will sometimes be more natural to denote it $\mathbb{1}$. So, remember that

$$\boxed{\mathbb{1} = \Sigma}.$$

5.1.2 *Dual dynamics*

A semigroup $\{P(t), t \ge 0\}$ of Markov operators in l^1 describes dynamics of distributions of the underlying Markov chain $X(t), t \ge 0$: if x is the initial distribution of the chain then $P(t)x$ is its distribution at time $t \ge 0$. In the theory of stochastic processes, along with dynamics of distributions, it is customary to consider dynamics of expected values; in fact, if the state-space of the process is not discrete, the latter approach is more common and useful. To explain, think of a bounded function $f : \mathbb{N} \to \mathbb{R}$ and of the expected value

$$g(i) = \mathbb{E}\,(f(X(t))|X(0) = i). \tag{5.3}$$

If $f(i)$ is the cost of staying at i (or the prize for being at i), then this quantity is an expected cost (expected prize) at time t provided we started at i.

What is the relation of the map $f \mapsto g$ to the Markov operators $\{P(t), t \ge 0\}$? In terms of transition probabilities,

$$g(i) = \sum_{j=1}^{\infty} p_{i,j}(t) f(j).$$

It follows that, treating g as a functional on l^1, for any $x = (\xi_i)_{i\ge 1} \in l^1$,

$$\begin{aligned} g(x) &= \sum_{i=1}^{\infty} g(i)\xi_i = \sum_{i=1}^{\infty}\sum_{j=1}^{\infty} p_{i,j}(t) f(j)\xi_i = \sum_{j=1}^{\infty} f(j) \sum_{i=1}^{\infty} p_{i,j}(t)\xi_i \\ &= f(P(t)x), \end{aligned}$$

which simply means that $g = P^*(t)f := [P(t)]^* f$. Thus, dynamics (5.3) is governed by the dual semigroup:

$$P^*(t)f(i) = \mathbb{E}\,(f(X(t))|X(0) = i) = \sum_{j=1}^{\infty} p_{i,j}(t) f(j). \tag{5.4}$$

5.1.3 A note

Here and in what follows we write $P^*(t)f(i)$ instead of more appropriate but less readable $([P(t)]^* f)(i)$ or $(P^*(t)f)(i)$, but it should be clear to the reader that the argument of $P^*(t)$ is the function f and not the number $f(i)$.

5.1.4 Williams's Theorem

From Section 1.5 we know that in general the dual of a strongly continuous semigroup need not be strongly continuous. As noted by D. Williams (see [75], Section 6; his remark was later greatly generalized by H.P. Lotz [65]), in the context of sub-Markov semigroups in l^1 this statement may be made more specific: condition $\lim_{t\to 0+} P^*(t)f = f, f \in l^\infty$ implies $\lim_{t\to 0+} \|P^*(t) - I\| = 0$. In other words, by assuming that the dual semigroup is strongly continuous in l^∞ we in fact rule out all semigroups with unbounded generators.

As we shall see now, this result is a consequence of Schur's Theorem discussed in Section 1.6.4. Suppose $A_n, n \ge 1$ are bounded linear operators in a Banach space $\mathbb{X}$ such that for all $x \in \mathbb{X}$,

$$\lim_{n\to\infty} A_n x = 0. \tag{5.5}$$

(Think of $A_n = P^*(t_n) - I_{l^\infty}$, where $\lim_{n\to\infty} t_n = 0$.) In general this condition does not imply

$$\lim_{n\to\infty} \|A_n\| = 0, \tag{5.6}$$

even if $\mathbb{X}$ is the dual of a Banach space $\mathbb{Y}$ ($\mathbb{X} = \mathbb{Y}^*$) and A_n's are duals of operators in $\mathbb{Y}$. For example, if $\mathbb{Y} = c_0$ and B is the shift to the right, then $A = B^*$ is the shift to the left in $l^1 = (c_0)^*$, and it is easy to see that for $A_n := A^n = (B^n)^*$ condition (5.5) holds whereas $\|A_n\| = 1$.

However, if $\mathbb{Y} = l^1$ and $A_n = B_n^*$ for certain $B_n \in \mathcal{L}(l^1)$ (in the case of interest to us, $B_n = P(t_n) - I_{l^1}$), then due to the curious property of l^1 described in Section 1.6.4, (5.5) implies (5.6).

For, if (5.6) does not hold, there is a $c > 0$ and infinitely many $n \in \mathbb{N}$ such that $\|B_n\| \ge c$. Therefore, there are $y_n \in l^1$ with $\|y_n\| = 1$ such that

$$\|B_n y_n\| \ge \frac{c}{2} \qquad \text{for infinitely many } n. \tag{5.7}$$

On the other hand, since

$$|f(B_n y_n)| = |(A_n f)(y_n)| \le \|A_n f\|,$$

for any $f \in l^\infty$, assumption (5.5) (the role of x is now played by f) implies that $(B_n y_n)_{n\geq 1}$, which is a sequence of elements of l^1, converges weakly to zero. By 1.6.4, it follows that $\lim_{n\to\infty} \|B_n y_n\| = 0$, which clearly contradicts (5.7). This shows that (5.6) is true.

In the case of interest to us we obtain that $\lim_{t\to 0+} P^*(t)f = f, f \in l^\infty$ implies

$$\lim_{t\to 0+} \|P^*(t) - I\| = 0,$$

as claimed (or, which is the same $\lim_{t\to 0+} \|P(t) - I\| = 0$).

5.1.5 Remark

Williams's Theorem should not be thought of as negative. This result simply reflects the somewhat strange nature of the space l^∞, inherited from l^1 (see also, e.g., Chapter 6 in [20]). In fact, Williams's Theorem opens new opportunities for us: in general, the dual dynamics is governed by a semigroup that is strongly continuous on a subspace of l^∞, which is the sun-dual $\mathbb{X}^\odot$ for the original semigroup $\{P(t), t \geq 0\}$ in l^1, and, as we shall see, along with the information on the process contained in the generator of the sun dual semigroup, *some information may be hidden in the 'size' and 'shape' of* $\mathbb{X}^\odot$. Thus, in a sense, working with l^∞ is like learning a new, mysterious language. We will be able to say about it more when commenting on particular cases of Markov chains, see in particular Sections 5.6–5.8.

5.2 Markov Chains with Feller Property

We start with Markov chains that possess Feller property: Let c_0 be the subspace of l^∞ composed of f such that $\lim_{i\to\infty} f(i) = 0$. A strongly continuous semigroup $\{P(t), t \geq 0\}$ of sub-Markov operators in l^1 is said to have the Feller property[1] if the dual semigroup maps c_0 into itself, and as restricted to c_0 is a strongly continuous semigroup there. In other words, we require that

$$P^*(t)f \in c_0 \qquad \text{whenever} \qquad f \in c_0 \tag{5.8}$$

and that

$$\lim_{t\to 0+} P^*(t)f = f, \qquad \text{for all} \qquad f \in c_0, \tag{5.9}$$

in the supremum norm of c_0.

[1] We will also say that the related chain has Feller property or that its transition probabilities have Feller property.

It is clear that these two conditions speak something about $\mathbb{X}^\odot$: the second requires that $c_0 \subset \mathbb{X}^\odot$ and the first says that c_0 is invariant for $\{P^*(t), t \geq 0\}$ (as is $\mathbb{X}^\odot$). It is perhaps as clear that c_0 may be a subspace of $\mathbb{X}^\odot$ without being its invariant subspace. Nevertheless, we note that an appropriate example is furnished in Section 5.6.2.

In the next two (sub-)sections we characterize (5.8) and (5.9) in terms of transition probabilities of the underlying Markov chain. In this analysis we will use the fact that the vectors e_j, $j \geq 1$ defined in (2.20) (now treated as elements of c_0) form a linearly dense subset of c_0: any element of c_0 can be approximated with arbitrary accuracy by (finite) linear combinations of e_j, $j \geq 1$. Indeed, for any $f \in c_0$, the distance between f and

$$f_n := \sum_{i=1}^{n} f(i) e_i \tag{5.10}$$

does not exceed $\sup_{i \geq n+1} |f(i)|$, which can be made as small as we wish by choosing n large enough.

5.2.1 *Strong continuity*

Condition (5.9) says that $c_0 \subset \mathbb{X}^\odot$, that is, that the sun dual space is appropriately 'large.' What does this mean for the transition probabilities of the underlying Markov chain? Taking $f = e_j$ in (5.9), we see that it implies that

$$\lim_{t \to 0+} \sup_{i \neq j} p_{i,j}(t) = 0 \qquad \text{for all} \qquad j \in \mathbb{N}. \tag{5.11}$$

Conversely, combined with (2.8) (which is a necessary and sufficient condition for $\{P(t), t \geq 0\}$ to be strongly continuous), (5.11) implies that (5.9) holds for all e_j, $j \geq 1$. Therefore, a three epsilon argument shows that (5.11) combined with (2.8) implies (5.9): given f and $\epsilon > 0$ we choose n so large that $\|f_n - f\| < \epsilon$ (see (5.10)), and then t so small that $\|P^*(t) f_n - f_n\| < \epsilon$. Then,

$$\begin{aligned} \|P^*(t) f - f\| &\leq \|P^*(t)(f - f_n)\| + \|P^*(t) f_n - f_n\| + \|f_n - f\| \\ &\leq 2\|f_n - f\| + \|P^*(t) f_n - f_n\| \leq 3\epsilon; \end{aligned}$$

the norm of $P^*(t)$ (even as an operator in l^∞) does not exceed one (since neither does the norm of $P(t)$). A take home message is thus:

$\mathbb{X}^\odot$ contains c_0 iff (2.8) and (5.11) are satisfied.

5.2.2 *Invariance of* c_0

Condition (5.8), in turn, says that c_0 is invariant for $\{P^*(t), t \ge 0\}$. Aiming at characterizing this condition in terms of transition probabilities we note that for each i (and fixed t),

$$x_i := \big(p_{i,j}(t)\big)_{j\in\mathbb{N}}$$

(the ith row in the matrix representing $P(t)$) is a member of l^1, and the latter space is the dual space for c_0: $c_0^* = l^1$. From this perspective, by (5.4), the requirement that $\lim_{i\to\infty} P^*(t)f(i) = \lim_{i\to\infty} x_i(f) = 0$ for all $f \in c_0$ is the requirement that $\lim_{i\to\infty} x_i = 0$ in the weak* topology. For a yet simpler and more practical characterization, taking $f = e_k$ we see that (5.8) implies that

$$\lim_{i\to\infty} p_{i,k}(t) = 0 \tag{5.12}$$

for all $t \ge 0$ and $k \in \mathbb{N}$. (This is the limit along the 'columns' of the matrix representing $P(t)$.) Hence, arguing as in the previous section (i.e., utilizing f_n's) we conclude that (5.12) is necessary and sufficient for c_0 to be invariant for $\{P^*(t), t \ge 0\}$.

5.2.3 *Corollary: A criterion for the Feller property*

Suppose that $\{P(t), t \ge 0\}$ and $\big\{\widetilde{P}(t), t \ge 0\big\}$ are two strongly continuous sub-Markov semigroups in l^1 such that

$$\widetilde{P}(t) \le P(t), \qquad t \ge 0.$$

If $\{P(t), t \ge 0\}$ has the Feller property, then so does $\big\{\widetilde{P}(t), t \ge 0\big\}$.

Proof In terms of the related transition probabilities, say, $p_{i,j}(t)$ and $\widetilde{p}_{i,j}(t)$, respectively, our assumption says that

$$\widetilde{p}_{i,j}(t) \le p_{i,j}(t), \qquad i, j \in \mathbb{N}, t \ge 0.$$

It follows that, if transition probabilities $p_{i,j}(t)$ satisfy (5.11) and (5.12), then so do the probabilities $\widetilde{p}_{i,j}(t)$. □

5.2.4 *Feller semigroups*

If $\{P(t), t \ge 0\}$ has the Feller property, that is, if the transition probabilities satisfy conditions (2.8), (5.11) and (5.12), the semigroup of operators $\{T(t), t \ge 0\}$ in c_0, where $T(t)$ is $P^*(t)$ restricted to c_0, is called the corresponding Feller semigroup. Any Feller semigroup $\{T(t), t \ge 0\}$ is thus a strongly continuous semigroup of contractions in c_0 such that $T(t)f \geq 0$ whenever $f \geq 0$. (Many authors add the requirement of honesty, but in this book it will be convenient to omit this condition.)

Since the space c_0 is sufficiently rich, we can reverse this procedure and build a strongly continuous semigroup $\{P(t), t \geq 0\}$ of sub-Markov operators from $\{T(t), t \geq 0\}$. To see this, suppose that $\{T(t), t \geq 0\}$ is a strongly continuous semigroup in c_0 satisfying the two conditions just given. Then, for each $t \geq 0$ and $i \in \mathbb{N}$, the formula $F_i(f) = T(t)f(i)$ defines a bounded linear functional on c_0. Therefore (see (5.10)),

$$T(t)f(i) = \lim_{n\to\infty} F_i(f_n) = \lim_{n\to\infty} \sum_{j=1}^{n} f(j)F_i(e_j) = \sum_{j=1}^{\infty} f(j)F_i(e_j).$$

Since $F_i(f) \geq 0$ provided $f \geq 0$, numbers $F_i(e_j)$ are nonnegative. Moreover, the norm of F_i not exceeding 1, we have for all n, $\sum_{j=1}^{n} F_i(e_j) = F_i(\sum_{j=1}^{n} e_j) \leq 1$ (since $\| \sum_{j=1}^{n} e_j \| = 1$) and this implies

$$\sum_{j=1}^{\infty} F_i(e_j) \leq 1$$

(we are simply rediscovering parts of the rich statement that the space of functionals on c_0 may be identified with l^1). Thus, for each i and t, defining $p_{i,j}(t) := F_i(e_j)$, we obtain nonnegative numbers with sum over j not exceeding 1 such that, for any $f \in c_0$,

$$T(t)f(i) = \sum_{j=1}^{\infty} p_{i,j}(t)f(j). \tag{5.13}$$

Since the dual of c_0 is l^1, along with $\{T(t), t \geq 0\}$ we also have the semigroup $\{T^*(t), t \geq 0\}$ of operators in l^1. Using (5.13) we easily see that $T^*(t)\,(\xi_i)_{i\geq 1} = (\eta_i)_{i\geq 1}$ where $\eta_i = \sum_{j=1}^{\infty} p_{j,i}(t)\xi_j$, that is, that $T^*(t)$ is the sub-Markov operator represented by the matrix $\left(p_{i,j}(t)\right)_{i,j\in\mathbb{N}}$. Moreover, since $\{T(t), t \geq 0\}$ is a strongly continuous semigroup in c_0, representation formula (5.13) shows (on considering $f = e_j$) that conditions (2.8) and (5.11) are satisfied. Similarly, we conclude that so is condition (5.12). Thus, as before, we may think of $\{T(t), t \geq 0\}$ as of (a part of) the sun-dual semigroup for a strongly continuous semigroup of sub-Markov operators in l^1 (possessing the special properties (5.11) and (5.12)). See also our Section 5.4 and Section 5.4.6 in particular.

5.2.5 *A comment*

Summarizing: if we restrict ourselves to transition probabilities satisfying (2.8), (5.11), and (5.12), there are two canonical descriptions of the underlying Markov chain: the description in l^1 through the semigroup of sub-Markov operators, and the description in c_0, through the Feller semigroup. In the general theory of Markov processes the situation is quite different. For example, if the state-space of the process is not countable but forms a compact metric space, say, S, constructing Markov semigroups in the space of signed measures on S is rather difficult. On the other hand, Feller semigroups in the space $C(S)$ of continuous functions on S can be constructed by classical means (see, e.g., [34, 43, 52, 53, 60, 86] and many other monographs).

Hence, while studying 'general' process in a more general state-space is quite difficult, Feller semigroups provide handy tools for studying Feller processes, that is, processes that satisfy certain additional regularity assumptions. Needless to say, Feller processes possess nicer properties than 'general' Markov processes.

We complete this section with a necessary and sufficient condition for the Feller property in terms of the resolvent.

5.2.6 *Another criterion for the Feller property*

Let $\{P(t), t \geq 0\}$ be a strongly continuous semigroup of sub-Markov operators, and let A be its generator. The semigroup $\{P(t), t \geq 0\}$ has the Feller property iff c_0 is invariant for R^*_λ, $\lambda > 0$:

$$R^*_\lambda f \in c_0 \qquad \text{whenever} \qquad f \in c_0.$$

Proof

(Necessity) Let $\mathbb{X}^\odot$ be the sun-dual space for $\{P(t), t \geq 0\}$, and let A^*_{p} be the generator of $\left\{P^*(t)_{|\mathbb{X}^\odot}, t \geq 0\right\}$ (see Section 1.5). In Sections 1.5.5 and 1.5.6 we have proved that $\left(\lambda - A^*_{\mathrm{p}}\right)^{-1} f = R^*_\lambda f$, $f \in \mathbb{X}^\odot$. By assumption, $c_0 \subset \mathbb{X}^\odot$, and thus for $f \in c_0$ the vector

$$R^*_\lambda f = \left(\lambda - A^*_{\mathrm{p}}\right)^{-1} f = \int_0^\infty \mathrm{e}^{-\lambda t} \mathrm{e}^{t A^*_{\mathrm{p}}} f \, \mathrm{d}t$$

belongs to c_0, because the Feller property ensures that $\mathrm{e}^{t A^*_{\mathrm{p}}} f \in c_0$ for all $t \geq 0$.

(Sufficiency) Let $U_\lambda := (R^*_\lambda)_{|c_0}$, and let A_0 be the part of A^* in c_0, that is, let A_0 be A^* restricted to the set of all $f \in \mathcal{D}(A^*) \cap c_0$ such that $A^* f \in c_0$. I claim that $U_\lambda, \lambda > 0$ is the resolvent of A_0. Indeed, for $f \in \mathcal{D}(A_0)$ and

$\lambda > 0$, the vector $\lambda f - A_0 f$ belongs to c_0 and thus it makes sense to calculate $U_\lambda(\lambda f - A_0 f) = R_\lambda^*(\lambda f - A^* f) = f$. Similarly, for any $\lambda > 0$ and $f \in c_0$, we have $A^* U_\lambda f = A^* R_\lambda^* f = \lambda R_\lambda^* f - f = \lambda U_\lambda f - f$, and $\lambda U_\lambda f - f$ belongs to c_0 by assumption. It follows that $U_\lambda f$ belongs to $\mathcal{D}(A_0)$ and we have $(\lambda - A_0)U_\lambda f = f$, completing the proof of the first claim.

Next, I claim that A_0 is densely defined. For, if this is not the case, there is a nonzero $x \in l^1 = (c_0)^*$ such that $f(R_\lambda x) = x(R_\lambda^* f) = x(U_\lambda f) = 0$ for $f \in c_0$. Taking $f = e_i, i \geq 1$ we see, however, that this implies $R_\lambda x = 0$, and this is impossible if $x \neq 0$. This contradiction proves the claim.

Since $\|\lambda R_\lambda\| \leq 1$, we have also $\|\lambda U_\lambda\| \leq 1$. Hence, A_0 satisfies all the conditions of the Hille–Yosida Theorem and thus is the generator of a contraction semigroup in c_0.

Since

$$\int_0^\infty e^{-\lambda t} e^{tA_0} f \, dt = (\lambda - A_0)^{-1} f = R_\lambda^* f, \qquad \lambda > 0, f \in c_0,$$

we have

$$\int_0^\infty e^{-\lambda t} e^{tA_0} f(x) \, dt = R_\lambda^* f(x), \qquad \lambda > 0, f \in c_0, x \in l^1.$$

On the other hand, $t \mapsto P^*(t) f(x) = f(P(t)x)$ is continuous with Laplace transform equal to, for $\lambda > 0$,

$$\int_0^\infty e^{-\lambda t} P^*(t) f(x) \, dt = f\left(\int_0^\infty e^{-\lambda t} P(t)x \, dt\right) = f(R_\lambda x) = R_\lambda^* f(x).$$

The function $t \mapsto e^{tA_0} f(x)$ being continuous as well, we infer that

$$P^*(t) f(x) = e^{tA_0} f(x), \qquad f \in c_0, t \geq 0, x \in l^1,$$

and this implies

$$P^*(t) f = e^{tA_0} f, \qquad f \in c_0, t \geq 0.$$

This, however, shows that $P^*(t)$ maps c_0 into c_0, because e^{tA_0} does. By the same token, $\lim_{t \to 0+} P^*(t) f = f, f \in c_0$. □

5.2.7 Example

We will check that the minimal pure birth chain of Sections 2.4.10 and 3.3.2 has the Feller property. To prove this, we recall formula (3.24) for the resolvent R_λ of the generator of the pure birth chain and, given $(\eta_i)_{i \geq 1} \in l^1$ and $g \in l^\infty$, calculate as follows:

$$g(R_\lambda\,(\eta_i)_{n\geq 1}) = \sum_{i=1}^{\infty} g(i) \frac{\pi_{i-1}}{\lambda + a_i} \sum_{j=1}^{i} \frac{\eta_j}{\pi_{j-1}} = \sum_{j=1}^{\infty} \eta_j \frac{1}{\pi_{j-1}} \sum_{i=j}^{\infty} \frac{\pi_{i-1}}{\lambda + a_i} g(i).$$

It follows that

$$R_\lambda^* g(i) = \frac{1}{\pi_{i-1}} \sum_{j=i}^{\infty} \frac{\pi_{j-1}}{\lambda + a_j} g(j). \tag{5.14}$$

Since $g \in l^\infty$ and the series $\sum_{j=1}^{\infty} \frac{\pi_{j-1}}{\lambda + a_j}$ converges (for a direct proof, see Section 2.4.10), so does the series $\sum_{j=i}^{\infty} \frac{\pi_{j-1}}{\lambda + a_j} g(j)$. Therefore its remainder converges to 0, and since $\lim_{i\to\infty} \pi_i$ is nonzero, we conclude that $\lim_{i\to\infty} R_\lambda^* g(i) = 0$, regardless of whether g is a member of c_0 or not. Since this is more than we need, the claim is proved.

5.3 Generators of Feller Semigroups

In the light of the preceding discussion, Feller semigroups in c_0 describe Markov chains that possess special, additional properties ((5.11) and (5.12)). This class of Markov chains is still rather large; hence, it is profitable to know the form of generators of Feller semigroups. In the analysis of Feller generators, as these generators are termed, the key role is played by the positive maximum principle.

5.3.1 *The positive maximum principle*

An operator $A\colon c_0 \supset \mathcal{D}(A) \to c_0$ is said to satisfy this principle if for any $f \in \mathcal{D}(A)$ and $i \in \mathbb{N}$, condition $f(i) = \sup_{j\geq 1} f(j)$ (which forces $f(i) \geq 0$) implies $Af(i) \leq 0$.

It is clear that Feller generators satisfy the positive maximum principle: if at a certain i, $f(i) = \sup_{j\geq 1} f(j)$, then by (5.13), $T(t)f(i) \leq f(i)$ implying that $Af(i) = \lim_{t\to 0+} t^{-1}(T(t)f(i) - f(i)) \leq 0$.

5.3.2 *The positive maximum principle: Examples*

Let a_n, $n \geq 1$ be positive numbers such that

$$\sum_{n=1}^{\infty} a_n^{-1} < \infty. \tag{5.15}$$

1. In the first example, the domain $\mathcal{D}(A)$ of our operator is composed of f such that $f(1) = 0$ and $\lim_{n\to\infty} a_n f(n) = 0$. By (5.15) it follows that for $f \in \mathcal{D}(A)$ the series $\sum_{n=1}^{\infty} |f(n)|$ converges, and in particular $\lim_{n\to\infty} f(n) = 0$, proving that $\mathcal{D}(A) \subset c_0$. We define A as follows:

$$Af(1) = \sum_{n=2}^{\infty} f(n), \qquad Af(i) = -a_i f(i), \quad i \geq 2.$$

It is easy to see that A satisfies the positive maximum principle since a maximum of an $f \in \mathcal{D}(A)$ is attained at $i = 1$ iff $f(n) \leq 0$ for $n \geq 2$.

2. In the second example, the domain $\mathcal{D}(A)$ is composed of $f \in c_0$ such that $\lim_{n\to\infty} a_n(f(n-1) - f(n)) = 0$. Operator A is then defined as follows:

$$\begin{aligned} Af(1) &= -f(1), \\ Af(2) &= 0, \\ Af(i) &= a_i[f(i-1) - f(i)], \qquad i \geq 3. \end{aligned}$$

Here it is also clear that the maximum principle is satisfied.

5.3.3 *Generators of Feller semigroups*

For Feller semigroups the Hille–Yosida Theorem takes the following form. An operator A is a Feller generator (i.e., the generator of a Feller semigroup) in c_0 iff the following three conditions are satisfied:

(a) A is densely defined,
(b) A satisfies the positive maximum principle,
(c) for any $\lambda > 0$ and $g \in c_0$, there is precisely one solution $f \in \mathcal{D}(A)$ to the resolvent equation $\lambda f - Af = g$.

Proof We need to check that for any $\lambda > 0$, $g \mapsto \lambda f$ where f is the solution to the resolvent equation and g is its right-hand side, is a positive contraction in c_0, for positivity of the resolvent operator implies positivity of the semigroup (by Yosida's or Hille's approximations).

Let, for given $\lambda > 0$ and $g \in c_0$, an f be such that $\lambda f - Af = g$. By the nature of c_0 there is an $i \in \mathbb{N}$ such that $\|f\| = |f(i)|$, and without loss of generality we may assume that $f(i) \geq 0$ (otherwise consider $-f$ and $-g$ instead of f and g). Then, by the positive maximum principle,

$$\|g\| \geq g(i) = \lambda f(i) - Af(i) \geq \lambda f(i) = \lambda \|f\|. \tag{5.16}$$

This shows that $g \mapsto \lambda f$ is a contraction.

Now suppose that for some $\lambda > 0$ and a nonnegative $g \in c_0$, the solution f to the resolvent equation is not nonnegative. Then, there is an i such that $f(i) \le f(j)$, $j \in \mathbb{N}$ and $f(i) < 0$. Hence, $-f$ attains a positive maximum at i, and thus we have $Af(i) \ge 0$. It follows that

$$g(i) = \lambda f(i) - Af(i) \le \lambda f(i) < 0,$$

a contradiction. This shows that $g \mapsto \lambda f$ is a positive contraction, completing the entire proof. □

5.3.4 Remark

As already remarked, a closer look at the proof of the (general form of the) Hille–Yosida Theorem reveals that to prove that A is a generator of a contraction semigroup it suffices to show (besides density of $\mathcal{D}(A)$) that the operators $\lambda(\lambda - A)^{-1}$ are contractions for all sufficiently large λ (and not for all $\lambda > 0$). Moreover, if A satisfies the positive maximum principle, it is in fact sufficient to check condition (c) from Section 5.3.3 for just one $\lambda > 0$ (see, e.g., [14,39]). We will not, however, use this result in what follows: usually (but not always) an argument establishing condition (c) is the same for all $\lambda > 0$ or at least for all sufficiently large $\lambda > 0$.

Moreover, we remark that, as we have seen in the proof of 5.3.3, the positive maximum principle implies condition (5.16) which says that

$$\|\lambda f - Af\| \ge \lambda \|f\| \qquad \text{for all} \qquad \lambda > 0,\ f \in \mathcal{D}(A). \tag{5.17}$$

It follows that there are no solutions of $\lambda f - Af = 0$ except for $f = 0$ and thus, by linearity, there is at most one solution to $\lambda f - Af = g$ for a $g \in c_0$. Hence, condition (c) of Section 5.3.3 may be modified as follows:

(c') for all (or all sufficiently large) $\lambda > 0$ and $g \in c_0$ there is a solution $f \in \mathcal{D}(A)$ to the resolvent equation $\lambda f - Af = g$.

The reader will also find that in many (but not all; see, e.g., our Example 5.4.10) cases the fact that f in the resolvent equation is nonnegative provided g is (shown above to be a consequence of the positive maximum principle) is clear from the argument establishing (c). Experience teaches that it is on (c) or (c') that the entire argument hinges and this is where usually the largest analytical difficulties are hidden (cf. the examples presented in Section 2.4). In particular, the reader should not be misled by the simplicity of Example 5.3.5, where (c) is checked almost directly. Two other examples of application of the Hille–Yosida Theorem (namely, the perturbations that lead to the Feynman–Kac and Volkonskii Formulae) will be discussed in Section 5.5.

5.3.5 Example

Let $a, b > 0$. We will show that the formula

$$\begin{aligned} Af(1) &= a[f(2) - f(1)], \\ Af(i) &= ai[f(i+1) - f(i)] + b[f(i-1) - f(i)], \qquad i \geq 2, \end{aligned}$$

defines a Feller generator in c_0; the domain of A is composed of f such that $\lim_{i\to\infty} i[f(i+1) - f(i)] = 0$. (The related birth and death process starting at a state $i \geq 2$ jumps, after an exponential time with parameter $ai + b$, either to $i - 1$, with probability $\frac{b}{ai+b}$, or to $i + 1$, with probability $\frac{ai}{ai+b}$. From $i = 1$ the chain jumps after an exponential time to 2.)

To this end, we write A as a sum of two operators: $A = A_0 + B$, where

$$A_0 f(i) = ai[f(i+1) - f(i)] \qquad i \geq 1,$$

(on $\mathcal{D}(A_0) = \mathcal{D}(A)$) and

$$Bf(1) = 0, \quad Bf(i) = b[f(i-1) - f(i)], \qquad i \geq 2.$$

The domain of B is equal to the entire c_0: in fact, B is a bounded linear operator.

We claim first that B is a Feller generator. Since it is clear that B satisfies the positive maximum principle, we need to check condition (c). To this end, we write B as $B = bR - bI_{c_0}$ where R defined by $Rf(i) = f(i-1), i \geq 1$ (with the proviso that $f(0) = 0$) is the shift to the right. Then, an f satisfies $\lambda f - Bf = g$ for given $\lambda > 0$ and $g \in c_0$, iff $(\lambda + b)f - bRf = g$. Since R has norm 1, however, the solution to the latter equation is unique and given by the Neumann series $f = \frac{1}{\lambda+b}\sum_{n=0}^{\infty}\left(\frac{b}{\lambda+b}\right)^n R^n g$, completing the proof of the claim (cf. Example 2.4.2).

Next, we claim that A_0 is also a Feller generator. By Exercise 1.1.17, we may, without loss of generality, assume that $a = 1$. It is clear that all $e_k, k \geq 1$ belong to $\mathcal{D}(A_0) = \mathcal{D}(A)$ and thus A_0 is densely defined. Moreover, the positive maximum principle is obviously satisfied. Hence, we only need to check condition (c).

In coordinates, the resolvent equation for A_0 (with $a = 1$) reads

$$(\lambda + i)f(i) - if(i+1) = g(i), \qquad i \geq 1$$

and thus renders the following recursion formula for $(f(i))_{i\geq 1}$:

$$f(i+1) = \frac{\lambda + i}{i} f(i) - \frac{g(i)}{i}, \qquad i \geq 1.$$

It follows that

$$f(i) = \pi_{i-1}\left[f(1) - \sum_{j=1}^{i-1} \frac{g(j)}{j\pi_j} \right], \qquad i \ge 1$$

where $\pi_i = \prod_{j=1}^{i} \frac{\lambda + j}{j}$, $i \ge 1$ and by convention $\pi_0 = 1$. By Criterion 2.4.6, $\lim_{i\to\infty} \pi_i = \infty$. Moreover, noting that $\frac{\lambda}{j\pi_j} = \frac{1}{\pi_{j-1}} - \frac{1}{\pi_j}$, $j \ge 1$, we see that the series $\sum_{j=1}^{\infty} \frac{1}{j\pi_j}$ converges (to λ^{-1}) and so does the series $\sum_{j=1}^{\infty} \frac{g(j)}{j\pi_j}$ (possibly to a different sum). Hence, if f is to belong to c_0, we must have $f(1) = \sum_{j=1}^{\infty} \frac{g(j)}{j\pi_j}$, and this in turn leads to

$$f(i) = \pi_{i-1} \sum_{j=i}^{\infty} \frac{g(j)}{j\pi_j}, \qquad i \ge 1. \tag{5.18}$$

It remains to check that f defined here belongs to $\mathcal{D}(A_0)$ and solves the resolvent equation. (By the way, it is clear that $f \ge 0$ provided $g \ge 0$; see Remark 5.3.4, above.) Given $\epsilon > 0$ we may find an i_0 such that $|g(i)| \le \epsilon\lambda$ for all $i \ge i_0$. For such i,

$$|f(i)| \le \pi_{i-1} \sum_{j=i}^{\infty} \frac{\epsilon\lambda}{j\pi_j} = \pi_{i-1}\epsilon \lim_{n\to\infty} (\pi_{i-1}^{-1} - \pi_n^{-1}) = \epsilon.$$

Hence, $f \in c_0$. Moreover, since $(\lambda + i)\pi_{i-1} = i\pi_i$, we have $(\lambda + i) f(i) - if(i+1) = i\pi_i \frac{g(i)}{i\pi_i} = g(i)$. This shows that $f \in \mathcal{D}(A_0)$ (because f and g belong to c_0) and that f solves the resolvent equation, thus completing the proof of the fact that A_0 is a Feller generator.

Finally, by Trotter's Product Formula, $A = A_0 + B$ is a Feller generator because both A_0 and B are Feller generators.

5.3.6 Exercise

Prove that $\sum_{i=1}^{\infty} \pi_i^{-1} < \infty$ (for π_i's of Section 5.3.5), as long as $\lambda > 1$, using Raabe's criterion of series convergence (see [69], p. 120).

5.4 More on Feller Semigroups

As mentioned before, Markov chains with Feller property are more regular than other chains. At the beginning of this section we discuss three results illustrating this rule. First of all, we show that in a chain with Feller property all states are stable. Next, we show that the generator of a Feller semigroup is a rather natural operator: the generator may in a sense be identified with the matrix of intensities. The third result is 'negative': restarting an explosive process after explosion is not a 'regular' procedure, and thus cannot be described within the framework of Feller semigroups. In other words, transition probabilities inscribed on a Feller semigroup are those of the minimal Markov chain. The section is completed with a result saying that in certain cases we may deduce that a chain has Feller property just by looking at its intensity matrix (Section 5.4.7).

5.4.1 *In a Markov chain with Feller property all states are stable*

As mentioned above, we start by showing that, in a chain with Feller property, $q_i < \infty$ for all $i \in \mathbb{N}$. To this end, let us fix $i \in \mathbb{N}$ and let $q(t) = 1 - p_{i,i}(t)$, $t \geq 0$. We claim that there is a $t > 0$ such that

$$q(h + ks) \geq q(h) + \frac{k}{2}q(s) \tag{5.19}$$

provided $h, s \geq 0$ and a natural number k are chosen so that $h + ks \leq t$.

If this property is established, the proof of the theorem is completed as follows: let $(s_n)_{n\geq 1}$ be a sequence of positive numbers converging to zero. For $k_n := \lfloor t/s_n \rfloor$ and $h_n := t - k_n s_n \geq 0$ we have $t = s_n k_n + h_n$. Then, since $q \geq 0$, (5.19) shows that $\frac{q(s_n)}{s_n} \leq \frac{2q(t)}{s_n k_n}$. Therefore, because of $\lim_{n\to\infty} s_n k_n = t$,

$$q_i = \lim_{n\to\infty} \frac{q(s_n)}{s_n} \leq \frac{2q(t)}{t} < \infty,$$

as desired.

To prove (5.19), it suffices to show that there is a $t > 0$ such that

$$q(h + s) \geq q(h) + \frac{1}{2}q(s)$$

as long as $s, h \geq 0$ and $s + h \leq t$; this implies (5.19) by an induction argument. Equivalently, in terms of $p_{i,i}$, it suffices to show that

$$p_{i,i}(h + s) \leq p_{i,i}(h) - \frac{1}{2}(1 - p_{i,i}(s)). \tag{5.20}$$

For t, we choose a number so small that $p_{i,i}(h) \geq \frac{3}{4}$ and $p_{j,i}(h) \leq \frac{1}{4}$ for $0 \leq h \leq t$ and $j \neq i$; such a t exists because for Feller semigroups condition (5.11) is satisfied. Then

$$\begin{aligned}p_{i,i}(h+s) &= p_{i,i}(s)p_{i,i}(h) + \sum_{j\neq i} p_{i,j}(s)p_{j,i}(h)\\ &\le p_{i,i}(s)p_{i,i}(h) + \frac{1}{4}\sum_{j\neq i} p_{i,j}(s)\\ &\le p_{i,i}(s)p_{i,i}(h) + \frac{1}{4}(1-p_{i,i}(s)).\end{aligned}$$

This completes the proof because, by $p_{i,i}(h) \ge \frac{3}{4}$, the last expression here is no greater than the right-hand side in (5.20).

5.4.2 *The form of a Feller generator*

Let's begin by recalling that, as we have seen in Section 5.2.4, a Feller semigroup $\{T(t), t \ge 0\}$ in c_0 contains information on certain transition probabilities $p_{i,j}(t), i, j \in \mathbb{N}$. By 5.4.1, all intensities on the diagonal in the related matrix of intensities are finite. Therefore, given $f \in c_0$, it is meaningful to consider the sequence g defined by

$$g(i) = \sum_{j=1}^{\infty} q_{i,j} f(j) = -q_i f(i) + \sum_{j\neq i} q_{i,j} f(j), \qquad i \in \mathbb{N};$$

the series on the right converging absolutely because f is bounded and $\sum_{j\neq i} q_{i,j} \le q_i$. It is then tempting to think (see also Section 5.7.1) that, if defined on a proper domain, an operator Q_0 mapping f to g is a good candidate for the generator of $\{T(t), t \ge 0\}$, and what domain could be more natural than that composed of all $f \in c_0$ such that g belongs to c_0?

As we shall now see, these hopes are not in vain. Formally, let Q_0 be defined by

$$\mathsf{Q}_0 f(i) = \sum_{j=1}^{\infty} q_{i,j} f(j), \tag{5.21}$$

on the domain composed of $f \in c_0$ such that $\mathsf{Q}_0 f \in c_0$. Then:

$$\text{The generator of } \{T(t), t \ge 0\} \text{ is } \mathsf{Q}_0. \tag{5.22}$$

For the proof we need the following result of independent interest.

5.4.3 Lemma

Transition probabilities $p_{i,j}(t), t \ge 0, i, j \in \mathbb{N}$ of a chain with Feller property are continuously differentiable in $t \ge 0$ and satisfy the **Kolmogorov backward equations**

$$p'_{i,j}(t) = \sum_{k=1}^{\infty} q_{i,k} p_{k,j}(t), \qquad i, j \in \mathbb{N}. \tag{5.23}$$

Proof Given $i, j \in \mathbb{N}, t \geq 0$ and $s > 0$, consider the difference quotient $s^{-1}(p_{i,j}(t+s) - p_{i,j}(t))$. Since (by the Chapman–Kolmogorov equation)

$$p_{i,j}(t+s) = \sum_{k=1}^{\infty} p_{i,k}(s) p_{k,j}(t),$$

the absolute value, say, $\psi(s)$, of the difference between this quotient and the right-hand side of (5.23) does not exceed

$$\begin{aligned}
&\sum_{k=1}^{\infty} \left| \frac{p_{i,k}(s) - \delta_{i,k}}{s} - q_{i,k} \right| p_{k,j}(t) \\
&\quad \leq \sum_{k=1}^{n} \left| \frac{p_{i,k}(s) - \delta_{i,k}}{s} - q_{i,k} \right| + \sum_{k=n+1}^{\infty} \left(\frac{p_{i,k}(s)}{s} + q_{i,k} \right) \sup_{k \geq n+1} p_{k,j}(t) \\
&\quad \leq \sum_{k=1}^{n} \left| \frac{p_{i,k}(s) - \delta_{i,k}}{s} - q_{i,k} \right| + \left(\frac{1 - p_{i,i}(s)}{s} + q_i \right) \sup_{k \geq n+1} p_{k,j}(t),
\end{aligned}$$

provided $n \geq i$. Therefore, $\limsup_{s \to 0+} \psi(s) \leq 2q_i \sup_{k \geq n+1} p_{k,j}(t)$.

On the other hand, by (5.12), $\sup_{k \geq n+1} p_{k,j}(t)$ can be made arbitrarily small by choosing n large enough. It follows that $t \mapsto p_{i,j}(t)$ has the right derivative at any t, and this derivative is $\sum_{k=1}^{\infty} q_{i,k} p_{k,j}(t)$. We note that the functions $t \mapsto p_{k,j}(t)$ are continuous, and the series $\sum_{k=1}^{\infty} q_{i,k} p_{k,j}(t)$ converges uniformly in $t \geq 0$. Hence, $t \mapsto \sum_{k=1}^{\infty} q_{i,k} p_{k,j}(t)$ is a continuous function, too.

Invoking the well-known fact of real analysis saying that a continuous function $x : [0, \infty) \to \mathbb{R}$ which has the right-hand derivative $x'_+(t)$ at all $t \geq 0$ with $t \mapsto x'_+(t)$ continuous, is necessarily (continuously) differentiable, we complete the proof. □

Proof of (5.22) Consider $\{P(t), t \geq 0\} := \{T^*(t), t \geq 0\}$ in l^1. As explained in 5.2.4, this is a strongly continuous semigroup of sub-Markov operators. Let A be its generator. We know, by Lemma 5.4.3, that the Kolmogorov backward differential equations are satisfied. Hence, see 2.5.6, all e_i's belong to $\mathcal{D}(A)$.

Let f belong to $\mathcal{D}(A^*) \subset l^\infty$. Then, $f(Ax) = (A^* f)(x)$ for all $x \in \mathcal{D}(A)$. In particular, in view of (2.45),

$$A^* f(i) = A^* f(e_i) = f(Ae_i) = f\left(q_{i,j}\right)_{j \geq 1} = \sum_{j=1}^{\infty} q_{i,j} f(j). \tag{5.24}$$

The fact that $\{T(t), t \geq 0\}$ is a Feller semigroup implies that the sun-dual space for $\{P(t), t \geq 0\}$ contains c_0 and that c_0 is an invariant subspace for the sun-dual semigroup. It follows that the generator of $\{T(t), t \geq 0\}$ is a part of the generator of the sun-dual semigroup, which in turn, as we know from Section 1.5, is a part of A^*. Hence, Q_0 is the part of A^* in c_0. In particular, its domain is composed of $f \in c_0$ such that $\mathsf{Q}_0 f (= A^* f)$ belongs to c_0. A glance at (5.21) and (5.24) completes the proof. □

5.4.4 Example

In 5.2.7 we have seen that the pure birth process of Section 2.4.10 has the Feller property. Therefore, the generator of the related Feller semigroup is Q_0 defined by

$$\mathsf{Q}_0 f(1) = a_i(f(i+1) - f(i)),$$

with domain composed of f such that $\lim_{i\to\infty} a_i(f(i+1) - f(i)) = 0$.

5.4.5 *A Feller semigroup describes the minimal chain*

Let us think again of the transition probabilities $p_{i,j}(t), i, j \in \mathbb{N}, t \geq 0$ inscribed on a Feller semigroup $\{T(t), t \geq 0\}$, and recall again that, by 5.4.1, the related intensity matrix has no infinities on the main diagonal. Hence, by Kato's Theorem, there is the minimal semigroup $\{S(t), t \geq 0\}$ in l^1, related to this intensity matrix. Let $\widetilde{p_{i,j}}(t), i, j \in \mathbb{N}, t \geq 0$ be the transition probabilities of the minimal chain. We claim that

$$p_{i,j}(t) = \widetilde{p_{i,j}}(t).$$

Proof Let $\{P(t), t \geq 0\} := \{T^*(t), t \geq 0\}$ be the adjoint semigroup for $\{T(t), t \geq 0\}$; recall that the matrix $(p_{i,j}(t))_{i,j\geq 1}$ represents both $T(t)$ and $P(t)$. Since $\{P(t), t \geq 0\}$ is a strongly continuous semigroup, the sun-dual space for $\{T(t), t \geq 0\}$ is the entire l^1 and it follows that the generator of $\{P(t), t \geq 0\}$ is the dual Q_0^* to Q_0, the generator of $\{T(t), t \geq 0\}$.

Consider $e_k \in l^1$ for a $k \in \mathbb{N}$, and think of this element as of a functional on c_0. Then

$$e_k(\mathsf{Q}_0 f) = (\mathsf{Q}_0 f)(k) = \sum_{j=1}^{\infty} q_{k,j} f(j), \qquad \text{for all } f \in \mathcal{D}(\mathsf{Q}_0).$$

Since $(q_{k,j})_{j\in\mathbb{N}}$ is a member of l^1, the map $\mathcal{D}(\mathsf{Q}_0) \ni f \mapsto e_k(\mathsf{Q}_0 f)$ can be extended to a continuous functional on l^1, represented by $(q_{k,j})_{j\in\mathbb{N}}$. It follows that e_k belongs to $\mathcal{D}(\mathsf{Q}_0^*)$ and that $\mathsf{Q}_0^* e_k = (q_{k,j})_{j\in\mathbb{N}}$. But this means that the generator of $\{P(t), t \ge 0\}$ is an extension of the operator which in Section 3.1.5 was denoted A_0. Since $P(t) \ge 0$, the latter section shows that $S(t) \le P(t), t \ge 0$. This in turn, by the criterion of 5.2.3, implies that the minimal semigroup $\{S(t), t \ge 0\}$ possesses the Feller property.

Let $\{\widetilde{T}(t), t \ge 0\}$ be the related strongly continuous semigroup in c_0, that is, let $\widetilde{T}(t)$ be the restriction of $S^*(t)$ to c_0:

$$\widetilde{T}(t) = [S^*(t)]_{|c_0}, \tag{5.25}$$

and let $\widetilde{p_{i,j}}(t)$ be the related transition probabilities. Since transition probabilities $p_{i,j}(t)$ and $\widetilde{p_{i,j}}(t)$ lead to the same Q-matrix, the generators of $\{\widetilde{T}(t), t \ge 0\}$ and $\{T(t), t \ge 0\}$ coincide (by (5.22)). This, however, implies that so do the semigroups, and this leads to the conclusion that $p_{i,j}(t) = \widetilde{p_{i,j}}(t)$. □

5.4.6 Corollary

In the proof presented above we have seen that the semigroup $\{P(t), t \ge 0\} := \{T^*(t), t \ge 0\}$ coincides in fact with the minimal semigroup $\{S(t), t \ge 0\}$. We have also seen that $\{\widetilde{T}(t), t \ge 0\}$ of (5.25) coincides with $T(t)$. Hence, if a Markov chain has the Feller property, then its minimal semigroup $\{S(t), t \ge 0\}$ is simply the dual of the Feller semigroup $\{T(t), t \ge 0\}$ defined in c_0; the semigroup $\{T(t), t \ge 0\}$ in turn may be recovered as the restriction of $\{S^*(t), t \ge 0\}$ to c_0:

$$\boxed{S(t) = T^*(t) \qquad \text{and} \qquad T(t) = [S^*(t)]_{|c_0}.}$$

Our next result is related to the following problem: given an intensity matrix Q can we say, just by examining Q itself, whether the related chain has the Feller property? By 5.4.5 this question makes sense only if by 'the related chain' we mean 'the minimal chain.' Below we describe one such situation.

5.4.7 *An intensity matrix leading to a Markov chain with Feller property*

Suppose Q is an intensity matrix such that

(i) $\lim_{i\to\infty} q_{i,j} = 0$ for all $j \ge 1$, and
(ii) for any $\lambda > 0$ there is no nonzero $(\xi_i)_{i\ge 1} \in l^1$ such that $\lambda\xi_i = \sum_{j=1}^{\infty} \xi_j q_{j,i}, i \ge 1$.

Then, the minimal chain related to Q has Feller property.

Proof Our strategy is to prove that Q_0 defined in (5.21) is closable and its closure generates a Feller semigroup. Then, we will show that this semigroup dominates the minimal semigroup: as in 5.4.5 this will allow us to deduce the Feller property for the minimal chain.

1. (Q_0 is closable.)

Since Q_0 satisfies the positive maximum principle, we have (see (5.16) or (5.17))

$$\|\lambda f - \mathsf{Q}_0 f\| \geq \lambda \|f\|, \qquad \text{for all } f \in \mathcal{D}(\mathsf{Q}_0). \tag{5.26}$$

Suppose $(f_n)_{n\geq 1}$ is a sequence of elements of $\mathcal{D}(\mathsf{Q}_0)$ such that $\lim_{n\to\infty} f_n = 0$ and $\lim_{n\to\infty} \mathsf{Q}_0 f_n = g \in c_0$. Then, for any $h \in \mathcal{D}(\mathsf{Q}_0)$ and $\lambda > 0$,

$$\|(\lambda - \mathsf{Q}_0)(\lambda f_n + h)\| \geq \lambda \|\lambda f_n + h\|.$$

Letting $n \to \infty$, we obtain

$$\|\lambda h - \lambda g - \mathsf{Q}_0 h\| \geq \lambda \|h\|.$$

Dividing by λ and letting $\lambda \to \infty$ yields $\|h - g\| \geq \|h\|$. Now, assumption (i) says that $e_j \in \mathcal{D}(\mathsf{Q}_0)$ for all $j \in \mathbb{N}$. Therefore, $\mathcal{D}(\mathsf{Q}_0)$ is a dense subset of c_0. Taking $h = h_n$ where $\lim_{n\to\infty} h_n = g$ and letting $n \to \infty$, we obtain $\|g\| \leq 0$, which is possible only if $g = 0$. This, by definition, means that Q_0 is closable.

2. (Definition of $\overline{\mathsf{Q}_0}$.)

If, for an $f \in c_0$, there is a sequence $(f_n)_{n\geq 1}$ of elements of $\mathcal{D}(\mathsf{Q}_0)$ such that $\lim_{n\to\infty} f_n = f$ and $\lim_{n\to\infty} \mathsf{Q}_0 f_n = g$, it makes sense to define $\overline{\mathsf{Q}_0} f = g$ because this definition does not depend on the choice of $(f_n)_{n\geq 1}$. Taking as the domain of $\mathcal{D}(\overline{\mathsf{Q}_0})$ the set of f with the property described above, we obtain a linear operator in c_0. It is easy to check that $\overline{\mathsf{Q}_0}$ extends Q_0 and is closed.

3. (The range of $\lambda - \overline{\mathsf{Q}_0}$ is c_0.)

Fix $\lambda > 0$. First, we claim that the range of $\lambda - \mathsf{Q}_0$ is dense in c_0. If this is not the case, there is a nonzero bounded linear functional on c_0, that is, a member $x = (\xi_i)_{i\geq 1}$ of l^1, such that $x(\lambda f - \mathsf{Q}_0 f) = 0$ for all $f \in \mathcal{D}(\mathsf{Q}_0)$. Taking $f = e_i$, we obtain that this implies $\lambda \xi_i = \sum_{j=1}^{\infty} \xi_j q_{j,i}$, $i \geq 1$. (By assumption (i), e_i's belong to $\mathcal{D}(\mathsf{Q}_0)$ and $\mathsf{Q}_0 e_i(j) = q_{j,i}$.) By assumption (ii), therefore, x cannot be nonzero. This contradiction proves the claim.

Next, we claim that the range of $\lambda - \overline{\mathsf{Q}_0}$ is closed. To this end, we first deduce from (5.26) that

$$\|\lambda f - \overline{\mathsf{Q}_0} f\| \geq \lambda \|f\|, \qquad \text{for all } f \in \mathcal{D}(\overline{\mathsf{Q}_0}). \tag{5.27}$$

Suppose now that g belongs to the the closure of the range of $\lambda - \overline{Q_0}$: there are $f_n \in \mathcal{D}(\overline{Q_0})$ such that $\lim_{n\to\infty}(\lambda f_n - \overline{Q_0} f_n) = g$. By (5.27), $(f_n)_{n\ge 1}$ is then a Cauchy sequence in c_0. Let $f = \lim_{n\to\infty} f_n$. Then $\lim_{n\to\infty} \overline{Q_0} f_n$ exists and equals $\lambda f - g$. Since $\overline{Q_0}$ is closed this means that $f \in \mathcal{D}(\overline{Q_0})$ and $\overline{Q_0} f = \lambda f - g$. Thus, $g = \lambda f - \overline{Q_0} f$, showing that g belongs to the range of $\lambda - \overline{Q_0}$.

Since the range of $\lambda - \overline{Q_0}$ contains the range of $\lambda - Q_0$, it follows that the former set is the entire c_0.

4. ($\overline{Q_0}$ generates a Feller semigroup.)

In point **3.** we have shown that the resolvent equation $\lambda f - \overline{Q_0} f = g$ has a solution $f \in \mathcal{D}(\overline{Q_0})$ for all $\lambda > 0$ and $g \in c_0$. Inequality (5.27) on the other hand, shows that such a solution is unique, and that the map $g \mapsto \lambda f$ is a contraction. Since $\overline{Q_0}$ is densely defined, the Hille–Yosida Theorem tells us that $\overline{Q_0}$ is the generator of a contraction semigroup, say $\{T(t), t \ge 0\}$ in c_0, but we still need to show that $T(t) \ge 0$.

It suffices to prove that $f \ge 0$ provided $g \ge 0$. To this end, suppose that $f(j) > 0$ for some j. Then there is an i such that $f(i) = \sup_{k\ge 1} f(k)$. Since $f \in \mathcal{D}(\overline{Q_0})$, there are $f_n \in \mathcal{D}(Q_0)$ such that $\lim_{n\to\infty} f_n = f$ and $\lim_{n\to\infty} Q_0 f_n = \overline{Q_0} f$. Then, defining $g_n = \lambda f_n - Q_0 f_n$ we have $\lim_{n\to\infty} g_n = g$. Also, for sufficiently large n, $f_n(i) > 0$, implying that there are $i_n \in \mathbb{N}$ such that $f_n(i_n) = \sup_{k\ge 1} f_n(k) > 0$. By the positive maximum principle, $Q_0 f_n(i_n) \le 0$, and by Exercise 5.4.11, we have $\lim_{n\to\infty} f_n(i_n) = f(i)$. Then, for sufficiently large n,

$$\|g_n - g\| < \frac{\lambda f(i)}{4} \qquad \text{and} \qquad f_n(i_n) > \frac{3}{4} f(i).$$

Therefore,

$$g(i_n) + \frac{\lambda f(i)}{4} > g_n(i_n) = \lambda f_n(i_n) - Q_0 f_n(i_n) \ge \lambda f_n(i_n) \ge \lambda \frac{3}{4} f(i),$$

that is, $g(i_n) > \frac{\lambda}{2} f(i) > 0$. It follows that $g \le 0$ implies $f \le 0$ or, which is the same, that $g \ge 0$ implies $f \ge 0$.

5. ($\{T^*(t), t \ge 0\}$ dominates in l^1 the minimal semigroup for Q.)

Let $\{T(t), t \ge 0\}$ be the semigroup generated by $\overline{Q_0}$. Its dual semigroup $\{T^*(t), t \ge 0\}$ is strongly continuous in l^1 and thus the generator of $\{T^*(t), t \ge 0\}$ is $(\overline{Q_0})^*$. As in 5.4.5, consider $e_k \in l^1$ for a $k \in \mathbb{N}$, and think of this element as of a functional on c_0. Then

$$e_k(Q_0 f) = (Q_0 f)(k) = \sum_{j=1}^{\infty} q_{k,j} f(j), \qquad \text{for all } f \in \mathcal{D}(Q_0). \tag{5.28}$$

For $f \in \mathcal{D}(\overline{\mathsf{Q}_0})$ there are $f_n \in \mathcal{D}(\mathsf{Q}_0)$ such that $\lim_{n\to\infty} f_n = f$ and $\lim_{n\to\infty} \mathsf{Q}_0 f_n = \overline{\mathsf{Q}_0} f$; without loss of generality we may assume that $\|f_n\| \le 2\|f\|, n \ge 1$. Then, by (5.28),

$$|e_k(\overline{\mathsf{Q}_0} f)| = |\lim_{n\to\infty} e_k(\mathsf{Q}_0 f_n)| \le 2q_k\|f_n\| \le 4q_k\|f\|, \qquad \text{for all } f \in \mathcal{D}(\overline{\mathsf{Q}_0}).$$

This shows that $e_k \in \mathcal{D}((\overline{\mathsf{Q}_0})^*)$. Moreover, by (5.28), the only possible value for $(\overline{\mathsf{Q}_0})^* e_k$ is $(q_{k,i})_{i\ge 1}$.

It follows that $\{T^*(t), t \ge 0\}$ may play the role of $\{P(t), t \ge 0\}$ of Section 3.1.5. Therefore, for the minimal semigroup $\{S(t), t \ge 0\}$ we have $S(t) \le T^*(t)$ and thus, by 5.2.3, the minimal chain has the Feller property.

5A. Here is another way the proof may be completed, communicated to me by E. Ratajczyk. Since $\{T(t), t \ge 0\}$ is a Feller semigroup, there is an intensity matrix $(\widetilde{q}_{i,j})_{i,j\ge 1}$ such that (a) transition probabilities of the related minimal chain possess the Feller property and (b) $\overline{\mathsf{Q}_0} f(i) = \sum_{k=1}^{\infty} \widetilde{q}_{i,k} f(k)$ for f in the domain of $\overline{\mathsf{Q}_0}$ (see Sections 5.4.2 and 5.4.5). However, e_j, $j \ge 1$ are members of $\mathcal{D}(\mathsf{Q}_0) \subset \mathcal{D}(\overline{\mathsf{Q}_0})$ and $(\mathsf{Q}_0 e_j)(i) = q_{i,j}$. It follows that $q_{i,j} = \widetilde{q}_{i,j}$, that is, that the minimal chain related to $(q_{i,j})_{i,j\ge 1}$ has the Feller property. □

5.4.8 Remark

In the proof presented above we have shown that Q_0 is closable and $\overline{\mathsf{Q}_0}$ is a Feller generator. This leads us to the conclusion that the related minimal chain has Feller property. This, however, when combined with 5.4.2, shows that Q_0 itself is a generator (see also part 5A in the proof above), and as such is closed. A direct proof of the fact that Q_0 is closed eludes me.

5.4.9 Corollary

As a by-product of the proof presented above we see that if, for a given intensity matrix Q, the operator Q_0 of (5.21) is a Feller generator, and $e_i \in \mathcal{D}(\mathsf{Q}_0)$ for $i \ge 1$, then the minimal chain related to Q has the Feller property.

5.4.10 Example

Consider the Kolmogorov matrix with the first row equal $(-24, 24, 0, 0, \dots)$ and the ith row equal

$$(0, \dots, 0, 3^i, -9 \cdot 3^i, 8 \cdot 3^i, 0, \dots), \qquad i \ge 2.$$

In this case, Q_0 is given by

$$\mathsf{Q}_0 f(1) = 24 f(2) - 24 f(1),$$
$$\mathsf{Q}_0 f(i) = 3^i (f(i-1) - 9 f(i) + 8 f(i+1)), \qquad i \ge 2,$$

on the domain composed of f such that $\lim_{i\to\infty} 3^i (f(i+1) - f(i)) = 0$ (this condition may be checked to be equivalent to $\lim_{i\to\infty} \mathsf{Q}_0 f(i) = 0$). It is clear that all $e_i, i \ge 1$ belong to this domain. We aim at proving that Q_0 is a Feller generator, and thus that the birth and death process related to Q has the Feller property. Since Q_0 clearly satisfies the positive maximum principle and is densely defined, it suffices to check that condition (c') of the Hille–Yosida Theorem is fulfilled. To this end, we introduce operators B (for 'birth') and D (for 'death') with $\mathcal{D}(B) = \mathcal{D}(D) = \mathcal{D}(\mathsf{Q}_0)$, given by

$$Bf(i) = 8 \cdot 3^i [f(i+1) - f(i)], \qquad i \ge 1,$$
$$Df(i) = 3^i [f(i-1) - f(i)], \qquad i \ge 2 \quad \text{and} \quad Df(1) = 0.$$

We have $\mathsf{Q}_0 = B + D$ and from Example 5.4.4 we know that B is a Feller generator. Also, for $f \in \mathcal{D}(B) = \mathcal{D}(D)$,

$$\|Df\| = \sup_{i\ge 2} 3^i |f(i-1) - f(i)| = 3 \sup_{i\ge 1} 3^i |f(i+1) - f(i)| = \frac{3}{8}\|Bf\|.$$

It follows that, for $g \in c_0$,

$$\|D(\lambda - B)^{-1} g\| \le \frac{3}{8}\|B(\lambda - B)^{-1} g\| = \frac{3}{8}\|\lambda(\lambda - B)^{-1} g - g\| \le \frac{3}{4}\|g\|.$$

Therefore, the series $(\lambda - B)^{-1} \sum_{k=0}^{\infty} \left[D(\lambda - B)^{-1}\right]^k g$ converges. It is easy to check that its sum is a solution to the resolvent equation for Q_0. This shows that condition (c') of the Hille–Yosida Theorem is satisfied, and our proof that Q_0 is a Feller generator is completed.

5.4.11 Exercise

Suppose for $f, g \in c_0$ there are $i, j \in \mathbb{N}$ such that $f(i) = \sup_{k\ge 1} f(k)$ and $g(j) = \sup_{k\ge 1} g(k)$. Then $|f(i) - g(j)| \le \|f - g\|$.

5.5 The Feynman–Kac and the Volkonskii Formulae

This section is devoted to two famous formulae for specifically perturbed Feller semigroups: the Feynman–Kac Formula and the (perhaps a bit less known) Volkonskii Formula.

5.5.1 *Increasing intensities on the main diagonal*

Let A be a Feller generator and $b \in l^\infty$ be a nonnegative sequence. Consider $B \in \mathcal{L}(c_0)$ given by

$$(Bf)(i) = b(i)f(i), \qquad f \in c_0, i \in \mathbb{N}. \tag{5.29}$$

We claim that $A - B$ with domain equal to $\mathcal{D}(A)$ is a Feller generator as well. Indeed, it is clear that this operator is densely defined and satisfies the positive maximum principle (nonnegativity of b is used here). Moreover (for the same reason), for $\beta := \|b\|$ the norm of the operator B' given by $B'f = \beta f - Bf$ does not exceed β. Hence, we have $\|B'(\lambda + \beta - A)^{-1}\| \le \frac{\beta}{\lambda+\beta} < 1$, and so the series

$$S_\lambda = (\lambda + \beta - A)^{-1} \sum_{n=0}^{\infty} \left[B'(\lambda + \beta - A)^{-1} \right]^n$$

converges in the operator norm for all $\lambda > 0$. Writing the resolvent equation $\lambda f - Af + Bf = g$ for the operator $A - B$ as

$$(\lambda + \beta)f - Af = B'f + g,$$

we easily check that $f := S_\lambda g$ solves this equation. This shows that condition (c') in Remark 5.3.4 is satisfied, and thus completes the proof. (Alternatively, we could use the Phillips Perturbation Theorem and the result discussed in the first part of Remark 5.3.4.)

5.5.2 *Increasing intensities on the main diagonal: Intuition*

How do the processes related to A and $A - B$ differ? The intensity with which the probability mass escapes from a state i in the first process is no greater than that intensity in the second process: if the first intensity is q_i then the second is $q_i + b(i)$. On the other hand, intensities $q_{i,j}$ with which the probability mass from the state i is collected at states j are in both cases the same. As a result, the process related to $A - B$ is dishonest even if the process related to A is honest (unless $b = 0$); in the former process after an exponential time spent at i a particle may jump to one of the other states or disappear (see Sections 2.5.8 and 2.5.9).

5.5.3 *The Feynman–Kac Formula*

As we shall see now, the information provided in the previous section can be made much more specific. Let $X(t), t \ge 0$ be a Markov chain related to A so that

$$\mathrm{e}^{tA} f(i) = \mathbb{E}_i\, f(X(t)), \qquad f \in c_0, i \in \mathbb{N}, t \geq 0,$$

where $\mathbb{E}_i f(X(t))$ is the expected value of $f(X(t))$ conditional on $X(0) = i$ (this notation will turn out to be more convenient than that used in (5.3)), and assume for safety's sake that the paths of the chain are right-continuous with left limits. Then the Riemann integral $\int_0^t b(X(s))\,\mathrm{d}s$ is well defined for all ω in the underlying probability space and for all $t \geq 0$, and thus the right-hand side in

$$\mathrm{e}^{t(A-B)} f(i) = \mathbb{E}_i\, \mathrm{e}^{-\int_0^t b(X(s))\,\mathrm{d}s} f(X(t)) \tag{5.30}$$

makes sense. Relation (5.30), which holds for all $f \in c_0, i \in \mathbb{N}$ and $t \geq 0$ is (a special case of) the famous Feynman–Kac Formula; we will prove it in Sections 5.5.4 and 5.5.5.

For now, we comment that this formula is a more quantitative version of the remarks made in Section 5.5.2. For, the difference between e^{tA} and $\mathrm{e}^{t(A-B)}$ lies of course in $\mathrm{e}^{-\int_0^t b(X(s))\,\mathrm{d}s}$. This factor gathers the information on how a particle might have disappeared along the path leading from a state at time 0 to a potential state at $t > 0$. Roughly speaking, the probability that the process related to $A - B$ is at a certain state at time t is that of the process related to A modified by that factor, but it should be kept in mind that this factor varies from path to path, and so we are effectively summing all such modified probabilities.

5.5.4 *Proof of the the Feynman–Kac Formula: Part I*

Turning to the proof of (5.30), we note that $-B$ is bounded and thus is a generator itself, and we have

$$\mathrm{e}^{-tB} f(i) = \mathrm{e}^{-b(i)t} f(i), \qquad f \in c_0, i \in \mathbb{N}, t \geq 0.$$

(Here, $\mathrm{e}^{-tB} := \mathrm{e}^{t(-B)}$.) This shows that $-B$ is in fact a Feller generator. It follows that the semigroup generated by $A - B$ may be obtained from Trotter's Product Formula.

This leads us to consider $H(t) = \mathrm{e}^{tA}\mathrm{e}^{-tB}, t \geq 0$; we note that

$$H(t) f(i) = \mathbb{E}_i\, \mathrm{e}^{-tb(X(t))} f(X(t)), \qquad f \in c_0, i \in \mathbb{N}, t \geq 0. \tag{5.31}$$

We claim, and this is the actual key to the proof of (5.30), that for any $t_1, \ldots, t_n > 0$, $f \in c_0$ and $i \in \mathbb{N}$,

$$H(t_1)H(t_2)\cdots H(t_n) f(i) = \mathbb{E}_i\, \mathrm{e}^{-\sum_{k=1}^n t_k b(X(s_k))} f(X(s_n)), \tag{5.32}$$

where $s_k = t_1 + \ldots t_k$. When expanded, the right-hand side here equals

$$\sum_{i_1,\ldots,i_n} \mathrm{e}^{-\sum_{k=1}^n t_k b(i_k)} f(i_n) P_{i,i_1,\ldots,i_n}(t_1, \ldots, t_n),$$

where summation is over all $i_1, \ldots, i_n \in \mathbb{N}$, and $P_{i,i_1,\ldots,i_n}(t_1, \ldots, t_n)$ is the probability of a path leading from $X(0) = i$ to $X(t_1+\cdots+t_n) = i_n$ via $X(t_1) = i_1$, $X(t_1 + t_2) = i_2$, and so on (provided we start at i). More specifically, that probability is

$$p_{i,i_1}(t_1) \prod_{k=2}^{n} \mathbb{P}[X(t_1 + \cdots + t_k) = i_k | X(t_1 + \cdots + t_{k-1}) = i_{k-1}].$$

To show (5.32), we use induction argument. For $n = 1$, (5.32) reduces to (5.31). Next, assuming that (5.32) holds for any n-tuple of times, we see that

$$h := H(t_1)H(t_2) \cdots H(t_{n+1})f$$

is $H(t_1)g$ where

$$g(j) = \sum_{i_2,\ldots,i_{n+1}} \mathrm{e}^{-\sum_{k=1}^{n} t_{k+1} b(i_{k+1})} f(i_{n+1}) P_{j,i_2,\ldots,i_{n+1}}(t_2, \ldots, t_{n+1}).$$

Therefore,

$$\begin{aligned} h(i) &= \sum_{i_1} \mathrm{e}^{-t_1 b(i_1)} g(i_1) p_{i,i_1}(t_1) \qquad (5.33) \\ &= \sum_{i_1,\ldots,i_{n+1}} \mathrm{e}^{-\sum_{k=1}^{n+1} t_k b(i_k)} f(i_{n+1}) P_{i_1,i_2,\ldots,i_{n+1}}(t_2, \ldots, t_{n+1}) p_{i,i_1}(t_1). \end{aligned}$$

Since $X(t), t \geq 0$ is time-homogeneous, the probability

$$\mathbb{P}[X(t_2 + \cdots + t_k) = i_k | X(t_2 + \cdots + t_{k-1}) = i_{k-1}]$$

is the same as

$$\mathbb{P}[X(t_1 + \cdots + t_k) = i_k | X(t_1 + \cdots + t_{k-1}) = i_{k-1}].$$

It follows that

$$P_{i_1,i_2,\ldots,i_{n+1}}(t_2, \ldots, t_{n+1}) p_{i,i_1}(t_1) = P_{i,i_1,i_2,\ldots,i_{n+1}}(t_1, \ldots, t_{n+1}).$$

This combined with (5.33) completes the induction step, and the entire proof of (5.32).

5.5.5 *Proof of the the Feynman–Kac Formula: Part II*

Having established (5.32) we specialize to the case where all t_k's are the same and equal t/n where $t > 0$ is fixed. Formula (5.32) then renders

$$[H(t/n)]^n f(i) = \mathbb{E}_i \, \mathrm{e}^{-\sum_{k=1}^{n} \frac{t}{n} b\left(X\left(\frac{kt}{n}\right)\right)} f(X(t)),$$

and we note that $\sum_{k=1}^{n} \frac{t}{n} b\left(X\left(\frac{kt}{n}\right)\right)$ is a partial sum of the Riemann integral $\int_0^t b(X(s))\,\mathrm{d}s$. As $n \to \infty$ this sum tends to the integral for all ω in the underlying probability space. By the Lebesgue Dominated Convergence Theorem $\left(\text{since } \left|\mathrm{e}^{-\sum_{k=1}^{n} \frac{t}{n} b\left(X\left(\frac{kt}{n}\right)\right)}\right| \le 1\right)$, $[H(t/n)]^n f(i)$ converges to the right-hand side of (5.30).

The convergence just established is point-wise: for all $i \in \mathbb{N}$, the limit $\lim_{n\to\infty}[H(t/n)]^n f(i)$ exists and equals to the right-hand side of (5.30). On the other hand, Trotter's Product Formula states that the limit $\lim_{n\to\infty}[H(t/n)]^n f$ exists in the norm of c_0 and equals $\mathrm{e}^{t(A-B)} f$. Since these two limits must coincide, we are done.

Our second subject in this section is the Volkonskii Formula. The Feynman–Kac Formula gives an explicit form of the semigroup generated by a (quite specific) additive perturbation of a generator of a given semigroup. The Volkonskii Formula plays a similar role for a multiplicative perturbation.

5.5.6 *A multiplicative perturbation of a generator*

Let A be a Feller generator, and let $b \in l^\infty$ be a positive sequence. We stress that in contrast to Section 5.5.1 we do assume that all $b(i)$ are positive (and not just nonnegative). In fact, we assume quite a bit more:

$$\inf_{i\in\mathbb{N}} b(i) =: \beta_0 > 0. \tag{5.34}$$

We claim that, under this assumption, the operator

$$BA$$

with domain equal to $\mathcal{D}(A)$ is a Feller generator; B featuring here is defined by (5.29).

As in Section 5.5.1, it is clear that BA is densely defined and satisfies the positive maximum principle. We are therefore left with showing that the resolvent equation

$$\lambda f - BAf = g \tag{5.35}$$

has a solution for all $\lambda > 0$ and $g \in c_0$ (see Remark 5.3.4, condition (c')). To this end, consider $c_\lambda = \lambda\frac{\beta - b}{b} \ge 0$ (i.e., $c_\lambda(i) = \lambda\frac{\beta - b(i)}{b(i)}$, $i \in \mathbb{N}$) where $\beta := \|b\|$. By assumption (5.34), $\sup_{i\in\mathbb{N}} c_\lambda(i) \le \lambda(\beta/\beta_0 - 1)$. Hence, c_λ belongs to l^∞ and, by 5.5.1, $\beta A - C_\lambda$ where C_λ is defined by $C_\lambda f(i) = c_\lambda(i) f(i)$, $f \in c_0, i \in \mathbb{N}$, is a Feller generator. It follows that $(\mu - \beta A + C_\lambda)^{-1}$ exists for all $\mu > 0$, and in particular we may use $\mu = \lambda$.

So prepared, we take

$$f := \beta(\lambda - \beta A + C_\lambda)^{-1} B^{-1} g. \tag{5.36}$$

Then f is a member of $\mathcal{D}(\beta A - C_\lambda) = \mathcal{D}(\beta A) = \mathcal{D}(A)$, and $\lambda f - \beta A f + C_\lambda f = \beta B^{-1} g$. On the other hand, $\lambda f + C_\lambda f = \lambda \beta B^{-1} f$. Therefore, $\lambda \beta B^{-1} f - \beta A f = \beta B^{-1} g$, showing that this $f \in \mathcal{D}(BA)$ solves (5.35), and thus completing the proof.

5.5.7 *A multiplicative perturbation of a generator: Intuition*

From the perspective of intensity matrices, the multiplicative perturbation by the operator B amounts to multiplying the ith row of the intensity matrix related to A by $b(i)$. Thus, if $b(i) > 1$, the modified process stays at i for a shorter time, and if $b(i) < 1$ it stays there for a longer time. Since, on the other hand, the entire row is multiplied by the same quantity, the probabilities of jumps from i to the other states are left intact. Therefore, we expect the process related to BA to be quite similar to that related to A: it merely runs faster through the states where $b(i) > 1$ and slows down at the states where $b(i) < 1$.

These intuitions are expressed more quantitatively in the Volkonskii Formula ([89] or [76], p. 278), presented in the next section. As we shall see the goal is achieved by modifying the time the process runs through its paths.

5.5.8 *The Volkonskii Formula*

Assume, as in Section 5.5.1 that the chain $X(t), t \geq 0$ related to A has right-continuous paths with left-hand limits, and consider (see Figure 5.1)

$$\theta(t) = \int_0^t \frac{\mathrm{d}s}{b(X(s))}, \qquad t \geq 0.$$

(More precisely, $\theta(t,\omega) = \int_0^t \frac{\mathrm{d}s}{b(X(s,\omega))}, t \geq 0$ for all ω in the underlying probability space.) This is a piece-wise linear, increasing function (for each ω). Its inverse τ, therefore, exists and is likewise piece-wise linear and increasing. In the time interval the chain is at a state i with $b(i) < 1$ the slope of θ is > 1, and the corresponding slope of τ is < 1. As a result, τ, which is designed as a modification of time, runs slower at such a state; at a state where $b(i) > 1$, τ runs faster. The Volkonskii Formula says that the chain related to BA is $X(\tau(t)), t \geq 0$:

$$\mathrm{e}^{tBA} f(i) = \mathbb{E}_i\, f(X(\tau(t))), \qquad f \in c_0, i \in \mathbb{N}, t \geq 0.$$

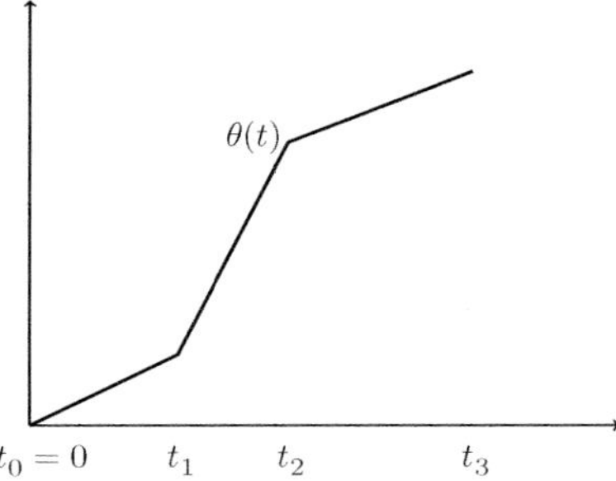

Figure 5.1 Function $\theta(t)$ for a path leading through states i_1, i_2, and i_3: between t_k and t_{k+1} the slope is $1/b(i_{k+1})$. The modified time τ runs though the ordinate axis.

5.5.9 *Proof of the Volkonskii Formula*

In Section 5.5.6, we have proved that $(\lambda - BA)^{-1} = \beta(\lambda - \beta A + C_\lambda)^{-1} B^{-1}$ (see (5.36)). Therefore, by the Feynman–Kac Formula and because convergence in c_0 implies convergence of coordinates, for any $f \in c_0$, $\lambda > 0$ and $i \in \mathbb{N}$,

$$\begin{aligned}(\lambda - BA)^{-1} f(i) &= \beta \int_0^\infty \mathrm{e}^{-\lambda t} \mathrm{e}^{t(\beta A - C_\lambda)} B^{-1} f(i)\,\mathrm{d}t \\ &= \beta \int_0^\infty \mathrm{e}^{-\lambda t} \mathbb{E}_i\, \mathrm{e}^{-\lambda \int_0^t \frac{(\beta - b(X(\beta s)))}{b(X(\beta s))}\,\mathrm{d}s} \frac{f(X(\beta t))}{b(X(\beta t))}\,\mathrm{d}t \\ &= \mathbb{E}_i\, \beta \int_0^\infty \mathrm{e}^{-\lambda\theta(\beta t)} \frac{f(X(\beta t))}{b(X(\beta t))}\,\mathrm{d}t \\ &= \mathbb{E}_i \int_0^\infty \mathrm{e}^{-\lambda\theta(t)} \frac{f(X(t))}{b(X(t))}\,\mathrm{d}t.\end{aligned}$$

To evaluate the inner integral, think of an ω in the underlying probability space and of a time interval $[t_1, t_2]$ in which $X(t, \omega) = j$ for some $j \in \mathbb{N}$. In this interval $f(X(t)) = f(j)$, $b(X(t)) = b(j)$ and θ is differentiable with $\theta'(t) = 1/b(j)$. Therefore,

$$\int_{t_1}^{t_2} \mathrm{e}^{-\lambda\theta(t)} \frac{f(X(t))}{b(X(t))}\,\mathrm{d}t = \int_{\tau(t_1)}^{\tau(t_2)} \mathrm{e}^{-\lambda s} f(j)\,\mathrm{d}s = \int_{\tau(t_1)}^{\tau(t_2)} \mathrm{e}^{-\lambda s} f(X(\tau(s)))\,\mathrm{d}s.$$

Dividing the time half-axis into such intervals (which *do* depend on ω) and summing, we obtain that the entire integral is $\int_0^\infty \mathrm{e}^{-\lambda t} f(X(\tau(t)))\,\mathrm{d}t$. It follows that

$$(\lambda - BA)^{-1} f(i) = \int_0^\infty \mathrm{e}^{-\lambda t}\, \mathbb{E}_i\, f(X(\tau(t)))\,\mathrm{d}t.$$

On the other hand, using again the fact that convergence in c_0 implies convergence of coordinates, we see that

$$(\lambda - BA)^{-1} f(i) = \int_0^\infty e^{-\lambda t} e^{tBA} f(i)\,dt.$$

Since the Laplace transform of a measurable function is identically zero iff this function is zero almost everywhere, we conclude that the Volkonskii Formula holds for each f and i save perhaps on a set of t of measure zero (depending on f and i). But $t \mapsto \mathbb{E}_i\, f(X(\tau(t)))$ is right-continuous, because the paths of $X(t), t \geq 0$ are right-continuous, and $t \mapsto e^{tBA} f(i)$ is continuous since the semigroup $\{e^{tBA}, t \geq 0\}$ is strongly continuous. This shows that the formula holds for all $t \geq 0$.

5.6 Kolmogorov, Kendall, and Reuter Revisited

Having gained some experience by learning about Feller semigroups, let us come back to the Kolmogorov–Kendall–Reuter examples. In this section we will find their sun-duals to see how differently the same information on these process is encrypted in l^1 and in l^∞ spaces. We start with the second example.

5.6.1 *Sun-dual of the second Kolmogorov–Kendall–Reuter semigroup*

Our aim is to characterize the dual of the operator A defined in Section 2.4.7. What do we need to assume about $f \in l^\infty$ to make sure that there is a constant $C = C(f)$ such that

$$\left|\sum_{i=1}^\infty (a_{i+1}\xi_{i+1} - a_i\xi_i) f(i)\right| \leq C(f)\|x\| \tag{5.37}$$

for all $x = (\xi_i)_{i\geq 1} \in \mathcal{D}(A) \subset l^1$? The left-hand side here is the limit, as $n \to \infty$, of

$$\left|\sum_{i=3}^n a_i\xi_i(f(i-1) - f(i)) + a_{n+1}\xi_{n+1} f(n) - \xi_1 f(1)\right| \tag{5.38}$$

(recall that $a_1 = 1$ and $a_2 = 0$). Since each $e_k, k \geq 3$ has norm 1 and belongs do $\mathcal{D}(A)$, and since for $x = e_k$ and $n \geq k$ the expression in (5.38) reduces to $|a_k(f(k-1) - f(k))|$, a necessary condition for (5.37) is

$$\sup_{k\geq 3} a_k|f(k-1) - f(k)| \leq C. \tag{5.39}$$

Conversely, if this condition is satisfied, the series $\sum_{k=2}^{\infty}(f(k-1)-f(k))$ converges absolutely. In particular, there exists the limit $f(\infty) := \lim_{k\to\infty} f(k)$. It follows that the expression under the absolute value sign in (5.38) converges to

$$\sum_{i=3}^{\infty} a_i\xi_i(f(i-1)-f(i)) + (f(\infty)-f(1))\xi_1$$

(recall that $\lim_{i\to\infty} a_i\xi_i = \xi_1$ for $(\xi_i)_{i\geq 1} \in \mathcal{D}(A)$), with the series converging absolutely. Also, the absolute value of the limit does not exceed $(C + |f(\infty) - f(1)|)\|x\|$.

We have thus proved that $\mathcal{D}(A^*)$ is composed of $f \in l^\infty$ such that (5.39) holds and, for such f,

$$\begin{aligned} A^*f(1) &= f(\infty) - f(1), \\ A^*f(2) &= 0, \\ A^*f(i) &= a_i(f(i-1)-f(i)), \qquad i \geq 3. \end{aligned}$$

In particular, as already remarked, $\mathcal{D}(A^*)$ is contained in c. In fact, $\mathcal{D}(A^*)$ is dense in c, because all linear combinations of $\mathbb{1}$ and $e_k, k \geq 1$ belong to $\mathcal{D}(A^*)$. Thus, for the second Kolmogorov–Kendall–Reuter semigroup

$$\mathbb{X}^{\odot} = c. \tag{5.40}$$

It follows also that A_{p}^* is A^* restricted to the domain composed of f such that $\lim_{i\to\infty} a_i(f(i-1)-f(i))$ exists (and is finite).

5.6.2 *Feller property?*

In spite of the fact that, by (5.40), c_0 is a subspace of $\mathbb{X}^{\odot}$, the second Kolmogorov–Kendall–Reuter semigroup does not possess the Feller property. For, recalling the formula for $(\lambda - A)^{-1}$ (see (2.33)) we check that

$$\begin{aligned} \left(\lambda - A^*\right)^{-1} f(i) &= \sum_{j=2}^{i} \frac{1}{\lambda + a_j}\left(\prod_{j<k\leq i} \frac{a_k}{\lambda + a_k}\right) f(j) \\ &= \widetilde{\pi}_i \sum_{j=2}^{i} \frac{1}{\lambda + a_j}\frac{f(j)}{\widetilde{\pi}_j}, \qquad i \geq 3, \end{aligned}$$

where $\widetilde{\pi}_i = \prod_{k=3}^{i} \frac{a_k}{\lambda + a_k}, i \geq 3$. Since $\widetilde{\pi}_i$ decreases when i increases, and $\widetilde{\pi}_\infty := \lim_{i\to\infty} \widetilde{\pi}_i$ exists and differs from 0, the series $\sum_{j=2}^{\infty} \frac{1}{\lambda + a_j}\frac{f(j)}{\widetilde{\pi}_j}$ converges, being dominated by $\sum_{j=2}^{\infty} \frac{1}{\lambda + a_j}\frac{\|f\|}{\widetilde{\pi}_\infty}$. Therefore,

$$\lim_{i\to\infty}\left(\lambda - A^*\right)^{-1} f(i) = \widetilde{\pi}_\infty \sum_{j=2}^{\infty} \frac{1}{\lambda + a_j} \frac{f(j)}{\widetilde{\pi}_j},$$

and this quantity need not be zero for $f \in c_0$.

5.6.3 *Learning new language*

Let's try to decipher the information just obtained. First of all, since $c_0 \subset c$, we see that transition probabilities of the second Kolmogorov–Kendall–Reuter example (these probabilities can be recovered from $\{e^{tA_p^*}, t \ge 0\}$ by $p_{i,j}(t) := e^{tA_p^*}e_j(i)$) satisfy condition (5.11), that is, are more regular than those in general Markov chains. But can the information we possess about the second Kolmogorov–Kendall–Reuter example be obtained from A_p^*?

The answer is in the affirmative, and, at least in this case, the message conveyed by A^* seems to be more clear than that of A. The reader might recall that we needed an approximation by simpler Markov chains to understand what is the meaning of the second assumption in the definition of the domain of A. Here, the situation is more transparent.

First of all, the fact that we need to work in c is an information in itself: $f(\infty)$ is well defined here and in fact appears as a key factor in the definition of A^*f. This tells us that something important is going on at infinity: there is a kind of additional point there that influences the way the chain behaves. Similarly to the third line of the definition of A^*f which tells us that after an exponential time with parameter a_i the chain starting at i jumps to $i-1$, the first line says that after an exponential time at the state 1 our chain jumps to the point at infinity. Moreover, the fact that $A^*f(\infty) = \lim_{i\to\infty} a_i(f(i-1) - f(i))$ tells us that ∞ is an instantaneous state: the chain starting at ∞ leaves this state immediately, with rate $\lim_{i\to\infty} a_i = \infty$ to jump, figuratively speaking, 'to the first rung of the infinite ladder next to infinity.'

5.6.4 *Intensity matrices versus generators*

Multiplying formally the intensity matrix of the second Kolmogorov–Kendall–Reuter chain (given by (2.53)) by an $f \in l^\infty$ from the right one obtains a vector that differs from A^*f only in the first coordinate. As we have commented above, it is precisely this missing factor, that is, $f(\infty)$, that tells the fate of the process starting at the state 1. Again, this information is not contained in the intensity matrix. But what is even more intriguing here is that the same information is conveyed by A and A_p^* in two radically different ways: it is skillfully hidden in the domain of A, and openly displayed in the way A^* acts.

5.6.5 A comparison with the related minimal chain

Since this is the first example we are examining (see cartoon preceding Preface), let's be even more specific about it.

In Example 3.2.3 we have shown that the intensity matrix Q of the second Kolmogorov–Kendall–Reuter process, when modified by replacing -1 in the first row by 0, is nonexplosive, and have found an explicit form of the Markov generator G related to this modified Q (see (3.14) and (3.15)).

Using G, let us define $G_\sharp$ on $\mathcal{D}(G_\sharp) = \mathcal{D}(G)$ by

$$G_\sharp x = Gx - \xi_1 e_1 = Gx + Px - x$$

where $x = (\xi_i)_{i\geq 1}$ and $Px = (0, \xi_2, \xi_3, \dots)$. Clearly, P is a sub-Markov operator and thus $P - I_{l^1}$ is a sub-Markov generator. Therefore, by Trotter's Product Formula, $G_\sharp$ is a sub-Markov generator as well. Since (a) the generator of the minimal semigroup is an extended limit of the operators $D + rO$ as $r \to 1$ (see Sections 3.1.4 and 1.4.3), (b) the off-diagonal operators O for the modified and unmodified Q coincide, (c) the diagonal operators D for the modified and unmodified Q have the same domain and their values on a $(\xi_i)_{i\geq 1}$ are also the same except for the first coordinate where they differ by $-\xi_1$, and (d) G is the generator of the minimal semigroup related to the modified Q, we may argue that $G_\sharp$ is the generator of the minimal semigroup for the unmodified Q.

It is perhaps worth recording explicitly that

$$G_\sharp\, (\xi_i)_{i\geq 1} = (a_{i+1}\xi_{i+1} - a_i\xi_i)_{i\geq 1}$$

where, in contrast to (3.14), a_1 is again 1, not 0. Thus, $G_\sharp$ is defined by the same formula as the generator A of the second Kolmogorov–Kendall–Reuter semigroup: the difference, however, lies in the domain. For $(\xi_i)_{i\geq 1}$ to belong to $\mathcal{D}(A)$, we require

$$\sum_{i=1}^{\infty} |a_{i+1}\xi_{i+1} - a_i\xi_i| < \infty \qquad \text{and} \qquad \lim_{i\to\infty} a_i\xi_i = \xi_1,$$

whereas for $(\xi_i)_{i\geq 1}$ to belong to $\mathcal{D}(G_\sharp)$ we require

$$\sum_{i=1}^{\infty} |a_{i+1}\xi_{i+1} - a_i\xi_i| < \infty \qquad \text{and} \qquad \lim_{i\to\infty} a_i\xi_i = 0.$$

Finding $G_\sharp^*$ is not difficult. In fact, the analysis we carried out in 5.6.1 may be repeated almost verbatim to see that $f \in \mathcal{D}(G_\sharp^*)$ iff condition (5.39) is satisfied, that is, that $\mathcal{D}(G_\sharp^*) = \mathcal{D}(A^*)$. Moreover, $A^* f$ and $G_\sharp^* f$ are the same except for the first coordinate:

$$G_\sharp^* f(1) = -f(1) \qquad \text{whereas} \qquad A^* f(1) = f(\infty) - f(1). \tag{5.41}$$

Now, we know that the chains related to $G_\sharp$ and A differ in the behavior starting at $i = 1$. The minimal chain (generated by $G_\sharp$), conditional on starting at $i = 1$, stays at $i = 1$ for an exponential time with parameter 1, and then is left undefined. By contrast, the process related to A, after the exponential time is over, jumps to ∞ (where it is immediately distributed across $\{i \geq 2\}$ according to an entrance law; see Exercise 4.4.9, cf. (2.55)). There is probably no clearer way of expressing this difference than by (5.41).

Intriguingly, in l^1 setting, the information on the fate of the process starting at $i = 1$ is hidden in the domain of the operators A and $G_\sharp$. The dual perspective, that is, that of l^∞, is dissimilar: A^* and $G_\sharp^*$ have the same domain but act differently. A boundary perturbation has become an additive perturbation of the generator.

5.6.6 *Sun-dual of the first Kolmogorov–Kendall–Reuter semigroup*

Let us turn to a characterization of the sun-dual in the first Kolmogorov–Kendall–Reuter example. What conditions on an $f \in l^\infty$ guarantee existence of a $C = C(f)$ such that

$$|f(Ax)| = \left| \sum_{j=2}^{\infty} (a_j \xi_j - \xi_1)[f(1) - f(j)] \right| \leq C(f)\|x\| \tag{5.42}$$

for all $x \in \mathcal{D}(A)$, where A is defined in Section 2.4.4? Rewriting the absolute value of the infinite series as

$$\lim_{n\to\infty} \left| \sum_{j=2}^{n} a_j \xi_j [f(1) - f(j)] - \xi_1 \sum_{j=2}^{n} [f(1) - f(j)] \right|,$$

and utilizing $x = e_k$, $k \geq 2$, we find, arguing as in Section 5.6.1, that we must have

$$\sup_{k\geq 2} a_k |f(1) - f(k)| \leq C. \tag{5.43}$$

Conversely, if the latter condition holds for some constant C, the series $\sum_{j=2}^{\infty} [f(1) - f(j)]$ converges absolutely, and $|f(Ax)|$ equals

$$\Big|\sum_{j=2}^{\infty} a_j\xi_j[f(1)-f(j)] - \xi_1\sum_{j=2}^{\infty}[f(1)-f(j)]\Big|$$

$$\le C\sum_{j=2}^{\infty}|\xi_j| + |\xi_1|\sum_{j=2}^{\infty}|f(1)-f(j)| \le (C+\sum_{j=2}^{\infty}|f(1)-f(j)|)\|x\|.$$

This analysis shows that f belongs to $\mathcal{D}(A^*)$ iff (5.43) holds and then

$$A^*f(1) = \sum_{j=2}^{\infty}[f(j)-f(1)],$$

$$A^*f(i) = a_i(f(1)-f(i)), \qquad i \ge 2. \tag{5.44}$$

Since (5.43) implies that $\lim_{k\to\infty} f(k) = f(1)$, $\mathcal{D}(A^*)$ is contained in the subspace $c_{1=\infty}$ of $c \subset l^\infty$ composed of $f \in c$ such that $f(\infty) = f(1)$. Since linear combinations of $\mathbb{1}$ and e_k, $k \ge 2$ belong to $\mathcal{D}(A^*)$, thus showing that $\mathcal{D}(A^*)$ is dense in $c_{1=\infty}$, we obtain that in the first Kolmogorov–Kendall–Reuter example,

$$\mathbb{X}^{\odot} = c_{1=\infty}.$$

Moreover, the domain of A^*_{p} is composed of $f \in \mathcal{D}(A^*)$ such that

$$\sum_{j=2}^{\infty}[f(j)-f(1)] = \lim_{i\to\infty} a_i(f(1)-f(i))$$

(with the series converging absolutely).

5.6.7 *What does it mean?*

Is this an elvish tongue or an indistinctive chatter? A glance at (5.44), especially if the series of $A^*f(1)$ is written as a limit of partial sums, convinces us that A^* tells the same story as A: A process starting at $i = 1$ escapes from this point 'immediately' and each of the states $i \ge 2$ has equal rights of being visited. Of course, A^* speaks in a dual language, as it describes infinitesimal changes in expected values.

Notably, the fact that the sun-dual space involves sequences that are convergent seems to point out (as in the previous example) to an additional point at infinity, and the fact that $A^*f(\infty) = \lim_{i\to\infty} a_i(f(1) - f(i))$ indicates that this point is instantaneous. However, $c_{1=\infty}$ is composed of f satisfying $f(1) = f(\infty)$, and so this additional point is identified with 1. In other words, this description warns us again that 1 is odd,[2] as if its strange nature was not clear from the form of $A^*f(1)$.

[2] I am glad it does not warn us 1 is even.

It is a matter of taste whether you find the language of (2.23) more appealing than that of (5.44) or vice versa. In fact, I must reluctantly admit, none of these descriptions even compares in depth to the detailed, revealing stochastic construction of Section 2.6.5.

5.7 Sun-Duals of Nonminimal Processes: A Case Study

As we have seen in the previous sections, the perspectives of l^1 and l^∞, although dual, are apparently dissimilar; the same facts are expressed in these spaces differently. For instance (as in the second Kolmogorov–Kendall–Reuter example), the same information on the fate of the process in the l^1 setting may be masterly hidden in the domain of the generator, and in the l^∞ setting may be displayed in the way the generator acts. Hence, by changing perspective, we transform perturbations of domains into additive perturbations of generators.

In this and the following two sections we continue to study these differences, by taking a closer look at nonminimal processes of Chapters 3 and 4. In particular, we will see that a reverse process to that described above is possible: by changing perspective from l^1 to l^∞, we may transform additive perturbations of generators into perturbations of their domains. More specifically, we know from Chapter 3 that the fate of a postexplosion process is described by additional terms perturbing the generator of the related minimal process. Now, in l^∞ all these generators may act in the same way, and differ only in their domains.

Furthermore, the reader might remember my warning from Section 3.4: generators of postexplosion processes are not obtained by restricting the maximal operator to suitable domains. Well, in l^1 they are not but, as we will see soon, in l^∞ they are.

In this section, we focus on the sun-dual for the semigroup of Section 3.5.1 (describing pure birth process starting afresh after explosion). In the next section, we will extend our result to a quite general class of explosive intensity matrices. In Section 5.9, this result will be further extended to examples where there is also entrance boundary.

5.7.1 *The master operator*

In this and in the following sections

$$Q = \left(q_{i,j}\right)_{i,j\geq 1}$$

is an explosive Kolmogorov matrix.

In the context of l^1, the natural operator related to Q is that of multiplying $x \in l^1$, seen as a row-vector, by Q *from the right* (see Section 3.4), so that

$x \mapsto x \cdot Q$. Its counterpart in l^∞ multiplies an $f \in l^\infty$, seen as a column-vector, by Q *from the left* (see also Section 5.4.2). We will denote this map by Q, so that[3]

$$\boxed{\mathsf{Q}f(i) = \sum_{j=1}^{\infty} q_{i,j} f(j), \qquad i \in \mathbb{N}} \tag{5.45}$$

whenever f is chosen so that $\sup_{i\geq 1} |\sum_{j=1}^{\infty} q_{i,j} f(j)| < \infty$; we recall that each series involved here converges absolutely for all $f \in l^\infty$.

Although this may come as a surprise, because in this aspect l^∞ setting differs from the l^1 setting, $e_i \in l^\infty$ (the unit vector with all but ith coordinate equal zero) in general belongs not to $\mathcal{D}(\mathsf{Q})$. For instance, for the Kolmogorov matrix of Example 3.2.6,

$$\mathsf{Q}e_i(j) = j - i, \qquad j > i,$$

showing that none of the e_i's belong to $\mathcal{D}(\mathsf{Q})$. On the other hand, the functional Σ of Chapter 2, which, to recall, we now denote $\mathbb{1}$, lies in $\mathcal{D}(\mathsf{Q})$ and we have $\mathsf{Q}\mathbb{1} = 0$.

We aim to show that the domain of Q is too large for Q itself to be a generator: by restricting Q to suitable sub-domains, though, we obtain various generators.

5.7.2 *A warning*

It is tempting to think that Q is the dual to $\mathfrak{Q}$ of Section 3.4. But, as we shall soon see in the pure birth process example, this is not the case.

5.7.3 *Trial and error*

To develop intuition, let us consider the case where Q is our old friend, the pure birth process matrix of Example 3.2.2. Here,

$$\mathsf{Q}f(i) = a_i(f(i+1) - f(i)), \qquad i \geq 1,$$

with domain composed of f such that $(a_i(f(i+1) - f(i)))_{i\geq 1} \in l^\infty$. In particular, $e_i \in \mathcal{D}(\mathsf{Q})$, $i \geq 1$ which considerably simplifies analysis of Q.

For $\lambda > 0$ and $g \in l^\infty$, the resolvent equation for the operator Q takes the form

$$(\lambda + a_i)f(i) - a_i f(i+1) = g(i), \qquad i \geq 1. \tag{5.46}$$

[3] In this notation, Q_0 of Section 5.4.2 is the part in c_0 of Q just defined.

This leads to the following recursion for the values of f:

$$f(i+1) = \frac{\lambda + a_i}{a_i} f(i) - \frac{1}{a_i} g(i), \qquad i \geq 1,$$

with solution (see Exercise 3.3.9, cf. Example 5.3.5)

$$\begin{aligned} f(i) &= \left(\prod_{j=1}^{i-1} \frac{\lambda + a_j}{a_j} \right) f(1) + \sum_{j=1}^{i-1} \left(\prod_{k=j+1}^{i-1} \frac{\lambda + a_k}{a_k} \right) \frac{-g(j)}{a_j} \\ &= \frac{1}{\pi_{i-1}} f(1) - \frac{1}{\pi_{i-1}} \sum_{j=1}^{i-1} \frac{\pi_j}{a_j} g(j) \\ &= \frac{1}{\pi_{i-1}} f(1) - \frac{1}{\pi_{i-1}} \sum_{j=1}^{i-1} \frac{\pi_{j-1}}{\lambda + a_j} g(j), \qquad i \geq 1, \end{aligned} \tag{5.47}$$

where $\pi_i = \prod_{j=1}^{i} \frac{a_j}{\lambda + a_j}$ and $\pi_0 = 1$. Since the series $\sum_{j=1}^{\infty} \frac{\pi_{j-1}}{\lambda + a_j}$ converges (see (3.34)) and $(g(i))_{i \geq 1}$ is bounded, $(f(i))_{i \geq 1}$ so defined is bounded regardless of the choice of $f(1)$, which may be treated as a parameter. In fact, $(f(i))_{i \geq 1}$ is a member of c, because

$$\lim_{i \to \infty} f(i) = \frac{1}{\pi_\infty} f(1) - \frac{1}{\pi_\infty} \sum_{j=1}^{\infty} \frac{\pi_{j-1}}{\lambda + a_j} g(j). \tag{5.48}$$

Since (5.46) shows that boundedness of $(f(i))_{i \geq 1}$ implies boundedness of $(a_i(f(i+1) - f(i)))_{i \geq 1}$, it follows that the resolvent equation for Q has infinitely many solutions. Thus, as we heralded before, the domain of Q is too large for Q to be a generator.

Let us, however, see what happens if Q is restricted to functions f with $\lim_{i \to \infty} f(i) = 0$. This condition forces $f(1) = \sum_{j=1}^{\infty} \frac{\pi_{j-1}}{\lambda + a_j} g(j)$ and then, by (5.47),

$$f(i) = \frac{1}{\pi_{i-1}} \sum_{j=i}^{\infty} \frac{\pi_{j-1}}{\lambda + a_j} g(j), \qquad i \geq 1. \tag{5.49}$$

This means in particular that the resolvent equation for Q restricted to $\mathcal{D}(\mathsf{Q}) \cap c_0$ has precisely one solution.

Similarly if, instead of $\lim_{i \to \infty} f(i) = 0$ we demand that $\lim_{i \to \infty} f(i) = f(1)$, formula (5.48) forces $f(1) = \frac{1}{1 - \pi_\infty} \sum_{j=1}^{\infty} \frac{\pi_{j-1}}{\lambda + a_j} g(j)$ and then (5.47) results in

$$f(i) = \frac{1}{\pi_{i-1}} \sum_{j=i}^{\infty} \frac{\pi_{j-1}}{\lambda + a_j} g(j) + \frac{\pi_\infty}{1 - \pi_\infty} \sum_{j=1}^{\infty} \frac{\pi_{j-1}}{\lambda + a_j} g(j), \qquad i \geq 1, \tag{5.50}$$

so that again the resolvent equation for Q restricted to the new domain has precisely one solution (as required for a generator).

These calculations probably would not tell us much, though, were it not for the following nice surprise: by (5.14), the map $g \mapsto f$ given by (5.49) coincides with the dual to the resolvent R_λ of the minimal pure birth chain of Section 3.3.2 (see (3.24)):

$$\left(\lambda - \mathsf{Q}_{|\mathcal{D}(\mathsf{Q})\cap c_0}\right)^{-1} = R_\lambda^*. \tag{5.51}$$

This reveals, by 1.5.5, that Q restricted to $\mathcal{D}(\mathsf{Q})\cap c_0$ is the dual to the generator G of the minimal pure birth process

$$G^* = \mathsf{Q}_{|\mathcal{D}(\mathsf{Q})\cap c_0}. \tag{5.52}$$

In particular, combining this with the fact established in Section 3.4 that in the pure birth process $\mathfrak{Q} = G$, we see that Q *is not* the dual to $\mathfrak{Q}$. Also, we see again that the minimality of the chain is reflected in the fact that the sun-dual space is c_0.

Similarly (see Exercise 5.7.6), the map $g \mapsto f$ given by (5.50) is the dual of the resolvent of the pure birth process which after explosion starts all over again at $i = 1$ (see Section 3.3.5 and formulae (3.30) and (3.32) there). Thus, Q with domain restricted to $f \in \mathcal{D}(\mathsf{Q})$ such that $f(1) = \lim_{i\to\infty} f(i)$ coincides with the dual of the generator G_1 of Section 3.3.6 (see in particular equation (3.36) there).

5.7.4 *Light begins to shine*

The discovery made at the end of the previous section may be generalized to any operator H of the form (3.55), that is, to the generator of the pure birth chain which after explosion starts all over again at i with probability equal to the ith coordinate of a $u \neq 0$ such that $\Sigma u \leq 1$; the case $u = 0$ was covered in Section 5.7.3. As a preparation for the generalization we have in mind, we introduce the following notations. First of all, we see the $u \in l^1$ as a functional on l^∞, thus a member of $(l^\infty)^*$, defined by $u(f) = f(u)$, $f \in l^\infty$. Also, we define

$$c_H = \{f \in c; u(f) = \ell(f)\}$$

where $\ell(f) := \lim_{i\to\infty} f(i)$. Since u and ℓ are bounded linear functionals on c, c_H is a closed subspace of c, hence a Banach space itself.

We will show that Q restricted to $\mathcal{D}(\mathsf{Q}) \cap c_H$ is the dual to H:

$$H^* = \mathsf{Q}_{|\mathcal{D}(\mathsf{Q})\cap c_H}. \tag{5.53}$$

Proof

Step 1. By (3.58) and Exercise 1.5.8, the resolvent of H^* is given by

$$\left(\lambda - H^*\right)^{-1} f = R_\lambda^* f + \frac{f(R_\lambda u)}{1 - \Sigma_\lambda(u)} \Sigma_\lambda, \qquad f \in l^\infty, \tag{5.54}$$

where $\Sigma_\lambda \in (l^1)^*$ is the functional given by

$$\Sigma_\lambda(x) = h(R_\lambda x) = \Sigma x - \Sigma \lambda R_\lambda x. \tag{5.55}$$

Here, R_λ is the resolvent of the minimal pure birth chain, and R_λ^* is its dual, which we know is given by (5.51).

Comparing (5.55) and (3.75) we see that Σ_λ is not only notationally identical with our old friend, which played so important a role in Section 3.6. Furthermore, in Section 3.6.9 we have shown that in the pure birth process example we are analyzing again now, Σ_λ may be identified with the sequence $\left(\frac{\pi_\infty}{\pi_{i-1}}\right)_{i \ge 1} \in l^\infty$.

Step 2. The previous step reveals that Σ_λ may be regarded as a member of c with $\ell(\Sigma_\lambda) = \lim_{i \to \infty} \frac{\pi_\infty}{\pi_{i-1}} = 1$. Since R_λ^* maps l^∞ into c_0, the image of $(\lambda - H^*)^{-1}$ is, by (5.54), contained in c. Moreover,

$$\ell(\left(\lambda - H^*\right)^{-1} f) = \frac{f(R_\lambda u)}{1 - \Sigma_\lambda(u)},$$

because, as we have already seen, $\ell(\Sigma_\lambda) = 1$. At the same time, equality $u(R_\lambda^* f) = R_\lambda^* f(u) = f(R_\lambda u)$ implies

$$\begin{aligned} u(\left(\lambda - H^*\right)^{-1} f) &= u(R_\lambda^* f) + \frac{f(R_\lambda u)}{1 - \Sigma_\lambda(u)} u(\Sigma_\lambda) \\ &= f(R_\lambda u) \left[1 + \frac{\Sigma_\lambda(u)}{1 - \Sigma_\lambda(u)}\right] = \frac{f(R_\lambda u)}{1 - \Sigma_\lambda(u)}. \end{aligned}$$

This shows that the domain of H^* is contained in c_H.

Next, since $a_i \left(\frac{1}{\pi_i} - \frac{1}{\pi_{i-1}}\right) = \frac{\lambda}{\pi_{i-1}}$, $i \ge 1$, we have $\Sigma_\lambda \in \mathcal{D}(\mathsf{Q})$ with $\mathsf{Q}\Sigma_\lambda = \lambda\Sigma_\lambda$. It follows, by (5.54), that $\mathcal{D}(H^*) \subset \mathcal{D}(\mathsf{Q}) \cap c_H$ and

$$\begin{aligned} (\lambda - \mathsf{Q}) \left(\lambda - H^*\right)^{-1} f &= (\lambda - \mathsf{Q}) R_\lambda^* f + \frac{f(R_\lambda u)}{1 - \Sigma_\lambda(u)} (\lambda - \mathsf{Q}) \Sigma_\lambda \\ &= (\lambda - G^*) R_\lambda^* f = f. \end{aligned}$$

Thus H^* and Q coincide on $\mathcal{D}(H^*)$.

Step 3. We are left with showing that $\mathcal{D}(\mathsf{Q}) \cap c_H \subset \mathcal{D}(H^*)$. Let f belong to $\mathcal{D}(\mathsf{Q}) \cap c_H$. Then, $\overline{f} := f - \ell(f)\Sigma_\lambda \in \mathcal{D}(G^*) = \mathcal{D}(\mathsf{Q}) \cap c_0$, and in view of (5.52), $\mathsf{Q}f = G^*\overline{f} + \ell(f)\lambda\Sigma_\lambda$. Therefore, $(\lambda - \mathsf{Q})f = (\lambda - G^*)\overline{f}$ and

$$\left(\lambda - H^*\right)^{-1}(\lambda - \mathsf{Q})f = \left(\lambda - H^*\right)^{-1}(\lambda - G^*)\overline{f}$$
$$= R_\lambda^*(\lambda - G^*)\overline{f} + \frac{(\lambda - G^*)\overline{f}(R_\lambda u)}{1 - \Sigma_\lambda(u)}\Sigma_\lambda.$$

Also,

$$(\lambda - G^*)\overline{f}(R_\lambda u) = R_\lambda^*(\lambda - G^*)\overline{f}(u) = \overline{f}(u) = f(u) - \ell(f)\Sigma_\lambda(u)$$
$$= \ell(f)(1 - \Sigma_\lambda(u))$$

(recall that $f(u) = u(f) = \ell(f)$ by assumption on f), implying

$$\left(\lambda - H^*\right)^{-1}(\lambda - \mathsf{Q})f = \overline{f} + \ell(f)\Sigma_\lambda = f.$$

This shows that each $f \in \mathcal{D}(\mathsf{Q}) \cap c_H$ is a member of $\mathcal{D}(H^*)$, completing the proof. □

Formula (5.53) is a stepping-stone for the following theorem, the main goal of this section.

5.7.5 Theorem

Let H be the generator defined by (3.55). Then the sun-dual space for H is c_H and the generator of the sun-dual semigroup is the part of Q in c_H.

Proof By (5.53) and the sun-dual theorem 1.5.6, it suffices to show that $\mathcal{D}(\mathsf{Q}) \cap c_H$ is dense in c_H. Let $f \in c_H$ be fixed. We will find $f_n \in \mathcal{D}(\mathsf{Q}) \cap c_H$ such that $\lim_{n\to\infty} f_n = f$.

Since $\mathbb{1}$ and e_i's are members of $\mathcal{D}(\mathsf{Q})$, so are $\mathbb{1}_n := \mathbb{1} - \sum_{i=1}^n e_i, n \ge 1$. It follows that also $g_n := \sum_{i=1}^n f(i)e_i + f(\infty)\mathbb{1}_n, n \ge 1$ belong to $\mathcal{D}(\mathsf{Q})$. Let $j \in \mathbb{N}$ be such that the jth coordinate of u is nonzero, ale let υ_j be this nonzero coordinate. Defining

$$f_n := g_n + \frac{u(f) - u(g_n)}{\upsilon_j}e_j, \qquad n \ge 1,$$

we see that $\ell(f_n) = \ell(g_n) = \ell(f)$ and $u(f_n) = u(g_n) + \frac{u(f)-u(g_n)}{\upsilon_j}\upsilon_j = u(f)$. Therefore, $f_n \in c_H$ because $f \in c_H$. Moreover, since $\lim_{n\to\infty} g_n = f$, $\lim_{n\to\infty} u(g_n) = \lim_{n\to\infty} u(f)$ and it follows that $\lim_{n\to\infty} f_n = f$, as desired. □

This theorem is a pleasing example of how different are the languages of l^1 and l^∞. As already mentioned, in l^1 to describe the fact that after explosion the process starts again at a random point with probability being equal to the ith coordinate of a vector u, we add an additional term, namely, $h(x)u$, to the minimal generator (see Section 3.5.5). Hence, the generator is the sum of the minimal semigroup generator G, responsible for the walk in the interior, and of this additional term, describing what happens after explosion, that is, on the boundary.

As exemplified by Theorem 5.7.5, in l^∞ all generators related to one intensity matrix, are restrictions of the single, large, maximal operator Q. This operator is devised to describe the walk in the interior. Moreover, the post explosion fate of the process is expressed in the shape of the sun-dual space $\mathbb{X}^\odot$. First of all, the information on existence of a boundary point is hidden in the fact that $\mathbb{X}^\odot \subset c$; for f in $\mathbb{X}^\odot$, $f(\infty)$ is well defined and so ∞ may be seen as an additional point of the state-space. Also, the space $\mathbb{X}^\odot$ is shaped so that if, for example, $u = e_i$, values of $f \in \mathbb{X}^\odot$ are the same at i and at ∞. This indicates that, from the viewpoint of our process, these two elements of the state-space should be identified, or, put otherwise, for the process, being at ∞ is the same as being at i. (This is of course just a figure of speech, because it is clear that, conversely, being at i is not the same as being at ∞; ∞ is instantaneous but i is not.) We are thus led to believe that right after the process reaches ∞ it jumps to i. In the general case, ∞ is 'identified' with the distribution u and we analogously 'see' that after reaching ∞ the process immediately is distributed according to u.

Again, this is simply a different language. Perhaps mysterious at the first encounter. Perhaps more beautiful.

5.7.6 Exercise

For $(\eta_i)_{i\ge1} \in l^1$ and $g \in l^\infty$, let $(\xi_i)_{i\ge1} \in l^1$ and $f \in l^\infty$ be given by (3.30) and (3.32), and (5.50), respectively. Then $f\,(\xi_i)_{i\ge1} = g\,(\eta_i)_{i\ge1}$.

5.8 Can We Generalize?

Here, we will generalize the findings of the previous section to a class of explosive intensity matrices: for this class, we want to find the dual for the operator defined by (3.92). Since the latter formula describes chains that may have several, different ways of exploding, our analysis needs to involve several 'infinities.' Each 'infinity' corresponds to one way an explosion may come about, that is, to one boundary point. Similarly as in the previous section, we will prove that the dual to H of (3.92) is the master operator Q of Section 5.7.1 restricted to the domain of $f \in l^\infty$ which (a) have limits at all those infinities and are such that (b) the ith of these limits coincides with the value of the corresponding functional u_i. We start by introducing the notion of convergence to boundary points. The main result of this section is formula (5.56).

5.8.1 *Topology in* $\mathbb{N} \cup \mathfrak{B}$

In the extended state-space $\mathbb{N} \cup \mathfrak{B}$ of Section 4.2.2 there is a natural topology. Although elements of $\mathbb{N}$ are naturally separated from other elements of $\mathbb{N}$, and elements of $\mathfrak{B}$ are separated from other elements of $\mathfrak{B}$, elements of $\mathbb{N}$ may converge to elements of $\mathfrak{B}$. To recall, $-j \in \mathfrak{B}$ is just an alias for the minimal active functional $f^j \in \mathfrak{B}$. In Section 4.1, we have discovered that this functional corresponds to a sojourn set, say, A, for the jump chain:

$$f^j = s_A,$$

or rather to a class of sojourn sets for which the above formula holds; our proof of 4.1.11 reveals that for A one can take any of the sets

$$A(\delta) = \{i \in \mathbb{N};\ f^j(i) > \delta\}$$

where δ is a number between 0 and 1. It is these sets that form natural neighborhoods of $-j$. Thus, it seems reasonable to say that a sequence of natural numbers $(i_n)_{n\geq 1}$ converges to $-j$ iff for any $\delta \in (0,1)$ all but finitely many elements of $(i_n)_{n\geq 1}$ belong to $A(\delta)$. Equivalently, $\lim_{n\to\infty} i_n = -j$ iff $\lim_{n\to\infty} f^j(i_n) = 1$.

The down-side of this definition is that there are sojourn sets such that $s_A(i) = 1$ for a number of i (see, e.g., 3.6.18). Hence, in terms of such a definition, a constant sequence $i_n = i, n \geq 1$ would converge to $-j$ for a certain j. To avoid such situations, we will say that $\lim_{n\to\infty} i_n = -j$ iff

$$\lim_{n\to\infty} i_n = \infty \qquad \text{and} \qquad \lim_{n\to\infty} f^j(i_n) = 1.$$

5.8.2 *Definition of* c_H

An $f \in l^\infty$ will be said to have the limit at the boundary point $-j$ iff for all sequences $(i_n)_{n\geq 1}$ converging to $-j$ the limit $\lim_{n\to\infty} f(i_n)$ exists, is finite, and does not depend on the choice of the sequence $(i_n)_{n\geq 1}$. This limit is then denoted $\ell_j(f)$. In other words,

$$\ell_j(f) = \lim_{i_n \to -j} f(i_n).$$

The definition of the operator H defined in (3.92) involves minimal functionals, or sojourn sets, $f^i, i = 1, \dots, k$, summing to $\Sigma_{\max}$, and nonnegative

$u_i, i = 1, \dots, \mathcal{k}$, members of l^1 such that $\Sigma u_i \leq 1$. The related space $c_H \subset l^\infty$, is defined as

$$c_H := \{f \in l^\infty, \ell_j(f) = u_j(f), j = 1, \dots, \mathcal{k}\}.$$

In other words, members of c_H have limits at all boundary points, that is, $\ell_j(f)$'s are well defined, and relations prescribed above hold. It is clear that c_H is a Banach space (with supremum norm) and that ℓ_j's are continuous functionals on c_H.

5.8.3 *The goal*

We aim at proving that, as in (5.53),

$$\boxed{H^* = \mathsf{Q}_{|\mathcal{D}(\mathsf{Q}) \cap c_H}.} \tag{5.56}$$

This relation will be established in 5.8.7; as a preparation, we need the result discussed in the next section.

5.8.4 *A borrowed result and its consequences*

Let $(i_n)_{n\geq 1}$ be a sequence converging to a boundary point $-j \in \mathfrak{B}$. By definition, we have then $\lim_{n\to\infty} f^j(i_n) = 1$. We will need a stronger result (see [42], p. 541), saying that we have also

$$\lim_{n\to\infty} f_\lambda^j(i_n) = 1$$

for all $\lambda > 0$ (where f_λ^j is the notation of Section 3.6.7); in terms of ℓ_j this says that $\ell_j(f_\lambda^j) = 1$. Since, by (3.81), $\mathbb{1} = \Sigma \geq (\Sigma_{\max})_\lambda = \Sigma_\lambda = \sum_{j=1}^{\mathcal{k}} f_\lambda^j$ and all summands are nonnegative, it follows that

$$\lim_{n\to\infty} f_\lambda^k(i_n) = 0, \qquad k \in \{1, \dots, \mathcal{k}\} \setminus \{j\}, \tag{5.57}$$

that is, that $\ell_k(f_\lambda^j) = \delta_{j,k}$, and this in turn shows that

$$\lim_{n\to\infty} \Sigma_\lambda(i_n) = 1.$$

In view of $\Sigma(i_n) = 1$, relation (3.76) proves now that $\lim_{n\to\infty} R_\lambda^* \Sigma(i_n) = 0$. Therefore, $\lim_{n\to\infty} R_\lambda^* f(i_n) = 0$ for all f satisfying $0 \leq f \leq \Sigma$, and thus for all nonnegative f. Finally, we obtain

$$\lim_{n\to\infty} R_\lambda^* f(i_n) = 0, \qquad f \in l^\infty \tag{5.58}$$

for any $(i_n)_{n\ge 1}$ converging to any boundary point. In other words,

$$\ell_j(R^*_\lambda f) = 0, \qquad f \in l^\infty, j = 1, \ldots, \kappa. \tag{5.59}$$

5.8.5 Remark

Formula (5.58) *does not* say that $\lim_{n\to\infty} R^*_\lambda f(i_n) = 0$ whenever $\lim_{n\to\infty} i_n = \infty$. For instance, there can be sequences $(i_n)_{n\ge 1}$ converging to infinity which approach no point of the exit boundary (i.e., sequences that are related to passive functionals): if the example of two infinite ladders (see Sections 3.5.7 and 3.5.8) is modified so that $a_{2i} = 1$ for all $i \in \mathbb{N}$, then arguing as in (5.47) we find that on the 'even ladder' $R^*_\lambda g$ must be of the form

$$R^*_\lambda g(2i) = (\lambda+1)^{i-1}\left(R^*_\lambda g(2) - \sum_{j=1}^{i-1}\frac{g(2j)}{(\lambda+1)^j}\right), \qquad i \in \mathbb{N}.$$

On the other hand, since Q restricted to l^1_{e} is bounded and generates a Markov semigroup there, the solution to the resolvent equation exists for $g \in l^1_{\mathrm{e}}$, and is uniquely determined. It follows that for this solution $R^*_\lambda g(2) = \sum_{j=1}^\infty \frac{g(2j)}{(\lambda+1)^j}$, which in turn implies

$$R^*_\lambda g(2i) = (\lambda+1)^{i-1}\sum_{j=i}^{\infty}\frac{g(2j)}{(\lambda+1)^j}.$$

In particular, for $g = (0,1,0,1,\ldots)$,

$$\lambda R^*_\lambda g(2i) = 1, \qquad i \ge 1,$$

and thus $\lim_{i\to\infty} R^*_\lambda g(2i)$ is not 0. (More generally, it may be proved that $\lim_{i\to\infty}\lambda R^*_\lambda g(2i) = \lim_{n\to\infty} g(2i)$, provided the latter limit exists.)

We are now ready to establish the first part of (5.56).

5.8.6 Proposition

Let H be given by (3.92). Then $\mathcal{D}(H^*) \subset c_H$.

Proof By 1.5.5, any element f of $\mathcal{D}(H^*)$ is of the form $f = [(\lambda - H)^{-1}]^* g$ for some $g \in l^\infty$, where $(\lambda - H)^{-1}$ is given by (3.99). The latter formula involves the $\kappa \times \kappa$ matrix $M_\infty = \left(m_{i,j}\right)_{i,j=1,\ldots,\kappa}$ defined by

$$M_\infty := \sum_{n=0}^{\infty} M^n$$

where $M = \left(h_j(R_\lambda u_i)\right)_{i,j=1,\ldots,\mathcal{k}}$ (vectors u change from row to row, functionals h change from column to column). More precisely, in terms of M_∞, (3.99) may be rewritten as

$$(\lambda - H)^{-1} y = R_\lambda y + \sum_{j=1}^{\mathcal{k}} \left(\sum_{i=1}^{\mathcal{k}} m_{i,j} h_i(R_\lambda y) \right) R_\lambda u_j, \qquad y \in l^1.$$

Therefore,

$$f(y) = g((\lambda - H)^{-1} y) = g(R_\lambda y) + \sum_{j=1}^{\mathcal{k}} \sum_{i=1}^{\mathcal{k}} m_{i,j} h_i(R_\lambda y) g(R_\lambda u_j).$$

Since (see (3.91)), $h_i(R_\lambda y) = f_\lambda^i(y)$, this means that

$$f = R_\lambda^* g + \sum_{i=1}^{\mathcal{k}} \left(\sum_{j=1}^{\mathcal{k}} m_{i,j} g(R_\lambda u_j) \right) f_\lambda^i. \tag{5.60}$$

All $\mathcal{k} + 1$ vectors in the linear combination on the right-hand side belong, by (5.57) and (5.59), to the intersection of the domains of the functionals $\ell_1, \ldots, \ell_{\mathcal{k}}$, and we have

$$\ell_k(f) = \sum_{j=1}^{\mathcal{k}} m_{k,j} g(R_\lambda u_j), \qquad k = 1, \ldots, \mathcal{k}.$$

On the other hand, using (3.91) again, we see that

$$u_k(f) = g(R_\lambda u_k) + \sum_{j=1}^{\mathcal{k}} \left(\sum_{i=1}^{\mathcal{k}} h_i(R_\lambda u_k) m_{i,j} \right) g(R_\lambda u_j).$$

Noting that $MM_\infty = M_\infty - I$, we realize that the sum in parentheses is $m_{k,j} - \delta_{k,j}$, and so the entire expression reduces to $\sum_{j=1}^{\mathcal{k}} m_{k,j} g(R_\lambda u_k)$. This shows that $u_k(f) = \ell_k(f)$ for all $f \in \mathcal{D}(H^*)$ and $k \in \{1, \ldots, \mathcal{k}\}$, that is, that $\mathcal{D}(H^*)$ is contained in c_H. □

Whereas the previous proposition is valid for all explosive matrices, the remaining results require some extra assumptions. Since we are not striving for generality but rather want our theorem to illustrate the idea well, in what follows we assume that

$$G^* = \mathsf{Q}_{|\mathcal{D}(\mathsf{Q}) \cap c_0}, \tag{5.61}$$

and that (see Remark 5.8.5), for an $f \in l^\infty$,

$$\text{conditions} \quad \ell_i(f) = 0, i = 1, \ldots, \mathcal{k} \qquad \text{imply} \qquad f \in c_0. \tag{5.62}$$

5.8.7 Proposition

Suppose (5.61) holds. Then, Q is an extension of H^*: $\mathcal{D}(H^*)$ is a subset of $\mathcal{D}(\mathsf{Q})$, and H^* coincides with Q on $\mathcal{D}(H^*)$.

Proof For an $f \in \mathcal{D}(H^*)$, let us consider representation (5.60) and take a closer look at the summands other than $R_\lambda^* g$. From Section 3.6.8 we know that $B_\lambda^* f_\lambda^i = f_\lambda^i, i = 1, \ldots, \Bbbk$. Recalling the definition of B_λ (see equation (3.4)) we easily see that

$$B_\lambda^* f(j) = \frac{1}{\lambda + q_j} \sum_{k \neq j} q_{j,k} f(k), \qquad f \in l^\infty.$$

Therefore, the result from Section 3.6.8 just recalled says that

$$(\lambda + q_j) f_\lambda^i(j) = \sum_{k \neq j} q_{j,k} f_\lambda^i(k), \qquad j \in \mathbb{N}.$$

It follows that

$$f_\lambda^i \in \mathcal{D}(\mathsf{Q}) \qquad \text{and} \qquad \mathsf{Q} f_\lambda^i = \lambda f_\lambda^i, \qquad i = 1, \ldots, \Bbbk. \tag{5.63}$$

Hence, by (5.61), the right-hand side of (5.60) belongs to $\mathcal{D}(\mathsf{Q})$, and

$$(\lambda - \mathsf{Q}) f = (\lambda - G^*) R_\lambda^* g = g.$$

Since, on the other hand, $(\lambda - H^*) f$ also equals g, we are done. □

5.8.8 Proposition

Suppose conditions (5.61) and (5.62) are satisfied. Then $\mathcal{D}(\mathsf{Q}) \cap c_H \subset \mathcal{D}(H^*)$.

Proof Suppose f belongs to $\mathcal{D}(\mathsf{Q}) \cap c_H$, and fix $\lambda > 0$. Our task will be completed if we succeed in proving that $(\lambda - H^*)^{-1} (\lambda - \mathsf{Q}) f = f$.

Let $\overline{f} = f - \sum_{k=1}^{\Bbbk} \ell_k(f) f_\lambda^k$. By (5.57), $\ell_k(\overline{f}) = 0, k = 1, \ldots, \Bbbk$. Hence, by assumption (5.62), $\overline{f}$ belongs to c_0. Therefore, because of (5.63) and (5.61), $(\lambda - \mathsf{Q}) f = (\lambda - \mathsf{Q}) \overline{f} = (\lambda - G^*) \overline{f}$, and thus $(\lambda - H^*)^{-1} (\lambda - \mathsf{Q}) f$ is the right-hand side of (5.60) with g replaced by $(\lambda - G^*) \overline{f}$. Since $R_\lambda^* (\lambda - G^*) \overline{f} = \overline{f}$ and $(\lambda - G^*) \overline{f} (R_\lambda u_j) = \overline{f}(u_j)$, we obtain

$$(\lambda - H^*)^{-1} (\lambda - \mathsf{Q}) f = \overline{f} + \sum_{i=1}^{\Bbbk} \left(\sum_{j=1}^{\Bbbk} m_{i,j} \overline{f}(u_j) \right) f_\lambda^i.$$

Thus, it suffices to show that the expression in parentheses is $\ell_i(f)$ and, by $\overline{f}(u_j) = f(u_j) - \sum_{k=1}^{\mathfrak{K}} \ell_k(f) f_\lambda^k(u_j)$, our task reduces to showing that

$$\sum_{j=1}^{\mathfrak{K}} m_{i,j}[f(u_j) - \sum_{k=1}^{\mathfrak{K}} \ell_k(f) f_\lambda^k(u_j)] = \ell_i(f), \qquad i = 1, \ldots, \mathfrak{K}. \quad (5.64)$$

However, recalling (3.91) and noting that $M_\infty M = M - I$ (consult Section 5.8.6), we see that

$$\sum_{j=1}^{\mathfrak{K}} m_{i,j} f_\lambda^k(u_j) = \sum_{j=1}^{\mathfrak{K}} m_{i,j} h_k(R_\lambda u_j) = m_{i,k} - \delta_{i,k}.$$

It follows that the left-hand side of (5.64) equals

$$\sum_{j=1}^{\mathfrak{K}} m_{i,j} f(u_j) - \sum_{k=1}^{\mathfrak{K}} \ell_k(f)[m_{i,k} - \delta_{i,k}].$$

This further reduces to $\ell_i(f)$, because f belongs to c_H. □

Combining Propositions 5.8.6, 5.8.7 and 5.8.8 we conclude that under assumptions (5.61) and (5.62) formula (5.56), the main subject of this section, holds. In this formula the fact that there are several ways the chain may explode, that is, there are several boundary exit points, is expressed in existence of limits $\ell_i(f)$, which are limits 'through sojourn sets' corresponding to these boundary points. Moreover, the fact that a boundary point i distributes paths of the process according to the distribution u_i is reflected in the equality $u_i(f) = \ell_i(f)$, $f \in \mathcal{D}(H^*)$. Again, we note that the same information on the fate of the process after explosion is expressed in one way in l^1 and in a very different way in l^∞.

5.9 Calling on P. Lévy Again

Before completing the book, we come back to the examples involving entrance laws. As already mentioned, it is intriguing that by introducing such laws one changes the domain of the 'master operator' in the l^1 setting (as seen in Sections 4.3 and 4.4) as well as in l^∞ setting (as will be seen now). Put otherwise, we will see again that duals of generators of postexplosion Markov chains are restrictions of the operator Q of Section 5.7.1 to very characteristic domains. It is in the shape of these domains, in the shape of the sun-dual space, that the information on postexplosion history is hidden.

5.9.1 *The dual to P. Lévy's flash generator*

Let us calculate the dual to the operator A of Section 4.4.1 (equation (4.39)); this operator includes, as a special case (for $p = 1$), the generator of P. Lévy's flash.

An $f \in l^\infty(\mathbb{Z})$ belongs to $\mathcal{D}(A^*)$ if there is a constant $C = C(f)$ such that

$$|f(Ax)| = \left|\lim_{n\to\infty} \sum_{i=-n}^{n} f(i)(a_{i-1}\xi_{i-1} - a_i\xi_i) + (1-p)\mathfrak{l}_+(x)f(u)\right|$$
$$\leq C(f)\|x\| \tag{5.65}$$

for all $x = (\xi_i)_{i\in\mathbb{Z}} \in \mathcal{D}(A)$. The sum featuring here may be rewritten as

$$\sum_{i=-n}^{n} a_i(f(i+1) - f(i))\xi_i - a_n\xi_n f(n+1) + f(-n)a_{-n-1}\xi_{-n-1}.$$

Moreover, $e_i \in \mathcal{D}(A)$ and $\mathfrak{l}_+(e_i) = 0$ for all $i \in \mathbb{Z}$, and for $x = e_i$ the sum above reduces to $a_i(f(i+1) - f(i))$ provided n is sufficiently large. Hence, a necessary condition for (5.65) is

$$\sup_{i\in\mathbb{Z}} a_i|f(i+1) - f(i)| < \infty. \tag{5.66}$$

It follows that for $f \in \mathcal{D}(A^*)$ the series $\sum_{i\in\mathbb{Z}} |f(i+1) - f(i)|$ converges (recall assumption (4.23)) and, thus, the limits $f(\infty) := \lim_{n\to\infty} f(n)$ and $f(-\infty) := \lim_{n\to\infty} f(-n)$ exist (and are finite).

However, (5.66) alone does not yet guarantee that $f \in \mathcal{D}(A^*)$; we claim that (5.65) forces also

$$f(\infty) = pf(-\infty) + (1-p)f(u), \tag{5.67}$$

and that (5.66) and (5.67) combined imply $f \in \mathcal{D}(A^*)$. For, if condition (5.66) is met, $f(Ax)$ equals (see the displayed formula following (5.65))

$$\sum_{i\in\mathbb{Z}} a_i(f(i+1) - f(i))\xi_i - \mathfrak{l}_+(x)f(\infty) + \mathfrak{l}_-(x)f(-\infty) + (1-p)\mathfrak{l}_+(x)f(u)$$

or, by the boundary condition,

$$\sum_{i\in\mathbb{Z}} a_i(f(i+1) - f(i))\xi_i + \mathfrak{l}_+(x)[pf(-\infty) + (1-p)f(u) - f(\infty)].$$

Thus, (5.65) implies (via (5.66)) that there is a $C' = C'(f)$ such that

$$|\mathfrak{l}_+(x)[pf(-\infty) + (1-p)f(u) - f(\infty)]| \leq C'(f)\|x\|, \qquad x \in \mathcal{D}(A).$$

Taking

$$x_n = (\ldots, pa^{-1}_{-n-1}, pa^{-1}_{-n}, 0, \ldots, 0, a_n^{-1}, a^{-1}_{n+1}, \ldots)$$

(nonzero elements end at the $(-n)$th coordinate and start again at the nth coordinate), we see that $x_n \in \mathcal{D}(\mathfrak{Q})$ (for the $\mathfrak{Q}$ defined in 4.3.1), $\mathfrak{l}_+(x_n) = 1$ and $\mathfrak{l}_-(x_n) = p$, implying $x_n \in \mathcal{D}(A)$ for all $n \ge 1$. Since $\lim_{n\to\infty} \|x_n\| = 0$, the last inequality is impossible unless the expression in brackets is zero. Conversely, if conditions (5.66) and (5.67) are satisfied, $f(Ax)$ equals $\sum_{i\in\mathbb{Z}} a_i(f(i+1) - f(i))\xi_i$ and thus $|f(Ax)|$ does not exceed

$$\|x\| \sup_{i\in\mathbb{Z}} a_i |f(i+1) - f(i)|.$$

This means that $f \in \mathcal{D}(A^*)$.

To summarize, $f \in l^\infty(\mathbb{Z})$ is a member of $\mathcal{D}(A^*)$ iff conditions (5.66) and (5.67) are satisfied and then $A^* f(i) = a_i(f(i+1) - f(i)), i \in \mathbb{Z}$. Put otherwise, A^* is the restriction of the operator Q to the domain where (5.67) holds; in our case, the operator Q of Section 5.7.1 is given by $\mathsf{Q}f(i) = a_i(f(i+1) - f(i)), i \in \mathbb{Z}$ on the domain composed of f satisfying (5.66).

5.9.2 *Sun-dual space for P. Lévy's flash*

I claim also that the sun-dual space here is identical to $c_{u,p} \subset l^\infty(\mathbb{Z})$ composed of f such that $f(\infty)$ and $f(-\infty)$ exist, and relation (5.67) holds. To prove this, it suffices, given $f \in c_{u,p}$, to find a sequence $(f_n)_{n\ge1}$ of elements of $\mathcal{D}(A^*) \subset c_{u,p}$ converging to f. To this end, we note that e_i's are members of $\mathcal{D}(\mathsf{Q})$ and that so are

$$\mathbb{1}_n := (\dots, 0, 0, 0, 1, 1, 1, \dots), \qquad n \in \mathbb{Z}$$

(the first nonzero element is on the $(n+1)$st coordinate). Therefore, also g_n defined by

$$g_n = \sum_{i=-n}^{n} f(i)e_i + f(\infty)\mathbb{1}_n + f(-\infty)(\mathbb{1} - \mathbb{1}_{-n-1}), \qquad n \ge 1$$

belongs to $\mathcal{D}(\mathsf{Q})$. Let $j \in \mathbb{Z}$ be such that the jth coordinate of u, denoted υ_j, is nonzero. Then

$$f_n = g_n + \frac{u(f) - u(g_n)}{\upsilon_j} e_j$$

belongs to $\mathcal{D}(\mathsf{Q})$, and

$$\begin{aligned} f_n(\infty) &= g_n(\infty) = f(\infty), \\ f_n(-\infty) &= g_n(-\infty) = f(-\infty), \\ u(f_n) &= u(g_n) + \frac{u(f) - u(g_n)}{\upsilon_j}\upsilon_j = u(f), \end{aligned}$$

proving that $f_n \in \mathcal{D}(A^*)$, because f belongs to $c_{u,p}$. Since $\lim_{n\to\infty} g_n = f$, we have also $\lim_{n\to\infty} u(g_n) = u(f)$, and this implies $\lim_{n\to\infty} f_n = f$, completing the proof.

5.9.3 *The dual to the generator of Section 4.4.7*

Arguing as in the previous section, one can find the dual to the generator A of Section 4.4.7, where two entrance laws were active. The analysis in this case involves four infinities: $-\infty$ and ∞ on $\mathbb{Z}$, and $-\infty'$ and ∞' on $\mathbb{Z}'$. Since the calculations are analogous, we will leave them as an exercise to the reader: the main result is that the dual to A is Q restricted to the domain where limits at infinities exist and satisfy the following two conditions

$$\boxed{f(\infty) = pf(-\infty) + qf(u) + (1-p-q)f(-\infty'),}$$

$$\boxed{f(\infty') = p'f(-\infty') + q'f(u') + (1-p'-q')f(-\infty).}$$

It is worth noting here that, in comparison to (4.43), the matrix of coefficients was transposed (as expected) and that new terms involving $f(u)$ and $f(u')$ are now present. But, most importantly, we see again that the shape of the domain of A^* contains the information on the postexplosion process: for example, the first part of the formula says that a particle starting at ∞ will after infinitesimal time be at $-\infty$ with probability p, at $-\infty'$ with probability $1-p-q$, or will be distributed according to the density u with probability q; interpretation of the second part is analogous.

5.10 Notes

According to Reuter and Riley [75], Sections 5.4.1, 5.4.2, and 5.4.5 are due to W. B. Jurkat who, however, published them merely in an internal research report and did not provide proofs. In the first of these sections I follow closely the exposition of [75], but the argument presented in 5.4.5 slightly differs from the original. Section 5.4.7 generalizes Theorem 8 in [75] (by removing the assumption that Q is nonexplosive); the proof that $\overline{A}$ is a Feller generator goes along more or less standard lines (see, e.g., [39]).

The fact that BA of Section 5.5.6 is a Feller generator is a simple case of the multiplicative perturbation theorem due to Dorroh ([31,37]). Remarkably, assumption (5.34) is heavily used in 5.5.6. Similarly, in the abstract setting, an analogue of condition (5.34) is to much extend indispensable; see discussion in [10, 11, 15, 26, 27].

To the best of my knowledge, Section 5.6 is new,[4] and so are the sections following it.

[4] An old joke tells about a review of a certain paper that allegedly read, 'The paper contains new and interesting results. New results are not interesting; interesting results are not new.'

Solutions and Hints to Selected Exercises

Hint to Exercise 1.2.13 Note that $\mathrm{e}^{tQ} = \mathrm{e}^{-t}\mathrm{e}^{tP}$ and expand e^{tP} as in (1.2).

Solution to Exercise 1.5.8 Since $(A^*f)(x) = f(Ax) = f_0(x)f(x_0)$, we see that $A^*f = f(x_0)f_0$.

Solution to Exercise 1.5.9 If, for a sequence $(f_n)_{n\ge1}$ of members of $\mathcal{D}(A^*)$, there are functionals f and g such that for all $x \in \mathbb{X}$, $\lim_{n\to\infty} f_n(x) = f(x)$ and $\lim_{n\to\infty}(A^*f_n)(x) = g(x)$, then $f(Ax) = \lim_{n\to\infty} f_n(Ax) = \lim_{n\to\infty}(A^*f_n)(x) = g(x)$.

Hint to Exercise 2.4.17 For nonnegative $x = (\xi_i)_{i\ge1} \in \mathcal{D}(A)$, $\Sigma Ax = (\alpha - 1)\xi_1 \le 0$.

Solution to Exercise 2.4.18 For $(\xi_i)_{i\ge1}$ in $\mathcal{D}(A)$,

$$\begin{aligned}\Sigma A\,(\xi_i)_{i\ge1} &= \eta_1 + \lim_{n\to\infty}\sum_{i=2}^{n}\eta_i\\ &= 9\xi_2 - 6\xi_1 + \lim_{n\to\infty}\left(\sum_{i=3}^{n+1}3^i\xi_i - 3\sum_{i=2}^{n}3^i\xi_i + 2\sum_{i=1}^{n-1}3^i\xi_i\right)\\ &= \lim_{n\to\infty} 3^n(3\xi_{n+1} - 2\xi_n).\end{aligned}$$

However, condition (2.37) implies $\lim_{n\to\infty} 3^n(3\xi_{n+1} - \xi_n) = 0$. Thus, if $(\xi_i)_{i\ge1}$ is nonnegative, $\Sigma A\,(\xi_i)_{i\ge1} = -\lim_{n\to\infty} 3^n\xi_n \le 0$.

Hint to Exercise 2.7.9 Recall that $p_{1,i}(0) = 0, i \ge 2$. Note also that $0 \le \lambda^2\pi_2 \le \frac{\lambda^2}{(\lambda+a_i)(\lambda+1)}\frac{1}{\lambda+a_{i+1}}$.

Hint to Exercise 3.1.9 Proceeed by induction: Write $A - B = A^{n-1}(A-B) + (A^{n-1} - B^{n-1})B$.

Solution to Exercise 3.2.8 We have $B_\lambda e_1 = 0$ and

$$B_\lambda e_i = \frac{a_i}{\lambda + 2a_i} e_1 + \frac{a_i}{\lambda + 2a_i} e_{i+1}, \qquad i \ge 2.$$

Thus, by induction,

$$B_\lambda^n e_i = \frac{\pi_{i+n}}{\pi_i}(e_1 + e_{i+n}), \qquad i \ge 2, n \ge 1,$$

where $\pi_i = \prod_{j=1}^{i-1} \frac{a_j}{\lambda+2a_j}$. Since $\lim_{n\to\infty} \frac{a_n}{\lambda+2a_n} = \frac{1}{2}$, we clearly have $\lim_{n\to\infty} \pi_{i+n} = 0$, and so $\lim_{n\to\infty} B_\lambda^n e_i = 0$.

Solution to Exercise 3.3.12 By (3.34),

$$\begin{aligned}\lambda \sum_{i=1}^\infty |\xi_i| &= \lambda \sum_{i=1}^\infty \left| \frac{\pi_{i-1}}{\lambda + a_i} \sum_{j=1}^i \frac{\eta_j}{\pi_{j-1}} \right| \le \sum_{j=1}^\infty \frac{|\eta_j|}{\pi_{j-1}} \lambda \sum_{i=j}^\infty \frac{\pi_{i-1}}{\lambda + a_i} \\ &= \sum_{j=1}^\infty \frac{|\eta_j|}{\pi_{j-1}} (\pi_{j-1} - \pi_\infty) = \sum_{j=1}^\infty |\eta_j| \left(1 - \frac{\pi_\infty}{\pi_{j-1}}\right) \\ &\le (1 - \pi_\infty) \sum_{j=1}^\infty |\eta_j|.\end{aligned}$$

Moreover, for $(\eta_i)_{i\ge1} = e_1$, all inequalities here turn out to be equalities.

Hint to Exercise 3.5.11 Let $(r_n)_{n\ge1}$ be a sequence of elements of $[0, 1)$ such that $\lim_{n\to\infty} r_n = 1$. By 3.1.4 and the Sova–Kurtz version of the approximation theorem (Section 1.4.3), the generator, say, H, of the minimal semigroup for the birth and death chain is the extended limit of the operators A_n defined on the common domain $\mathcal{D}(A_n) = \{(\xi_i)_{i\ge1}; \sum_{i\ge1}(a_i + b_i)|\xi_i| < \infty\}$ by the formulae

$$A_n (\xi_i)_{i\ge1} = -(a_i\xi_i)_{i\ge1} - (b_i\xi_i)_{i\ge1} + r_n (a_{i-1}\xi_{i-1})_{i\ge1} + r_n (b_{i+1}\xi_{i+1})_{i\ge1},$$

where b_1 is modified to be equal to 0 (note that b_1 in fact does not feature in the intensity matrix of the chain). However, since $(b_n)_{n\ge1}$ is bounded, $\mathcal{D}(A_n) = \{(\xi_i)_{i\ge1}; \sum_{i\ge1} a_i|\xi_i| < \infty\}$, and sequences $x_n, n \ge 1$ (with $x_n = (\xi_{n,i})_{i\ge1} \in \mathcal{D}(A_n)$) and $A_n x_n, n \ge 1$ converge simultaneously iff so do $x_n, n \ge 1$ and

$$(-a_i\xi_{n,i})_{i\ge1} + r_n (a_{i-1}\xi_{n,i-1})_{i\ge1}, n \ge 1.$$

By 3.3.2 and the Sova–Kurtz version of the approximation theorem, it follows that the domain of H coincides with $\mathcal{D}(G)$ defined in (3.25) and that

$$H (\xi_i)_{i\ge1} = G (\xi_i)_{i\ge1} + (b_{i+1}\xi_{i+1} - b_i\xi_i)_{i\ge1}.$$

Solution to Exercise 3.6.19 The functional f_λ is the largest of g satisfying $B_\lambda^* g = g$ and $0 \le g \le f$. In particular, $0 \le f_\lambda \le f$. On the other hand, $(f_\lambda)^\diamond$ is the smallest of g satisfying $f_\lambda \le g \le \Sigma$ and $\Pi^* g = g$. Since, for $f \in \mathfrak{B}$, $\Pi^* f = f$, f is a member of the latter set and thus $(f_\lambda)^\diamond \le f$.

Solution to Exercise 3.6.21 By (3.84), $f(\lambda R_\lambda x) = f(x)$, for all $x \in l^1$, $\lambda > 0$. On the other hand, by the Hille approximation, $S(t)x = \lim_{n\to\infty}\left(\frac{n}{t}R_{\frac{n}{t}}\right)^n x$. Since f is continuous,

$$f(S(t)x) = \lim_{n\to\infty} f\left(\left(\frac{n}{t}R_{\frac{n}{t}}\right)^n x\right) = f(x).$$

Solution to Exercise 3.6.22 We have

$$B_\lambda e_1 = p_\lambda e_2 + p'_\lambda e_3 \qquad \text{and} \qquad B_\lambda e_2 = q_\lambda e_1 + q'_\lambda e_4.$$

It follows that $B_\lambda^2 e_1 = p_\lambda q_\lambda e_1 + p'_\lambda B_\lambda e_3 + p_\lambda q'_\lambda e_4$, and so

$$B_\lambda^{n+2} e_1 = p_\lambda q_\lambda B_\lambda^n e_1 + p'_\lambda B_\lambda^{n+1} e_3 + p_\lambda q'_\lambda B_\lambda^n e_4.$$

Thus, $\Sigma_\lambda(1) = p_\lambda q_\lambda \Sigma_\lambda(1) + p'_\lambda \Sigma_\lambda(3) + p_\lambda q'_\lambda \Sigma_\lambda(4)$, or

$$\Sigma_\lambda(1) = \frac{p'_\lambda}{1 - p_\lambda q_\lambda}\Sigma_\lambda(3) + \frac{p_\lambda q'_\lambda}{1 - p_\lambda q_\lambda}\Sigma_\lambda(4) = \frac{p'_\lambda}{1 - p_\lambda q_\lambda}\Sigma_\lambda(3),$$

and, similarly,

$$\Sigma_\lambda(2) = q_\lambda \Sigma_\lambda(1) + q'_\lambda \Sigma_\lambda(4) = q_\lambda \Sigma_\lambda(1) = \frac{p'_\lambda q_\lambda}{1 - p_\lambda q_\lambda}\Sigma_\lambda(3).$$

Solution to Exercise 3.8.5 Suppose $f \cap g \ne 0$. Then, by 3.8.2, there are $\alpha, \beta \in (0,1]$ such that $f \cap g = \alpha f = \beta g$. If $\alpha \ge \beta$ (the other case is analogous), $g = \frac{\alpha}{\beta} f \ge f$. Thus, $f \cap g = f$, and in particular, $f \cap g$ is extremal. Since g is minimal, we must have $g = f \cap g = f$.

Solution to Exercise 4.2.6 An $x \in l^1(\mathbb{I})$ is in $\mathcal{D}(G_{\text{stop}})$ iff Lx is in $\mathcal{D}(G)$; since this places no restrictions on the coordinates $\xi_{-k}, \dots, \xi_{-1}$ of x and since $\mathcal{D}(G)$ is dense in $l^1(\mathbb{N})$, $\mathcal{D}(G_{\text{stop}})$ is seen to be dense in $l^1(\mathbb{I})$. Next, for nonnegative $x \in \mathcal{D}(G_{\text{stop}})$,

$$\begin{aligned}\Sigma G_{\text{stop}} x &= \sum_{i=1}^{k} h_i(Lx) + \Sigma G L x = -\sum_{i=1}^{k} f^i(GLx) + \Sigma G L x \\ &= -\Sigma_{\max} G L x + \Sigma G L x = 0\end{aligned}$$

(for the last step, see Theorem 3.7.1 (d)). Hence, it suffices to show that for each nonnegative $y \in l^1(\mathbb{I})$, there is precisely one $x \in \mathcal{D}(G_{\text{stop}})$ solving the resolvent equation

$$\lambda x - G_{\text{stop}} x = y,$$

and this x is nonnegative. Applying L to both sides of this equation and noting that $Le_{-i} = 0$ for $i = 1, \dots, \mathcal{k}$, we see that Lx, that is, the part of x lying in $l^1(\mathbb{N})$, is determined by $Lx - GLx = Ly$. Also, noting that the vectors $e_{-\mathcal{k}}, \dots, e_{-1}$ are linearly independent and independent of any $x \in l^1(\mathbb{N})$, and comparing the coefficients of $e_{-i}, i = 1, \dots, \mathcal{k}$, we see that x solves the resolvent equation if $Lx = (\lambda - G)^{-1} Ly$ and

$$\lambda \xi_i - h_i(Lx) = \eta_i, \qquad i = -k, \dots, -1$$

(where η_i's are coordinates of y). Since h_i's are nonnegative functionals, this completes the proof.

Hint to Exercise 4.2.7 Let

$$T(t)x = \sum_{i=-\mathcal{k}}^{-1} (\xi_i + f^i Lx - f^i S(t)Lx)e_i + S(t)Lx.$$

(To guess this formula, either use probabilistic intuitions similar to those employed in defining $\{S_a(t), t \geq 0\}$ or recall that $(\lambda - G_{\text{stop}})^{-1}$ obtained explicitly in the previous exercise is the Laplace transform of the semigroup $\{e^{tG_{\text{stop}}}, t \geq 0\}$.) Check to see that this is a strongly continuous semigroup. To show that each $T(t)$ is a Markov operator, recall that $\Sigma_{\max} = \sum_{i=1}^{\mathcal{k}} f^i$ and use Exercise 3.6.21 with $f = \Sigma_{\text{pass}}$. Finally, check that the generator of this semigroup is G_{stop}.

Hint to Exercise 4.4.10 *(by A. Gregosiewicz)* For $n \geq 1$, let O_n be the operator in l^1 that changes all the coordinates of an $x \in l^1$ with indices $< -n$ to zeros and leaves the remaining coordinates intact. Clearly, $\lim_{n\to\infty} O_n x = x, x \in l^1$. For $x \in \mathcal{D}(A)$, check to see that $x_n := O_n x$ belongs to $\mathcal{D}(G)$, and $Gx_n = \mathfrak{Q}x_n = -a_{-n-1}\xi_{-n-1}e_{-n} + O_n\mathfrak{Q}x$.

Solution to Exercise 5.4.11 We have $f(i) - g(j) \leq f(i) - g(i) \leq \|f - g\|$. By symmetry, $g(j) - f(i) \leq \|g - f\|$.

Solution to Exercise 5.7.6 Since the first term on the right-hand side of (5.50) is identical to the right-hand side of (5.49), and the first term defining ξ_i in (3.30) and (3.32) is identical to the right-hand side of (3.24), the calculation presented in Example 5.2.7 reduces the task to showing that

$$\sum_{i=1}^{\infty} \eta_i \frac{1}{\pi_{i-1}} \sum_{j=1}^{\infty} \frac{\pi_{j-1}}{\lambda + a_j} g(j) = \sum_{j=1}^{\infty} g(j) \frac{\pi_{j-1}}{\lambda + a_j} \sum_{i=1}^{\infty} \frac{\eta_i}{\pi_{i-1}}.$$

This, however, results from the obvious change of the order of summation.

Commonly Used Notation

Entries are arranged by topic, not alphabetically.

$\mathbb{I}$ ———— (countable) set of indices

$\mathbb{P}, \mathbb{E}$ ———— probability and expected value

l^1, l^∞, c, c_0 ———— the spaces of summable sequences, bounded sequences, convergent sequences, and sequences covering to 0, respectively

x, y, z ———— elements of a Banach space, possibly of l^1

f, g, h ———— functionals, probably members of $(l^1)^*$

$f_\lambda, f^\diamond$ ———— see Sections 3.6.7 and 3.6.11

ξ_i, η_i ———— coordinates of $x = (\xi_i)_{i\in\mathbb{I}}$ and $y = (\eta_i)_{i\in\mathbb{I}}$

e_i ———— the vector with 1 at the ith coordinate and zeros on all the other coordinates, seen as a member of l^1 in Chapters 2–4 and as a member of l^∞ in Chapter 5

$\delta_{i,j}$ ———— this number equals 1 iff $i = j$ and is zero otherwise

A, B, C, P, R, L ———— operators – P is probably a (sub)-Markov operator; however, A and B may also be sets, and C might be a real constant

$A_{|S}$ ———— operator A with domain restricted to S

$I_{\mathbb{X}}$ or I ———— identity operator in a Banach space $\mathbb{X}$, mapping $x \in \mathbb{X}$ to itself

A^* ———— the dual to an operator A

$\mathcal{D}(A)$ ———— domain of an operator A

G ———— the generator of Kato's minimal semigroup

H ———— the generator of a semigroup dominating Kato's minimal semigroup

$\{S(t), t \geq 0\}$ ———— Kato's minimal semigroup

$\{P(t), t \geq 0\}$ ——— a semigroup; probably a (sub-)Markov semigroup in l^1

$\{T(t), t \geq 0\}$ ——— a semigroup; probably a Feller semigroup in c_0

$\{e^{tA}, t \geq 0\}$ ——— a semigroup generated by A

$\Sigma, \mathbb{1}$ ——— the functional mapping $(\xi_i)_{i\in\mathbb{I}} \in l^1$ to $\sum_{i\in\mathbb{I}} \xi_i \in \mathbb{R}$; I tend to write Σ when I see it as a functional on l^1 and $\mathbb{1}$ when I see it as a member of l^∞

$\Sigma_{\text{pass}}, \Sigma_{\max}$ ——— see Section 3.6.14

$p_{i,j}(t)$ ——— transition probabilities of a Markov chain

$q_{i,j}, Q$ ——— intensities for a Markov chain (see Section 2.2) and the matrix of intensities; sometimes, though, Q may denote an operator

$\mathfrak{Q}, \mathsf{Q}$ ——— see Sections 3.4 and 5.7.1, respectively

λ, μ ——— real, positive numbers, perhaps arguments of the Laplace transform; also, often I write λ instead of the more proper $\lambda I_{\mathbb{X}}$ – in the latter case, λ is an operator

$\mathfrak{l}_+, \mathfrak{l}_-, \ell, \ell_j$ ——— see Sections 4.3.1, 5.7.4, and 5.8.2, respectively

R_λ ——— a shorthand for $(\lambda - A)^{-1}$ if an operator A is clear from the context; in the latter part of Chapter 3 and in Chapter 4, this symbol is almost exclusively reserved for $(\lambda - G)^{-1}$ (a rather reserved resolvent?)

$\mathfrak{B}, \mathfrak{k}$ ——— $\mathfrak{B}$ is the discrete part of the exit boundary, $\mathfrak{k} := \#\mathfrak{B}$

References

[1] *Mathematically Speaking: A Dictionary of Quotations*, selected and arranged by Carl C. Gaither and Alma E. Cavazos-Gaither, Institute of Physics, Bristol, 1998. (Cited on page 157.)

[2] W. J. Anderson, *Continuous-Time Markov Chains: An Applications-Oriented Approach*, Springer Series in Statistics: Probability and Its Applications, Springer, 1991. (Cited on page 88.)

[3] W. Arendt, C. J. K. Batty, M. Hieber, and F. Neubrander, *Vector-Valued Laplace Transforms and Cauchy Problems*, Birkhäuser, 2001. (Cited on page 38.)

[4] L. Arlotti and J. Banasiak, Strictly substochastic semigroups with application to conservative and shattering solutions to fragmentation equations with mass loss, *J. Math. Anal. Appl.* **293**, no. 2 (2004), 693–720. (Cited on page 156.)

[5] J. Banasiak, A complete description of dynamics generated by birth-and-death problem: A semigroup approach, in *Mathematical Modelling of Population Dynamics*, Banach Center Publication 63, Polish Academy of Sciences Institute of Mathematics, 2004, 165–176. (Cited on page 155.)

[6] J. Banasiak and L. Arlotti, *Perturbations of Positive Semigroups with Applications*, Springer, 2006. (Cited on pages 116, 118, 155, and 156.)

[7] J. Banasiak and M. Lachowicz, Around the Kato generation theorem for semigroups, *Stud. Math.* **179**, no. 3 (2007), 217–238. (Cited on page 155.)

[8] J. Banasiak and W. Lamb, On the application of substochastic semigroup theory to fragmentation models with mass loss, *J. Math. Anal. Appl.* **284**, no. 1 (2003), 9–30. (Cited on page 156.)

[9] J. Banasiak, W. Lamb, and P. Laurencot, *Analytic Methods for Coagulation–Fragmentation Models*, CRC Press, 2019. (Cited on page 156.)

[10] C. J. K. Batty, Derivations on compact spaces, *Proc. London Math. Soc.* (3) **42**, no. 2 (1981), 299–330. (Cited on page 244.)

[11] Derivations on the line and flows along orbits, *Pac. J. Math.* **126**, no. 2 (1987), 209–225. (Cited on page 244.)

[12] N. H. Bingham, A conversation with David Kendall, *Stat. Sci.* **11**, no. 3 (1996), 159–188. (Cited on page xiv.)

[13] D. Blackwell, Another countable Markov process with only instantaneous states, *Ann. Math. Stat.* **29** (1958), 313–316. (Cited on page 82.)

[14] A. Bobrowski, *Functional Analysis for Probability and Stochastic Processes: An Introduction*, Cambridge University Press, 2005. (Cited on pages 8, 30, 31, 38, 51, 66, 76, 193, and 204.)

[15] On a semigroup generated by a convex combination of two Feller generators, *J. Evol. Equ.* **7**, no. 3 (2007), 555–565. (Cited on page 244.)

[16] *Convergence of One-Parameter Operator Semigroups*; *In Models of Mathematical Biology and Elsewhere*, New Mathematical Monographs 30, Cambridge University Press, 2016. (Cited on pages xii, 23, 38, 155, and 193.)

[17] A. Bobrowski and R. Bogucki, Semigroups generated by convex combinations of several Feller generators in models of mathematical biology, *Stud. Math.* **189** (2008), 287–300. (Cited on pages xii and 193.)

[18] A. Bobrowski and M. Kimmel, Dynamics of the life history of a DNA-repeat sequence, *Arch. Control Sci.* **9(45)**, no. 1–2 (1999), 57–67. (Cited on page 101.)

[19] H. Brezis, *Functional Analysis, Sobolev Spaces and Partial Differential Equations*, Universitext, Springer, 2011. (Cited on page 38.)

[20] N. L. Carothers, *A Short Course on Banach Space Theory*, London Mathematical Society Student Texts 64, Cambridge University Press, 2004. (Cited on pages 36, 38, and 196.)

[21] K. L. Chung, *Markov Chains with Stationary Transition Probabilities*, Springer, 1960. (Cited on pages 53, 82, and 87.)

[22] *Lectures on Boundary Theory for Markov Chains*, with Paul-André Meyer. Annals of Mathematics Studies 65, Princeton University Press, 1970. (Cited on pages xiii, 148, and 156.)

[23] B. Collins, T. Kousha, R. Kulik, T. Szarek, and K. Życzkowski, The accessibility of convex bodies and derandomization of the hit and run algorithm, *J. Convex Anal.* **24**, no. 3 (2017), 903–916. (Cited on page 192.)

[24] R. Dautray and J.-L. Lions, *Mathematical Analysis and Numerical Methods for Science and Technology, Vol. 5*, with Michel Artola, Michel Cessenat, and Hélène Lanchon, Evolution Problems I, Springer, Berlin, 1992. (Cited on page 38.)

[25] E. B. Davies, *One-Parameter Semigroups*, Academic Press, London, 1980. (Cited on page 38.)

[26] R. deLaubenfels, Well-behaved derivations on $C[0, 1]$, *Pac. J. Math.* **115**, no. 1 (1984), 73–80. (Cited on page 244.)

[27] Correction to: 'Well-behaved derivations on $C[0, 1]$' [Pacific J. Math. **115** (1984), no. 1, 73–80, *Pac. J. Math.* **130**, no. 2 (1987), 395–396. (Cited on page 244.)

[28] R. L. Dobrushin, An example of a countable homogeneous Markov process all states of which are instantaneous, *Teor. Veroyatnost. Primenen.* **1** (1956), 481–485. (Cited on page 82.)

[29] W. Doeblin, Sur les propriétés asymptotiques de mouvement régis par certains types de chaînes simples, *Bull. Math. Soc. Roumaine Sci.* **39**, no. 1 (1937), 57–115. (Cited on page 192.)

[30] J. L. Doob, *Stochastic Processes*, Wiley, 1953. (Cited on page 88.)

[31] B. Dorroh, Contraction semigroups in a function space, *Pac. J. Math.* **19** (1966), 35–38. (Cited on page 243.)

[32] N. Dunford and J. T. Schwartz, *Linear Operators, Part I, General Theory*, with William G. Bade and Robert G. Bartle, reprint of the 1958 original, Wiley Classics Library, John Wiley, 1988. (Cited on page 38.)

[33] R. Durrett and S. Kruglyak, A new stochastic model of microsatellite evolution, *J. Appl. Prob.*, **36** (1999), 621–631. (Cited on page 101.)

[34] E. B. Dynkin, *Markov Processes, Vols. I and II*, translated by J. Fabius, V. Greenberg, A. Maitra, and G. Majone, Die Grundlehren der Mathematischen Wissenschaften 121, vol. 122, Academic Press 1965. (Cited on pages 38 and 200.)

[35] E. B. Dynkin and A. A. Yushkevich, *Markov Processes: Theorems and Problems*, Plenum Press, 1969. (Cited on page xv.)

[36] E. Yu. Emel′yanov, *Non-Spectral Asymptotic Analysis of One-Parameter Operator Semigroups*, Operator Theory: Advances and Applications 173, Birkhäuser, 2007. (Cited on page 38.)

[37] K.-J. Engel and R. Nagel, *One-Parameter Semigroups for Linear Evolution Equations*, Springer, 2000. (Cited on pages 38 and 243.)

[38] *A Short Course on Operator Semigroups*, Springer, 2006. (Cited on page 38.)

[39] S. N. Ethier and T. G. Kurtz, *Markov Processes: Characterization and Convergence*, Wiley, 1986. (Cited on pages xii, 38, 204, and 243.)

[40] W. Feller, *An Introduction to Probability Theory and Its Applications*, Vol. 1, 3rd ed., Wiley, 1970. (Cited on page 88.)

[41] Boundaries induced by non-negative matrices, *Trans. Am. Math. Soc.* **83** (1956), 19–54. (Cited on pages xiii, 132, 148, 156, and 190.)

[42] On boundaries and lateral conditions for the Kolmogorov differential equations, *Ann. Math.* (2) **65** (1957), 527–570. (Cited on pages xiii, 132, 155, 156, 180, 182, and 236.)

[43] *An Introduction to Probability Theory and Its Applications*, Vol. 2, 2nd ed., Wiley, 1971. (Cited on page 200.)

[44] W. Feller and H. P. McKean, Jr., A diffusion equivalent to a countable Markov chain, *Proc. Natl. Acad. Sci. U.S.A.* **42** (1956), 351–354. (Cited on page 82.)

[45] D. Freedman, *Markov Chains*, Holden-Day, 1971. (Cited on pages 41, 82, and 87.)

[46] *Approximating Countable Markov Chains*, 2nd ed., Springer, 1983. (Cited on page 88.)

[47] J. A. Goldstein, *Semigroups of Linear Operators and Applications*, Oxford University Press, 1985. (Cited on pages 38 and 59.)

[48] G. R. Grimmett and D. R. Stirzaker, *Probability and Random Processes*, 3rd ed., Oxford University Press, 2001. (Cited on page 88.)

[49] M. Hairer and J. Mattingly, The strong Feller property for singular stochastic PDEs, *Ann. Inst. Henri Poincaré Probab. Stat.* **54**, no. 3 (2018), 1314–1340. (Cited on page 192.)

[50] B. Hayes, First Links in the Markov Chain, *Am. Sci.* **101(2)** (March 2013), 92–97. (Cited on page iv.)

[51] E. Hille and R. S. Phillips, *Functional Analysis and Semi-groups*, Colloquium Publication 31, American Mathematical Society, 1957. (Cited on pages 38, 89, and 155.)

[52] N. Jacob, *Pseudo Differential Operators and Markov Processes: Vol. I. Fourier Analysis and Semigroups*, Imperial College Press, 2001. (Cited on page 200.)

[53] O. Kallenberg, *Foundations of Modern Probability*, Springer, 1997. (Cited on pages 38 and 200.)

[54] T. Kato, On the semi-groups generated by Kolmogoroff's differential equations, *J. Math. Soc. Jpn.* **6** (1954), 1–15. (Cited on pages 89 and 155.)

[55] *Perturbation Theory for Linear Operators*, Classics in Mathematics, Springer, 1995. (Cited on page 38.)

[56] D. G. Kendall, Some further pathological examples in the theory of denumerable Markov processes, *Q. J. Math. Oxford Ser. (2)* **7** (1956), 39–56. (Cited on pages 53, 185, and 190.)

[57] D. G. Kendall and G. E. H. Reuter, Some pathological Markov processes with a denumerable infinity of states and the associated semigroups of operators on l, in *Proceedings of the International Congress of Mathematicians, 1954, Amsterdam*, vol. III, Erven P. Noordhoff, 1956, pp. 377–415. (Cited on pages 53, 72, 73, 75, and 88.)

[58] J. Kisyński, *Semi-groups of Operators and Some of Their Applications to Partial Differential Equations*, Control Theory and Topics in Functional Analysis III, International Atomic Energy Agency 1976, pp. 305–405. (Cited on page 38.)

[59] A. N. Kolmogorov, On the differentiability of the transition probabilities in stationary Markov processes with a denumberable number of states, *Moskov. Gos. Univ. Učenye Zapiski Matematika* **148(4)** (1951), 53–59. (Cited on page 53.)

[60] V. N. Kolokoltsov, *Markov Processes, Semigroups and Generators*, De Gruyter Studies in Mathematics 38, Walter de Gruyter, 2011. (Cited on page 200.)

[61] T. G. Kurtz, Extensions of Trotter's operator semigroup approximation theorems, *J. Funct. Anal.* **3** (1969), 354–375. (Cited on page 24.)

[62] A. Lasota and M. C. Mackey, *Chaos, Fractals, and Noise: Stochastic Aspects of Dynamics*, Springer, 1994. (Cited on page 192.)

[63] P. Lévy, Systèmes markoviens et stationnaires: Cas dénombrable, *Ann. Sci. École Norm. Sup.* (3) **68** (1951), 327–381. (Cited on page 190.)

[64] T. M. Liggett, *Continuous Time Markov Processes: An Introduction*, American Mathematical Society, 2010. (Cited on pages 82 and 87.)

[65] H. P. Lotz, Uniform convergence of operators on L^∞ and similar spaces, *Math. Z.* **190**, no. 2 (1985), 207–220. (Cited on page 195.)

[66] A. C. McBride, *Semigroups of Linear Operators: An Introduction*, Pitman Research Notes in Mathematics Series 156, Longman Scientific & Technical, 1987. (Cited on page 38.)

[67] M. A. McKibben, *Discovering Evolution Equations with Applications: Vol. 1: Deterministic Equations*, CRC Press, 2011. (Cited on page 38.)

[68] M. Mokhtar-Kharroubi and J. Voigt, On honesty of perturbed substochastic C_0-semigroups in L^1-spaces, *J. Operator Theory* **64**, no. 1 (2010), 131–147. (Cited on page 156.)

[69] M. Mureşan, *A Concrete Approach to Classical Analysis*, CMS Books in Mathematics/Ouvrages de Mathématiques de la SMC, Springer, 2009. (Cited on page 206.)

[70] J. Neveu, Théorie des semi-groupes de Markov, *Univ. Calif. Publ. Stat.* **2** (1958), 319–394. (Cited on page 180.)

[71] Lattice methods and submarkovian processes, in *Proceedings of the 4th Berkeley Symposium of Mathematics, Statistics, and and Probability*, Vol. II, University of California Press, 1961, pp. 347–391. (Cited on page 190.)

[72] J. R. Norris, *Markov Chains*, Cambridge University Press, 1997. (Cited on pages xvi, 68, 87, 107, and 171.)

[73] A. Pazy, *Semigroups of Linear Operators and Applications to Partial Differential Equations*, Springer, 1983. (Cited on page 38.)

[74] T. Pratchett, *Equal Rites*, The Discworld Series, HarperCollins, 1987. (Cited on page 72.)

[75] G. E. H. Reuter and P. W. Riley, The Feller property for Markov semigroups on a countable state space, *J. London Math. Soc.* (2) **5** (1972), 267–275. (Cited on pages 195 and 243.)

[76] L. C. G. Rogers and D. Williams, *Diffusions, Markov Processes and Martingales: Vol. 1. Foundations*, Cambridge University Press, Cambridge, 2000. (Cited on page 220.)

[77] R. Rudnicki, *Models and Methods of Mathematical Biology* (in Polish), Institute of Mathematics of the Polish Academy of Sciences, 2014. (Cited on page 100.)

[78] L. Saloff-Coste, *Lectures on Finite Markov Chains*, Lectures on Probability Theory and Statistics, Lecture Notes in Mathematics 1665, Springer, 1997. (Cited on page xvi.)

[79] J. Schur, Über lineare Transformationen in der Theorie der unendlichen Reihen, *J. reine angewandte Math.* **151** (1921), 79–111. (Cited on page 36.)

[80] W. Sierpiński, *Infinite Operations* (in Polish), Mathematical Monographs, Czytelnik, 1948. (Cited on page 55.)

[81] M. Sova, Convergence d'opérations linéaires non bornées, *Rev. Roumaine Math. Pures Appl.* **12** (1967), 373–389. (Cited on page 24.)

[82] J. M. O. Speakman, Two Markov chains with a common skeleton, *Z. Wahrscheinlichkeitstheorie Verw. Gebiete* **7** (1967), 224. (Cited on page 41.)

[83] K. Taira, *Semigroups, Boundary Value Problems and Markov Processes*, Springer Monographs in Mathematics, Springer, 2004. (Cited on page 38.)

[84] H. R. Thieme and J. Voigt, Stochastic semigroups: Their construction by perturbation and approximation, in *Positivity IV—Theory and Applications*, Technical University Dresden, 2006, pp. 135–146. (Cited on page 155.)

[85] J. Tiuryn, R. Rudnicki, and D. Wójtowicz, *A Case Study of Genome Evolution: From Continuous to Discrete Time Model*, Mathematical Foundations of Computer Science 2004, Lecture Notes in Computer Science, 3153, Springer, 2004, pp. 1–24. (Cited on page 100.)

[86] J. A. van Casteren, *Markov Processes, Feller Semigroups and Evolution Equations*, Series on Concrete and Applicable Mathematics 12, World Scientific, 2011. (Cited on page 200.)

[87] J. van Neerven, *The Adjoint of a Semigroup of Linear Operators*, Lecture Notes in Mathematics 1529, Springer, 1992. (Cited on page 38.)

[88] J. Voigt, On substochastic C_0-semigroups and their generators, in *Proceedings of the Conference on Mathematical Methods Applied to Kinetic Equations* (Paris, 1985), vol. 16, 1987, pp. 453–466. (Cited on page 155.)

[89] V. A. Volkonskiĭ, Random substitution of time in strong Markov processes, *Teor. Veroyatnost. Primenen* **3** (1958), 332–350. (Cited on page 220.)

[90] I. I. Vrabie, *C_0-semigroups and Applications*, North-Holland Mathematics Studies 191, North-Holland, 2003. (Cited on page 38.)

[91] A. D. Wentzell, *A Course in the Theory of Stochastic Processes*, translated from the Russian by S. Chomet, with a foreword by K. L. Chung, McGraw-Hill. (Cited on page 38.)

[92] Chin Pin Wong, *Kato's perturbation theorem and honesty theory*, PhD thesis, University of Oxford, 2015. (Cited on page 156.)

[93] New approaches to honesty theory and applications in quantum dynamical semigroups, *J. Oper. Th.* **75**, no. 2 (2016), 443–474. (Cited on pages 155 and 156.)

[94] D. Worm, *Semigroups on spaces of measures*, PhD thesis, Thomas Stieltjes Institute for Mathematics, Leiden University, 2010. (Cited on page 192.)

[95] K. Yosida, *Functional Analysis*, Springer, 1965. (Cited on page 38.)

Index

Printed in the United States
By Bookmasters